NAMED ORGANIC REACTIONS

NAMED ORGANIC REACTIONS

RONALD C. DENNEY
B.Sc., Ph.D., A.R.I.C.
*Lecturer in Chemistry,
Woolwich Polytechnic, London*

NEW YORK
PLENUM PRESS

LONDON
BUTTERWORTHS

Published in the U.S.A. by
PLENUM PRESS
a division of
PLENUM PUBLISHING CORPORATION
227 West 17th Street, New York, N.Y. 10011

First published by
Butterworth & Co. (Publishers) Ltd

©
Butterworth & Co. (Publishers) Ltd.
1969

Suggested U.D.C. *No.* 541·124/·127:547
Library of Congress Catalog Card Number: 79–84057

Printed offset in Great Britain by
The Camelot Press Ltd., London and Southampton

PREFACE

For many years it has been a common practice in organic chemistry to associate well established reactions with the name of the scientist who developed the procedure. The student who is unable to differentiate between a Wittig reaction and a Claisen condensation is likely to find chemical discussion rather difficult. Although many chemists disapprove of this practice, it continues unabated in spite of modern teaching methods, as a name automatically conjures up a mental picture of the reaction and is less time-consuming in communication.

In order to cater for the needs of the student, the author has tried to weld together the historical, theoretical and practical aspects of each reaction. Recent years have shown an increase of interest in the historical growth of chemistry as a science in its own right. It was felt that the reader would find the historical background to each reaction of some interest and perhaps become more appreciative of the modern facilities available for chemical investigations.

Owing to limitations of space, the standard organic textbook has little room to spare for details of reaction conditions and applications. An attempt has been made to overcome this deficiency by drawing attention to well-established procedures and to syntheses of natural products and novel compounds.

In order to assist and encourage the student to look more deeply into his chemistry, an extensive list of references has been included with each reaction. In addition to this, all non-English language journal references have been coupled, as far as possible, with an English language abstract, either from *Chemical Abstracts* or, prior to 1907, to those published as part of the *Journal of the Chemical Society* (marked *A*).

Use has been made of numerous publications for the details contained in this book. In addition to the standard journals, the volumes of *Organic Syntheses* and *Organic Reactions* have been of particular value.

The author is indebted to Duke University, North Carolina, for the library facilities afforded him during his two-year stay in the United States of America. He wishes to express his appreciation to his colleague, Dallas Pinion, for reading the manuscript and making many helpful suggestions which have improved the quality of the text; also to his wife for her understanding and consideration during many working evenings. A general word of thanks is necessary for

PREFACE

many people who have made suggestions, supplied references and found books. Without all of them this book would still be a dream.

Undoubtedly, a book of this type must lead to questions on why a particular reaction or reference has been omitted. The reactions included have been those which occur most frequently in standard undergraduate textbooks. The references have been selected widely enough to enable a person seeking further details to trace other pertinent work on the subject.

The author has sought to produce a book which he hopes will be of value to the user.

R. C. D.

CONTENTS

		PAGE
PREFACE		v
1	Arndt–Eistert Synthesis and Wolff Rearrangement	1
2	Beckmann Rearrangement	4
3	Curtius Reaction (Curtius Rearrangement)	8
4	Hofmann Rearrangement (Hofmann Reaction)	11
5	Lossen Rearrangement	15
6	Schmidt Reaction	19
7	Bouveault–Blanc Reduction	23
8	Clemmensen Reduction	25
9	Rosenmund (Rosenmund–Saytzeff) Reduction	28
10	Wolff–Kishner Reduction	31
11	Cannizzaro Reaction	34
12	Tishchenko Reaction	37
13	Claisen Condensation	40
14	Michael Condensation (Addition)	43
15	Claisen–Schmidt Condensation (Claisen Reaction)	47
16	Knoevenagel Condensation	50
17	Perkin Reaction	54
18	Stobbe Condensation	57
19	Wittig Reaction	61
20	Claisen Rearrangement	65
21	Fries Rearrangement	68
22	Crum-Brown–Walker Reaction	72
23	Kolbe Reaction	74
24	Hell–Volhard–Zelinsky Halogenation	77
25	Wohl–Ziegler Reaction	80
26	Hinsberg's Reaction (Separation of Amines)	83
27	Schotten–Baumann Reaction	86
28	Kiliani–Fischer Synthesis	89
29	Lobry De Bruyn–Alberda Van Ekenstein Transformation	93
30	Ruff Degradation	97
31	Weerman Degradation	100
32	Wohl Degradation	104
33	Meerwein–Ponndorf–Verley Reduction	107
34	Oppenauer Oxidation	111
35	Barbier–Wieland Degradation	114
36	Elbs Persulphate Oxidation	118

CONTENTS

37	Étard Reaction	122
38	Blanc Reaction (Blanc's Rule)	125
39	Dieckmann Reaction	127
40	Ruzicka Ring Synthesis	130
41	Ziegler–Thorpe Large-Ring Synthesis	133
42	Fischer Indole Synthesis	137
43	Skraup Synthesis	140
44	Friedel–Crafts Reaction	143
45	Gattermann Aldehyde Synthesis	148
46	Gattermann–Koch Reaction	151
47	Grignard Reagents (Grignard Reactions)	154
48	Reformatsky Reaction	159
49	Gattermann Reaction	163
50	Sandmeyer Reaction	166
51	Baekeland Polymerization Process	170
52	Bucherer Reaction	174
53	Dakin Reaction	177
54	Darzens' Procedure	180
55	Diels–Alder Reaction	183
56	Fischer–Speier Esterification	187
57	Gabriel Synthesis	190
58	Hofmann Exhaustive Methylation	193
59	Hofmann–Martius Rearrangement	197
60	Hunsdiecker (Borodine–Hunsdiecker) Reaction	200
61	Kolbe–Schmitt Reaction	204
62	Reimer–Tiemann Synthesis	207
63	Sommelet Reaction	210
64	Stephen Reaction	214
65	Strecker Synthesis	217
66	Ullmann Reaction	220
67	Von Richter Reaction	223
68	Wagner–Meerwein Rearrangement	226
69	Walden Inversion	230
70	Willgerodt–Kindler Reaction	234
71	Williamson Synthesis	238
72	Wurtz–Fittig Reaction (Wurtz Reaction)	241
	Index	245

1

ARNDT–EISTERT SYNTHESIS AND WOLFF REARRANGEMENT

Nature of the reaction
An organic acid is converted into a derivative of the next higher homologue by the action of diazomethane on the acid chloride, followed by a Wolff rearrangement of the resulting diazoketone

$$RCOOH + SOCl_2 \longrightarrow RCOCl + HCl + SO_2$$
$$RCOCl + 2CH_2N_2 \longrightarrow RCOCHN_2 + CH_3Cl + N_2$$
$$RCOCHN_2 + R'OH \xrightarrow{Ag_2O} RCH_2CO_2R' + N_2$$

Historical development
In 1912, Wolff[1] reported that the action of silver oxide on diazoketones in ammoniacal solution led to the formation of amides as a result of a rearrangement taking place. General application of the rearrangement to achieve chain lengthening was made after Arndt and Eistert[2] used diazomethane to prepare diazoketones which had previously been obtained only by involved processes

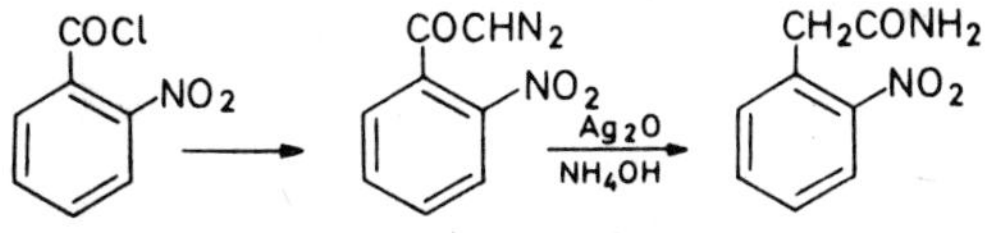

Many uses of the procedure have since been reported. Bachmann and Struve[3] reviewed the early work on the Arndt–Eistert synthesis, and a separate review of the Wolff rearrangement has also been published[4].

The nature of the final product depends upon the medium in which the diazoketone is treated: with water, an acid is formed, with alcohol an ester and with ammonia an amide

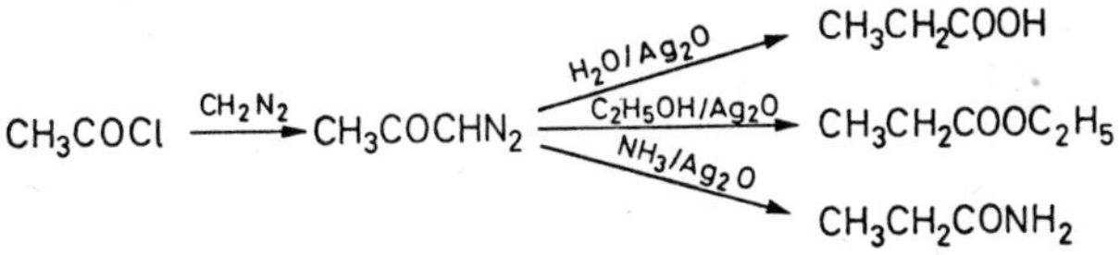

$$CH_3COCl \xrightarrow{CH_2N_2} CH_3COCHN_2 \begin{array}{c} \xrightarrow{H_2O/Ag_2O} CH_3CH_2COOH \\ \xrightarrow{C_2H_5OH/Ag_2O} CH_3CH_2COOC_2H_5 \\ \xrightarrow{NH_3/Ag_2O} CH_3CH_2CONH_2 \end{array}$$

Overall yields for the three steps are good, usually between 50 and 80 per cent.

Mechanism

A number of studies[5-7] on the Wolff rearrangement have led to the proposed mechanism[8]

$$O=\overset{R}{\underset{|}{C}}-\overset{..}{C}H-\overset{+}{N}\equiv N \longrightarrow O=\overset{R}{\underset{|}{C}}-\overset{..}{C}H \longrightarrow O=\overset{+}{C}-\overset{..}{C}HR$$

$$\xrightarrow{OH^-} \quad O=\underset{\underset{O=C-\overset{..}{C}HR}{\overset{|}{OH}}}{} \quad \xrightarrow{H^+} \quad O=\underset{\underset{O=C-CH_2R}{\overset{|}{OH}}}{}$$

Smith[9] has discussed the mechanism of the Wolff rearrangement in a review of rearrangements to electron-deficient nitrogen and oxygen atoms. This embraces the related Curtius, Hofmann, Lossen and Schmidt rearrangements.

General reaction conditions

The synthesis is applicable to both aliphatic and aromatic systems.

Conversion of the carboxylic acid starting material to the acid chloride is carried out by one of the standard procedures to give as pure a product as possible. An ethereal (or benzene) solution of the acid chloride (1 mol) is gradually added with stirring to a solution of diazomethane (3 mol) in the same solvent. The temperature in the solution is maintained below 5° for several hours prior to removing the solvent under reduced pressure at about 25°.

Rearrangement of the diazoketone depends upon the required product. Esters are obtained by dissolving it in the appropriate anhydrous alcohol, followed by heating with fresh silver oxide until nitrogen is no longer evolved.

For carboxylic acids, the diazoketone is dissolved in an inert solvent, e.g. dioxan, and added to a warmed suspension of silver oxide in dilute sodium thiosulphate.

Amides are formed if a dioxan solution of the diazoketone is warmed with aqueous ammonia and silver nitrate.

Modification

Wilds and Meader[10] have reported the successful use of diazoethane, CH_3CHN_2, in place of the conventional diazomethane as a general method for obtaining α-methyl carboxylic acids. The rearrangement of the diazoketones in these cases is accomplished in the presence of a base such as dimethylaniline[11] at 180°

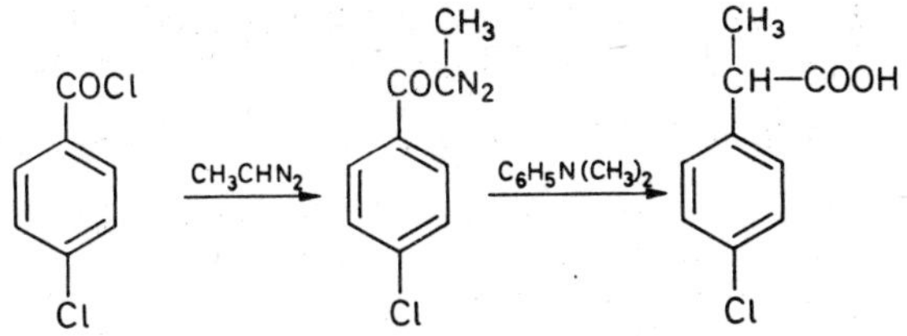

Applications

The Arndt–Eistert reaction has been employed in steroid and alkaloid syntheses. Harley-Mason and Laird[12] used the method to prepare 3,4-dihydroxy-5-methoxyphenylacetic acid which is a human metabolite of the alkaloid mescaline

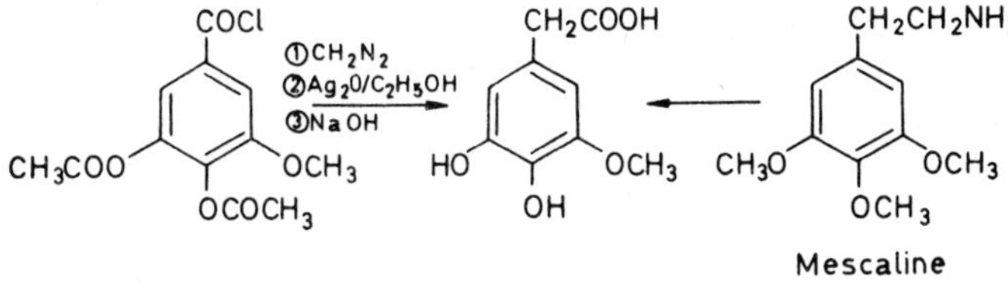

The use of the diazoethane procedure enabled Adelfang and Daub[13] to convert α-(1,2,3,11b-tetrahydro-7H-meso-benzanthrenyl-3) propionyl chloride to α-methyl-β-(1,2,3,11b-tetrahydro-7H-meso-benzanthrenyl-3) butyric acid, with 80 per cent yield

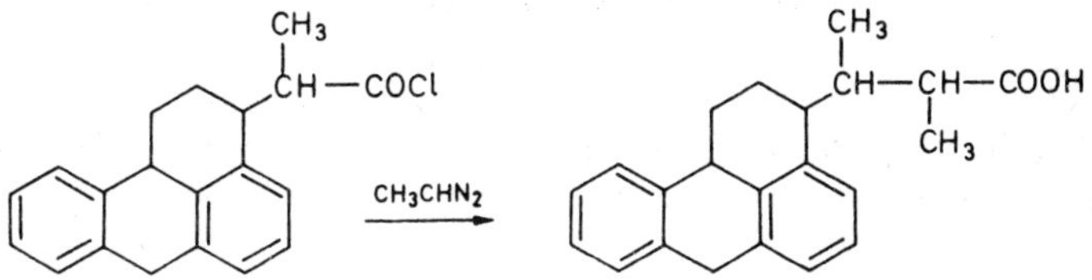

REFERENCES

[1] Wolff, L. *Justus Liebigs Annln Chem.* 394 (1912) 25 [*Chem. Abstr.* 7 (1913) 787]
[2] Arndt, F. and Eistert, B. *Ber. dt. chem. Ges.* 68B (1935) 200; Eistert, B. 69B (1936) 1074; Arndt, F. and Eistert, B. *ibid.* 1805 [*Chem. Abstr.* 30 (1936) 5214, 6326]

[3] Bachmann, W. E. and Struve, W. S. *Org. React.* 1 (1942) 38
[4] Rodina, L. L. and Korobitsyna, I. K. *Usp. Khim.* 36 (1967) 611; *Russ. chem. Revs* (1967) 260
[5] Eistert, B. *Ber. dt. chem. Ges.* 68B (1935) 208 [*Chem. Abstr.* 29 (1935) 3224]
[6] Huggett, C., Arnold, R. T. and Taylor, T. I. *J. Am. chem. Soc.* 64 (1942) 3043
[7] Lane, J. F., Willenz, J., Weissberger, A. and Wallis, E. S. *J. org. Chem.* 5 (1940) 276
[8] Finar, I. L. *Organic Chemistry*, 4th edn., 1, 328, London (Longmans) 1963
[9] Smith, P. A. S. *Molec. Rearrangements* 1 (1963) 457
[10] Wilds, A. L. and Meader, A. L. *J. org. Chem.* 13 (1948) 763
[11] Kägi, H. *Helv. chim. Acta* 24 (1941) 141E [*Chem. Abstr.* 36 (1942) 5176]
[12] Harley-Mason, J. and Laird, A. *J. chem. Soc.* (1959) 2629
[13] Adelfang, J. L. and Daub, G. H. *J. org. Chem.* 23 (1958) 749

2

BECKMANN REARRANGEMENT

Nature of the reaction

Ketoximes and aldoximes are converted into substituted acid amides by the action of acidic reagents

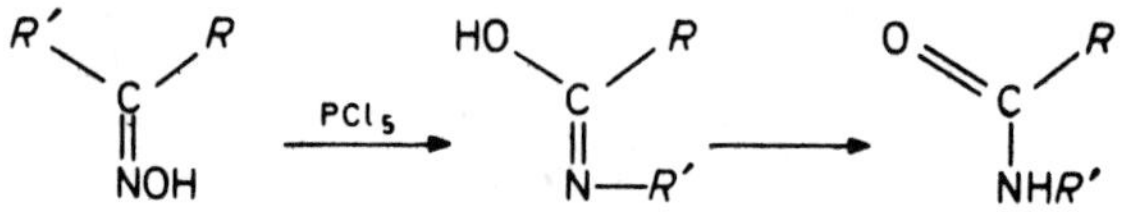

$$\underset{\text{NOH}}{\overset{R'\diagdown\diagup R}{\underset{\|}{C}}} \xrightarrow{PCl_5} \underset{\underset{\|}{N-R'}}{\overset{HO\diagdown\diagup R}{C}} \longrightarrow \underset{NHR'}{\overset{O=\diagup R}{C}}$$

Historical development

This reaction was first reported by Beckmann[1] in 1886 with reference to the rearrangement of benzophenone oxime

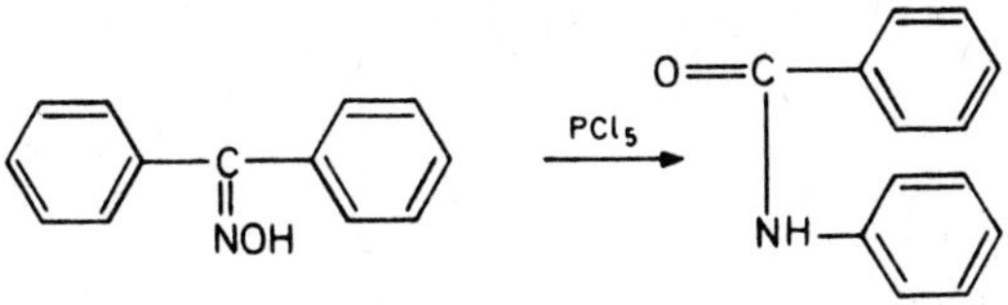

BECKMANN REARRANGEMENT

Rearrangements occur under the influence of Lewis acids as well as concentrated sulphuric acid and phosphorus pentoxide. With unsymmetrical ketones, two isomeric oximes can exist that may rearrange to give different amides

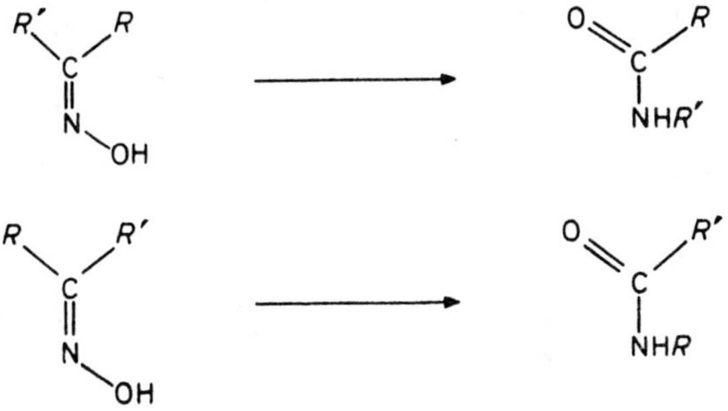

Yields from the reaction vary a great deal, depending upon the nature of the catalyst and substituents, but are usually well above 50 per cent and often greater than 90 per cent.

Blatt[2] and Jones[3] have reviewed the early work and development of the Beckmann rearrangement. It is often compared with the Curtius and Hofmann and related reactions which also involve migration to an electron-deficient nitrogen atom.

Mechanism

Theoretically an unsymmetrical ketoxime may rearrange to give both possible amides. In one case the migration is termed *syn*, in the other *anti*:

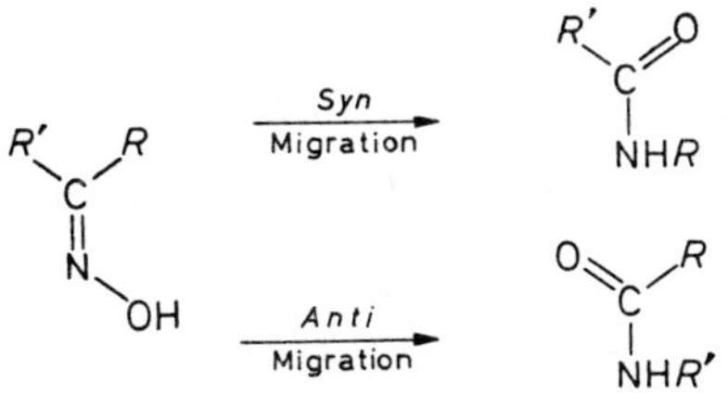

On the basis of rearrangements carried out with phosphorus pentachloride in ether, Meisenheimer[4] concluded that the process occurred by *anti*-migration. *Syn*-migration, however, does occur. It has been suggested that this is due to partial rearrangement of the unsymmetrical oxime to the other isomer prior to rearrangement to the amide. Donaruma and Heldt[5] have reviewed the current aspects of the mechanism.

Anti-migration proceeds via an intramolecular mechanism, initially by partial ionization of the oxygen–nitrogen bond of the oxime

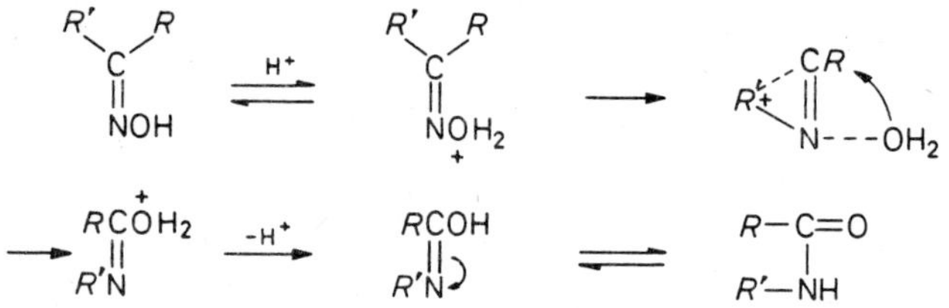

$$\underset{NOH}{\overset{R' \ R}{C}} \underset{H^+}{\rightleftharpoons} \underset{\overset{+}{NOH_2}}{\overset{R' \ R}{C}} \longrightarrow \underset{N---OH_2}{\overset{CR}{\underset{\|}{C}}} R'_+$$

$$\longrightarrow \underset{\underset{R'N}{\|}}{\overset{R\overset{+}{COH_2}}{}} \underset{-H^+}{\longrightarrow} \underset{\underset{R'N}{\|}}{\overset{RCOH}{}} \rightleftharpoons \underset{R'-NH}{\overset{R-C=O}{|}}$$

Evidence for the intramolecular nature of the rearrangement was found by Kenyon and co-workers[6]. They showed that if R' was an optically active group, the activity was retained after the rearrangement had taken place.

More recently, it has been suggested that with some reagents intermolecular rearrangements also occur. By using a mixture of two unsymmetrical ketoximes in polyphosphoric acid, Conley[7] obtained four different products:

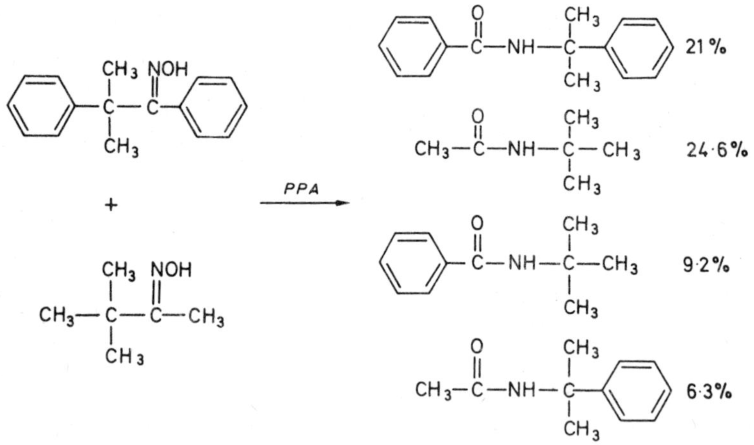

[Ph–C(CH₃)(NOH)–C(CH₃)–Ph structure]

+

CH₃–C(CH₃)(NOH)=C(CH₃)–CH₃

→ (PPA)

Ph–C(=O)–NH–C(CH₃)₂–Ph   21%

CH₃–C(=O)–NH–C(CH₃)₂–CH₃   24·6%

Ph–C(=O)–NH–C(CH₃)₂–CH₃   9·2%

CH₃–C(=O)–NH–C(CH₃)₂–Ph   6·3%

BECKMANN REARRANGEMENT

The kinetics and mechanisms of Beckmann rearrangements of alicyclic ketoximes in sulphuric acid and oleum have been reviewed by Vinnik and Zarakhani[8].

General reaction conditions

The procedure is commonly carried out in ethereal solution, the oxime being dissolved in specially dried ether and cooled in an ice bath before the catalyst is added. Further cooling is often necessary as reactions may become vigorous. After standing at room temperature for several hours, the mixture is poured onto crushed ice from which the ether can be evaporated. Solid products are then filtered off.

In the case of polyphosphoric acid (PPA) reactions[9], the oxime is added to the PPA and the mixture heated above 100° for 30 min before being poured into the iced water from which the product is isolated.

Applications

Oximes of cyclic ketones may also be rearranged by the Beckmann procedure to give lactones. Marvel and Eck[10] obtained a 60–65 per cent yield of ε-caprolactam by rearranging cyclohexanone oxime in concentrated sulphuric acid

Similarly anthraquinone dioxime can be rearranged to the dianthranilide[11]

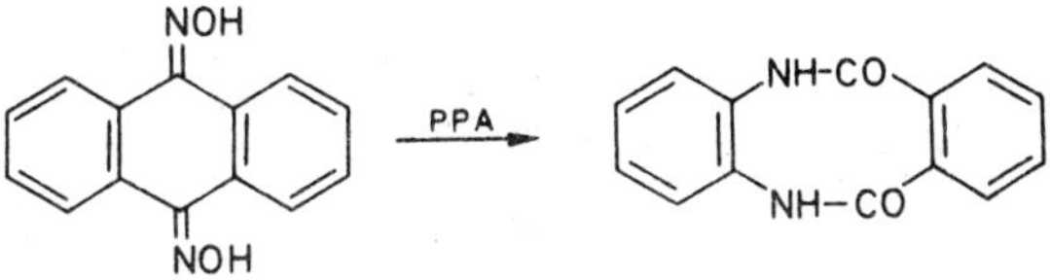

REFERENCES

[1] Beckmann, E. *Ber. dt. chem. Ges.* 19 (1886) 988; 20 (1887) 1507 [*J. chem. Soc.* (A) 50 (1886) 618; 52 (1887) 826]
[2] Blatt, A. H. *Chem. Rev.* 12 (1933) 215
[3] Jones, B. *Chem. Rev.* 35 (1944) 335
[4] Meisenheimer, J. *Ber. dt. chem. Ges.* 54B (1921) 3206 [*Chem. Abstr.* 16 (1922) 2105]
[5] Donaruma, L. G. and Heldt, W. Z. *Org. React.* 11 (1960) 1

[6] Campbell, A. and Kenyon, J. *J. chem. Soc.* (1946) 25; Kenyon, J. and Young, D. P. (1941) 263
[7] Conley, R. T. *J. org. Chem.* 28 (1963) 278
[8] Vinnik, M. I. and Zarakhani, N. G. *Usp. Khim.* 36 (1967) 167; *Russ. chem. Revs* (1967) 51
[9] Popp, F. D. and McEwen, W. E. *Chem. Rev.* 58 (1958) 370
[10] Marvel, C. S. and Eck, J. C. *Org. Synth., Coll. Vol.* 2 (1943) 371
[11] Rydon, H. N., Smith, N. H. P. and Williams, D. *J. chem. Soc.* (1957) 1900

3

CURTIUS REACTION
(CURTIUS REARRANGEMENT)

Nature of the reaction

Acids are converted to primary amines, possessing one carbon atom less, via the formation and rearrangement of an acid azide

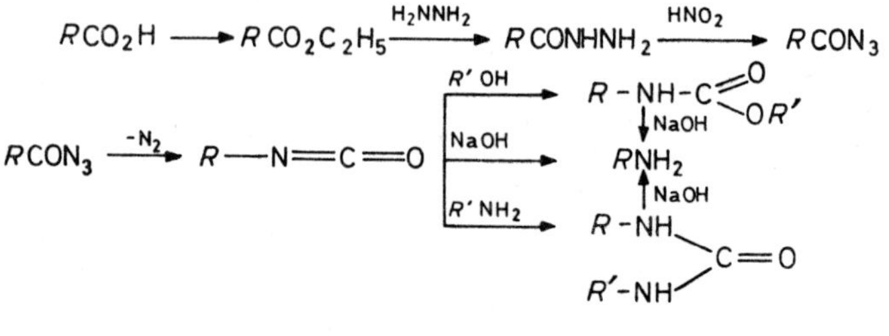

$$RCO_2H \longrightarrow RCO_2C_2H_5 \xrightarrow{H_2NNH_2} RCONHNH_2 \xrightarrow{HNO_2} RCON_3$$

$$RCON_3 \xrightarrow{-N_2} R{-}N{=}C{=}O \begin{cases} R'OH \longrightarrow R{-}NH{-}C{\overset{O}{\underset{OR'}{\Longleftrightarrow}}} \\ \quad\quad\quad\quad\quad\quad \downarrow NaOH \\ NaOH \longrightarrow RNH_2 \\ \quad\quad\quad\quad\quad\quad \uparrow NaOH \\ R'NH_2 \longrightarrow R{-}NH{\underset{R'{-}NH}{\Longleftrightarrow}}C{=}O \end{cases}$$

Historical development

Curtius' early work on organic azides[1] led him to investigate their interaction with alcohols and other compounds. From this he developed[2] the method for forming amines, involving chain shortening, that bears his name. As shown in the above equations, the route is also of value for preparing substituted ureas and urethans.

8

CURTIUS REACTION (CURTIUS REARRANGEMENT)

Benzoylazide was one of the first substances to be converted in this manner, via the urethan, to aniline

$$\text{PhCON}_3 \xrightarrow{C_2H_5OH} \text{PhNHCO}_2C_2H_5 \xrightarrow{HCl} \text{PhNH}_2$$

A long list of yields for the various steps of the reaction has been given in a review by Smith[3]. Formation of the isocyanates occurs with yields >50 per cent, and conversion to the amine >70 per cent.

Mechanism

The formation of an isocyanate is a common feature between this and the Hofmann (No. 4) and Lossen (No. 5) reactions, and due to this they are often discussed together[4]. In the case of the Curtius rearrangement, the sequence is normally written[5] to show the existence of an intermediate in which the nitrogen atom possesses only six electrons

$$R-\overset{O}{\overset{\|}{C}}-\overset{-}{N}-\overset{+}{N}\equiv N \longrightarrow N_2 + R-\overset{O}{\overset{\|}{C}}-\ddot{N}: \longrightarrow \overset{O}{\overset{\|}{C}}-\overset{-}{\underset{R}{N}} \longleftrightarrow R-N=C=O$$

Hauser and Kantor[6], however, have concluded that elimination of molecular nitrogen and migration of the group R occur simultaneously rather than as two separate steps.

General reaction conditions

There are two general methods for forming the acid azides required for rearrangement:

(*a*) by refluxing the ester with hydrazine to give an acid hydrazide which is treated with nitrous acid;

(*b*) by refluxing an acid chloride with sodium azide in benzene; in many cases the latter is superior and preferable[7]:

$$R\,COCl + NaN_3 \longrightarrow R\,CON_3 + NaCl$$

The decomposition and rearrangement are carried out by refluxing the azide in dry benzene until the evolution of nitrogen is complete. The product is then isolated by distilling off the benzene.

The preparation of undecyl isocyanate from lauroyl chloride has been described in detail[8] using the sodium azide route and isolating the isocyanate in 81–86 per cent yield

$$n-C_{11}H_{23}COCl + NaN_3 \longrightarrow n-C_{11}H_{23}CON_3 \longrightarrow n-C_{11}H_{23}NCO + N_2$$

Isocyanates are hydrolysed by warming with 50 per cent aqueous potassium hydroxide to give primary amines in almost quantitative yields.

Urethans, in yields of 40–70 per cent, are obtained by refluxing the isocyanate with excess alcohol. In some cases a high-boiling solvent, such as xylene, is also used.

Applications

A great deal of interest has been shown in the application of the Curtius reaction to the benzenesulphonyl system. Lwowski and Scheiffele[9] have studied benzenesulphonyl azide in methanol solution irradiated with ultra-violet light. In addition to a 15 per cent yield of *N*-methoxybenzene sulphonamide, they recorded the first formation of a phenylsulphamic ester, with a yield of 23 per cent

$$\text{Ph–SO}_2N_3 \xrightarrow{h\nu} [\text{Ph–NSO}_2] \xrightarrow{CH_3OH} \text{Ph–NH–SO}_2OCH_3$$

Curtius reactions on tetrahydropyran-α,α-dicarboxylic acids have been found[10] to give lactones

$$\text{(tetrahydropyran-}\alpha,\alpha\text{-(CO}_2H)_2) \longrightarrow \text{(tetrahydropyran-}\alpha,\alpha\text{-(NCO)}_2) \longrightarrow \delta\text{-Caprolactone}$$

δ-Caprolactone

These results were anticipated, as the Curtius reaction with geminal dicarboxylic acids is known to give aldehydes[3].

REFERENCES

[1] Curtius, T. *Ber. dt. chem. Ges.* 23 (1890) 3023 [*J. chem. Soc.* (A) 60 (1891) 56]

[2] Curtius, T. *Ber. dt. chem. Soc.* 27 (1894) 778 [*J. chem. Soc.* (A) 66 (1894) [1] 331]; *J. prakt. Chem.* 50 (1894) [2] 275 [*J. chem. Soc.* (A) 68 (1895) [1] 32]

[3] Smith, P. A. S. *Org. React.* 3 (1946) 337

[4] Wallis, E. S. and Lane, J. F. *Org. React.* 3 (1946) 268; Smith, P. A. S. *Molec. Rearrangements* 1 (1963) 457
[5] Hine, J. *Physical Organic Chemistry*, 2nd edn., p. 336, New York (McGraw-Hill) 1962
[6] Hauser, C. R. and Kantor, S. W. *J. Am. chem. Soc.* 72 (1950) 4284
[7] Naegeli, C., with Gruntuck, L. and Lendorff, P. *Helv. chim. Acta* 12 (1929) 227; with Lendorff, P. 15 (1932) 49 [*Chem. Abstr.* 23 (1929) 2418; 26 (1932) 2430]
[8] Allen, C. F. H. and Bell, A. *Org. Synth., Coll. Vol.* 3 (1955) 846
[9] Lwowski, W. and Scheiffele, E. *J. Am. chem. Soc.* 87 (1965) 4359
[10] Zamojski, A. and Jankowski, K. *Roczn. Chem.* 38 (1964) 707 [*Chem. Abstr.* 62 (1965) 1560]

4

HOFMANN REARRANGEMENT (HOFMANN REACTION)

(The name 'Hofmann degradation' should be avoided because of its ambiguity with the Hofmann exhaustive methylation [No. 58]).

Nature of the reaction

Amides are decomposed by the action of bromine or chlorine dissolved in alkali to give isocyanates which are rearranged to primary amines possessing one less carbon atom than the original amide

$$R-\underset{NH_2}{\overset{O}{C}} + Br_2 + 4NaOH \longrightarrow [R-NCO] \longrightarrow RNH_2 + 2NaBr + 2H_2O + Na_2CO_3$$

Historical development

Hofmann[1] reported in 1881 that the action of sodium hydroxide on a solution of acetamide and bromine led to the formation of *N*-bromoacetamide which could be decomposed on heating in the alkaline solution to give methylamine. On this basis he made a variety of *N*-bromoamides[2] and showed that isocyanates were intermediates in the formation of the amines

$$CH_3CH_2CONH_2 \underset{\text{Propionamide}}{\longrightarrow} CH_3CH_2CONHBr \xrightarrow{OH^-} [CH_3CH_2CONBr]^-$$

$$\longrightarrow CH_3CH_2NCO \xrightarrow{OH^-} \underset{\text{Ethylamine}}{CH_3CH_2NH_2}$$

Isocyanates formed in alcoholic solution are directly converted to urethans, as in the Curtius rearrangement

$$R-NCO + R'OH \longrightarrow RNHCO_2R'$$

The relationship of the Hofmann rearrangement to those of Beckmann, Curtius and Lossen has been dealt with in a review by Franklin[3]. Wallis and Lane[4] have given a more detailed treatment of the scope of, and conditions for, the reaction.

Yields from the rearrangements are generally above 60 per cent and frequently above 80 per cent, but lower when aliphatic amides possessing more than 8 carbon atoms are used under standard Hofmann conditions.

Mechanism

Any acceptable mechanism has to account for the formation of *N*-haloamides and isocyanates, since these have been isolated from reaction media. The mechanism given by Finar[5] is generally accepted and satisfies the experimental data:

$$R-\underset{\underset{}{\|}}{\overset{O}{C}}-NH_2 \xrightarrow[OH^-]{X_2} R-\underset{\underset{}{\|}}{\overset{O}{C}}-\underset{\underset{}{|}}{\overset{H}{N}}-X \xrightarrow{KOH} \left[R-\underset{\underset{}{\|}}{\overset{O}{C}}-\ddot{N}-Br\right]^- K^+$$

$$\xrightarrow{-KBr} \left[R-\underset{\underset{}{\|}}{\overset{O}{C}}-N:\right] \longrightarrow R-N=\underset{\underset{}{\|}}{\overset{O}{C}} \xrightarrow{H_2O} RNH_2 + CO_2$$
$$A$$

Hauser and Kantor[6] have stated that the existence of the intermediate, *A*, with a nitrogen atom possessing 6 electrons is unlikely. The loss of the halide ion must then be accompanied by simultaneous transfer of the group *R*.

Studies employing optically active groups have shown retention of optical rotation[7], supplying evidence that the reaction is intra-

HOFMANN REARRANGEMENT (HOFMANN REACTION)

molecular. Optically pure (+)2-amino-3-phenylpropane was obtained from (+)2-methyl-3-phenylpropionamide.

Where strong electronegative groups exist adjacent to the carbonyl group in the amide, it has been observed[8] that a second mechanism comes into play. In the case of trifluoroacetamide this leads to the formation of bromotrifluoromethane and bromoheptafluoroethane. Patterson and co-workers[9] have extended this study and suggest three simultaneous reaction paths to account for the range of products obtained from similar starting materials.

General reaction conditions

The reaction is straightforward and easy to operate. Sodium hypobromite is freshly prepared by adding bromine to a 10 per cent solution of sodium hydroxide cooled by iced water. The amide is added and the solution maintained at 0° and stirred until all the solid has dissolved. The rearrangement is effected by warming the solution to 50–80° for 30 min or longer. If the required amine is a liquid, this may be steam-distilled off; in the case of a solid product, isolation and purification may often be achieved by forming the hydrochloride.

A good example of the preparation is the formation of 3-amino-pyridine in 70 per cent yields from nicotinamide[10]

[pyridine-3-carboxamide] + 4NaOH + Br$_2$ ⟶ [3-aminopyridine] + Na$_2$CO$_3$ + 2NaBr + 2H$_2$O

Buck and Ide[11] employed alkaline sodium hypochlorite to convert veratric amide to 4-amino veratrole with an 80 per cent yield

[3,4-dimethoxybenzamide] + NaOCl + 2NaOH ⟶ [4-amino-1,2-dimethoxybenzene] + NaCl + Na$_2$CO$_3$ + H$_2$O

Modifications

The application of the Hofmann rearrangement conditions to imides leads to the formation of an amino acid. Clarke and Behr[12]

used this route to prepare β-alanine, a 45 per cent yield of pure amino acid being obtained

$$\begin{matrix} CH_2-CO \\ NH \\ CH_2-CO \end{matrix} + KOBr + 2KOH \longrightarrow \begin{matrix} CH_2NH_2 \\ | \\ CH_2COOH \end{matrix} + KBr + K_2CO_3$$

In the cases of long-chain aliphatic amides, low yields of amines are obtained by the standard procedure. To overcome this limitation, the amides may be converted to carbamates by treating with bromine in a methanolic solution of sodium methoxide[13]. The amine is obtained by refluxing the carbamate with sodium hydroxide

$$RCONH_2 \xrightarrow{Br_2/NaOCH_3} RNHCO_2CH_3 \xrightarrow{OH^-} RNH_2$$

REFERENCES

[1] Hofmann, A. W. *Ber. dt. chem. Ges.* 14 (1881) 2725; 15 (1882) 407 [*J. chem. Soc. (A)* 42 (1882) 822, 950]
[2] Hofmann, A. W. *Ber. dt. chem. Ges.* 15 (1882) 762; 18 (1885) 2734 [*J. chem. Soc. (A)* 42 (1882) 1052; 50 (1886) 45]
[3] Franklin, E. C. *Chem. Rev.* 14 (1934) 219
[4] Wallis, E. S. and Lane, J. F. *Org. React.* 3 (1946) 267
[5] Finar, I. L. *Organic Chemistry*, 4th edn., 1, 206, London (Longmans) 1963
[6] Hauser, C. R. and Kantor, S. W. *J. Am. chem. Soc.* 72 (1950) 4284
[7] Wallis, E. S., with Nagel, S. C. *J. Am. chem. Soc.* 53 (1931) 2787; with Moyer, W. W. 55 (1933) 2598
[8] Barr, D. A. and Haszeldine, R. N. *Chemy Ind.* (1956) 1050; *J. chem. Soc.* (1957) 30
[9] Patterson, J. M., Wilson, D. D. and Trimnell, D. *J. org. Chem.* 27 (1962) 3148
[10] Allen, C. F. H. and Wolf, C. N. *Org. Synth., Coll. Vol.* 4 (1963) 45
[11] Buck, J. S. and Ide, W. S. *Org. Synth., Coll. Vol.* 2 (1943) 44
[12] Clarke, H. T. and Behr, L. D. *Org. Synth., Coll. Vol.* 2 (1943) 19
[13] Jeffreys, E. *Am. chem. J.* 22 (1899) 14; *Ber. dt. chem. Ges.* 30 (1897) 898 [*J. chem. Soc. (A)* 72 (1897) [1] 315]

5
LOSSEN REARRANGEMENT

Nature of the reaction
Hydroxamic acids and their acyl derivatives are converted to isocyanates either by thermal decomposition or by treatment with thionyl chloride

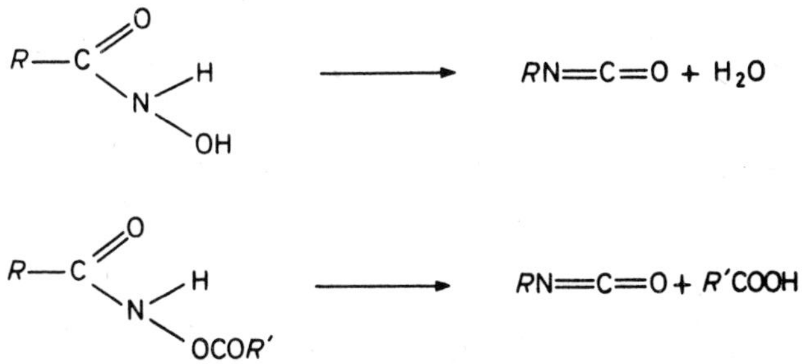

$$R-C(=O)-N(H)(OH) \longrightarrow RN=C=O + H_2O$$

$$R-C(=O)-N(H)(OCOR') \longrightarrow RN=C=O + R'COOH$$

Historical development
The work by Lossen[1] on reactions of hydroxylamine with acid chlorides was reported in 1872 and led eventually, over several years, to this method of preparing isocyanates. He found that benzoylbenzohydroxamate formed from benzoyl chloride and hydroxylamine decomposed on heating to give phenyl isocyanate and benzoic acid

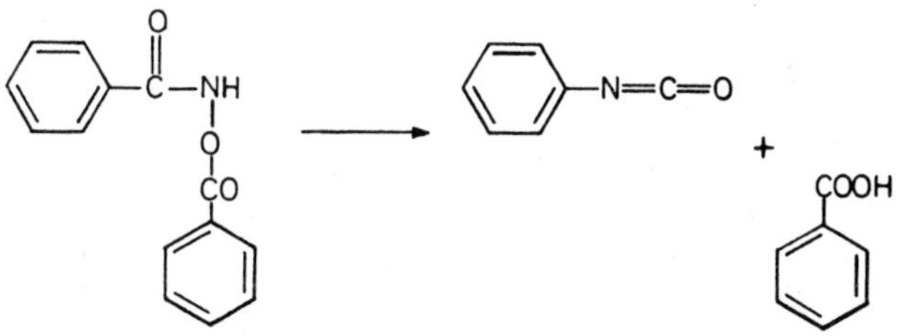

$$PhC(=O)NH-O-C(=O)Ph \longrightarrow Ph-N=C=O + PhCOOH$$

Similar to other reactions involving the formation of isocyanates

(Hofmann, Curtius), direct hydrolysis affords the corresponding amine

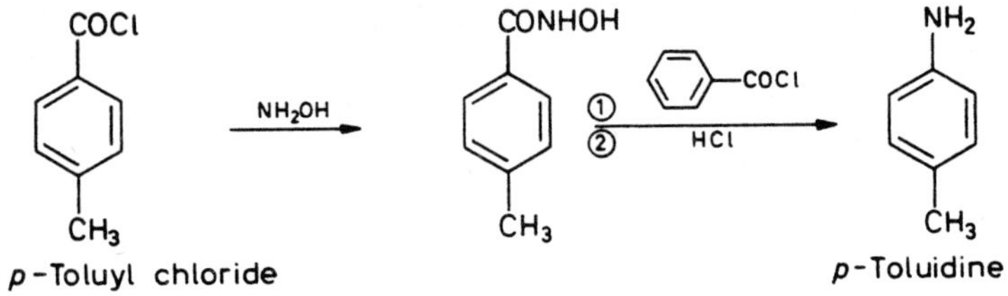

$$\text{p-Toluyl chloride} \xrightarrow{NH_2OH} \text{p-CH}_3\text{C}_6\text{H}_4\text{CONHOH} \xrightarrow[\text{HCl}]{\text{① COCl, ②}} \text{p-Toluidine}$$

Yale[2] has reviewed the development of the reaction in which it is accepted that two tautomeric forms of the monohydroxamic acids can exist

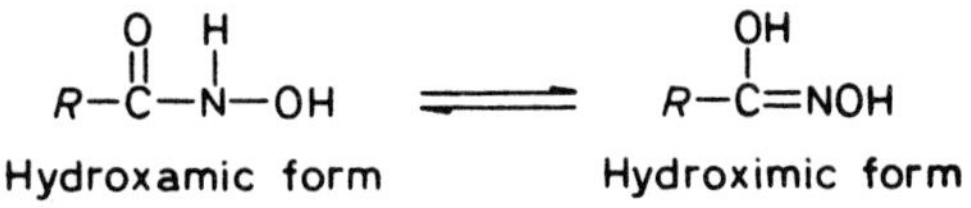

$$\underset{\text{Hydroxamic form}}{R-\underset{\underset{O}{\|}}{C}-\underset{\underset{H}{|}}{N}-OH} \rightleftharpoons \underset{\text{Hydroximic form}}{R-\underset{\underset{OH}{|}}{C}=NOH}$$

Overall yields of isocyanates or the hydrolysis products vary greatly. High yields in the region 60–80 per cent are common, but unexpectedly low ones frequently occur.

The relationship of this reaction to similar routes has been dealt with by Franklin and by Smith[3].

Mechanism

The mechanism proposed by Stieglitz and co-workers[4] has been investigated by a number of research groups.

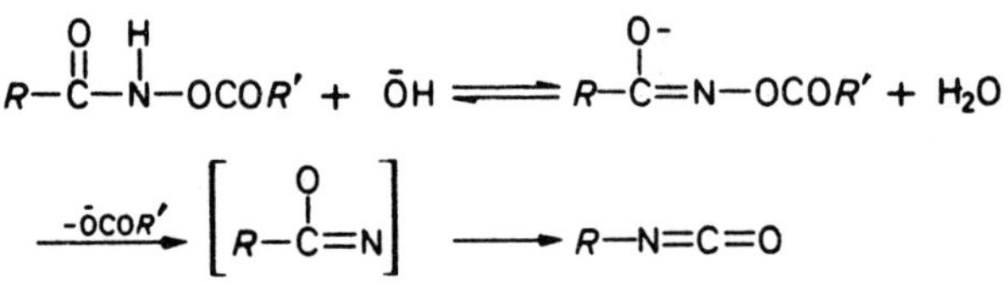

$$R-\underset{\underset{O}{\|}}{C}-\underset{\underset{H}{|}}{N}-OCOR' + \bar{O}H \rightleftharpoons R-\underset{\underset{O^-}{|}}{C}=N-OCOR' + H_2O$$

$$\xrightarrow{-\bar{O}COR'} \left[R-\underset{\underset{O}{\|}}{C}=N \right] \longrightarrow R-N=C=O$$

Renfrow and Hauser[5] showed that electron-donating groups in R increase the reactivity of the molecule and the rate-determining step is the release of the group –OCOR′.

Berndt and Shechter[6] found the kinetics for the reaction to be first-order for sodium benzoyl acyl hydroxamates where the group R

varied from methyl to t-butyl. They have expressed the currently accepted form of the mechanism as

$$R-\underset{\underset{O}{\|}}{C}-\underset{\underset{H}{|}}{N}-OCOR' \xrightleftharpoons{OH^-} R-\underset{\underset{O}{\|}}{C}-\underset{\underset{}{|}}{\bar{N}}-OCOR' \longrightarrow$$

$$\left[\begin{array}{c} O \\ \| \\ C \\ \diagup \diagdown \\ R\text{-----}N\text{----}O\text{---}COR' \end{array} \right] \longrightarrow R-N=C=O + R'CO_2^-$$

General reaction conditions

Many methods exist for the formation of monohydroxamic acids[2]. The most common is the reaction between an ester and hydroxylamine which occurs normally at room temperature.

Hydroxylamine hydrochloride is dissolved in absolute alcohol and converted to the free base by the addition of sodium ethoxide. Sodium chloride is filtered off and the filtrate added to an ethanolic solution of the organic ester, followed by an alcoholic solution of sodium ethoxide. After standing for a few minutes, evaporation of the solution gives a crystalline sodium salt of the hydroxamic acid[7]

$$\text{C}_6\text{H}_5-CH_2-\underset{\underset{CH_3}{|}}{CH}-\underset{\underset{O}{\|}}{C}-OCH_3 + H_2NOH \xrightarrow{NaOC_2H_5}$$

$$\text{C}_6\text{H}_5-CH_2-\underset{\underset{CH_3}{|}}{CH}-\underset{\underset{O}{\|}}{C}-\underset{\underset{H}{|}}{N}-OH$$

Esters of hydroxamic acids are prepared by the addition of the appropriate acid chloride to a cold aqueous alkaline solution of the hydroxamic acid. After agitation the product crystallizes from solution within a short time and may be filtered off[8].

Thermal decomposition of the solid hydroxamic acid or its esters above their melting points leads to the formation of the isocyanate. Usually this is carried out under acid conditions in order to form the amine immediately.

Since the formation of the isocyanates from hydroxamic acids is essentially a dehydration reaction, the same result can be obtained by heating the acid in the presence of thionyl chloride or acetic anhydride.

Modification

Polyphosphoric acid has frequently been employed to prepare amines by this method[9]. A more convenient procedure is to heat the carboxylic acid with hydroxylamine hydrochloride in PPA at 150° for 5-10 min. After dilution and neutralization the amine may be extracted[10]

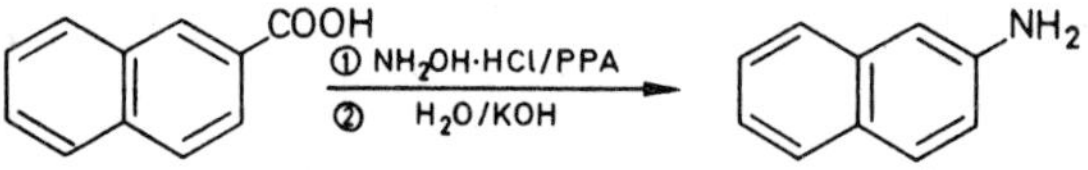

Applications

The Lossen rearrangement was used by Bredt and Perkin[11] in their transformation of methyl-D-bornylene-3-carboxylate to epicamphor

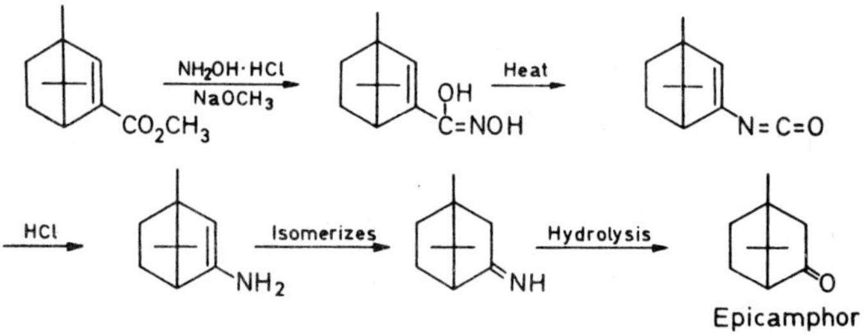

Bredt[12] had earlier prepared methyl-D-bornylene-3-carboxylate from camphor, so that this process completed the conversion of camphor to its isomer.

REFERENCES

[1] Lossen, W. *Justus Liebigs Annln Chem.* 161 (1872) 347; 175 (1875) 271, 313 [*J. chem. Soc.* (A) 25 (1872) 414; 28 (1875) 634, 769]
[2] Yale, H. L. *Chem. Rev.* 33 (1943) 209
[3] Franklin, E. C. *Chem. Rev.* 14 (1934) 219; Smith, P. A. S. *Molec. Rearrangements* 1 (1963) 457
[4] Stieglitz, J., with Leech, P. N. *J. Am. chem. Soc.* 36 (1914) 272; with Stagner, B. A. 38 (1916) 2064
[5] Renfrow, W. B. and Hauser, C. R. *J. Am. chem. Soc.* 59 (1937) 2308
[6] Berndt, D. C. and Shechter, H. *J. org. Chem.* 29 (1964) 916
[7] Jones, L. W. and Wallis, E. S. *J. Am. chem. Soc.* 48 (1926) 175
[8] Jones, L. W. and Neuffer, L. *J. Am. chem. Soc.* 39 (1917) 663

[9] Popp, F. D. and McEwen, W. E. *Chem. Rev.* 58 (1958) 374
[10] Snyder, H. R., Elston, C. T. and Kellom, D. B. *J. Am. chem. Soc.* 75 (1953) 2014
[11] Bredt, J. and Perkin, W. H. *J. chem. Soc.* 103 (1913) 2182
[12] Bredt, J. *Justus Liebigs Annln Chem.* 366 (1909) 1 [*J. chem. Soc.* (*A*) 96 (1909) [1] 498]

6

SCHMIDT REACTION

Nature of the reaction

Degradations of acids, aldehydes and ketones are brought about by the action of hydrazoic acid in concentrated mineral acid. Free carboxylic acids are converted to amines, aldehydes to nitriles, and ketones to amides

$$RCOOH + HN_3 \longrightarrow RNH_2 + CO_2 + N_2$$
$$RCHO + HN_3 \longrightarrow RCN \text{ and } RNHCHO$$
$$RCOR' + HN_3 \longrightarrow RCONHR' + N_2$$

Historical development

As a result of investigations with hydrazoic acid and sulphuric acid, Schmidt[1] was led to the conclusion that a free imide radical, NH, was formed. On the basis of this result, a reaction in the presence of benzophenone was performed and resulted in the preparation of benzanilide

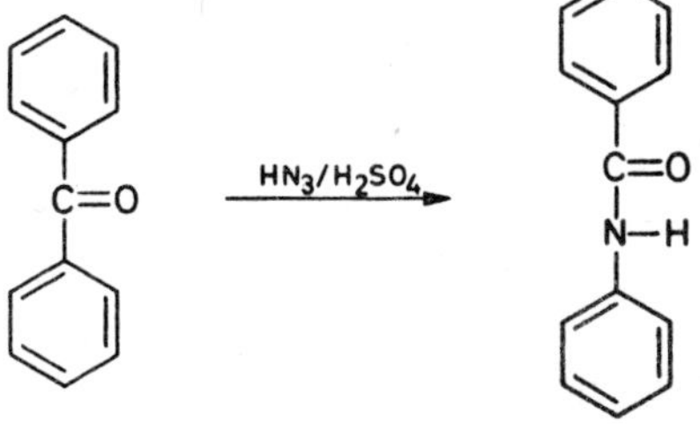

NAMED ORGANIC REACTIONS

The significance of this reaction was quickly recognized by Schmidt and extended to other carbonyl compounds, demonstrating its value as a synthetic procedure of wide application.

The reaction, which has been reviewed by Wolff[2], gives high yields, in most cases 65–95 per cent. The greatest application has been in the preparation of amines from both aromatic and aliphatic acids

$$CH_3(CH_2)_6 CO_2H \xrightarrow{HN_3} CH_3(CH_2)_6 NH_2 + CO_2 + N_2$$

Caprylic acid → n-Heptylamine

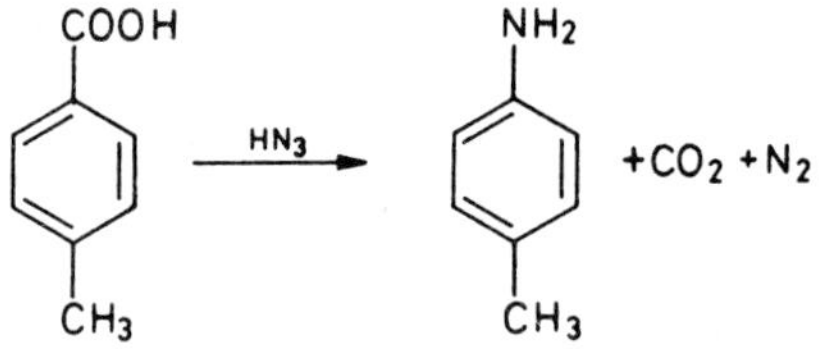

p-Toluic acid → p-Toluidine (+ CO_2 + N_2)

The value and efficacy of the Schmidt reaction have been compared[3] with those of the related Curtius (No. 3) and Hofmann (No. 4) reactions. The main limitation is in the danger of working with quantities of hydrazoic acid or azides.

Mechanism

A number of mechanisms have been proposed, and that by Smith[4] has been widely accepted. In this, the conjugate acid of the carbonyl is believed to react with molecular hydrogen azide, not with an imide radical as believed by Schmidt:

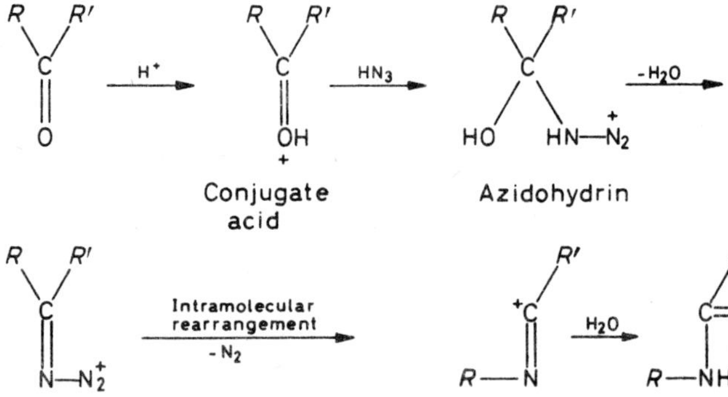

Conjugate acid → Azidohydrin

Iminodiazonium ion → (Intramolecular rearrangement, $-N_2$) → H_2O

SCHMIDT REACTION

In the case of unsymmetrical ketones, two possible isomers may be formed. It has been assumed that the migrating group lies *anti* to the diazonium nitrogens. This mechanism has managed to withstand a certain amount of criticism based upon observations with substituted benzophenones[5]. Steric factors and conjugation appear to play a greater role in determining which group migrates than was originally believed[6].

General reaction conditions

Reactions have been carried out with either hydrazoic acid or sodium azide. In the former case, a chloroform or benzene solution of hydrazoic acid is gradually added to a stirred solution of the carbonyl compound in sulphuric acid. A slight excess of hydrazoic acid is used and the rate of reaction followed by the evolution of nitrogen. In the latter case, sodium azide is added portionwise to a stirred solution of the carbonyl compound in an organic solvent (e.g. acetic acid) and sulphuric acid. Stirring is continued until evolution of nitrogen ceases and the product is isolated by neutralizing the solution, followed by extraction or steam distillation.

Detailed procedures for the preparation of 3,5-dinitroaniline and acetanilide have been given by Vogel[7]

3,5-dinitrobenzoic acid $\xrightarrow{NaN_3/H_2SO_4}$ 3,5-dinitroaniline

McNamara and Stothers[8] found that yields could be improved by employing 20 per cent oleum in place of concentrated sulphuric acid.

Modifications

Conley[9] observed that the Schmidt reaction could be effected by using molar proportions of ketone and sodium azide in polyphosphoric acid. High yields were obtained in a number of reactions including 89 per cent of ε-caprolactam from cyclohexanone

cyclohexanone $\xrightarrow{NaN_3/PPA}$ ε-caprolactam

The application of this modification has been extended[10] to a large variety of compounds, and a number of abnormal reactions were noted.

Applications

The method has been of value in the preparation of amino acids from aryl and alkylacetoacetic esters. Schmidt prepared both phenyl alanine and aspartic acid in high yields

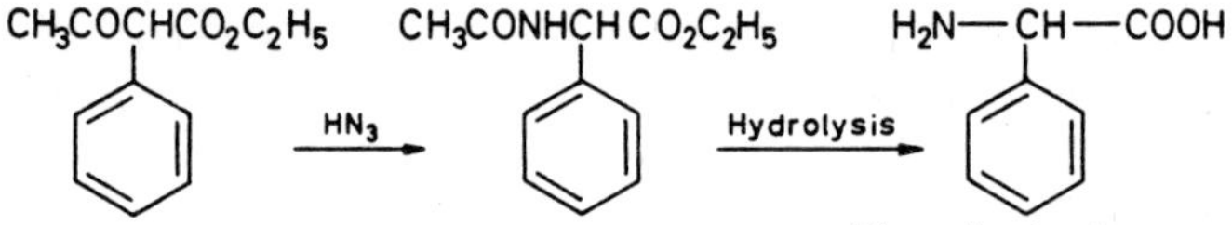

$CH_3COCHCO_2C_2H_5$ (phenyl) $\xrightarrow{HN_3}$ $CH_3CONHCHCO_2C_2H_5$ (phenyl) $\xrightarrow{Hydrolysis}$ $H_2N-CH-COOH$ (phenyl)

Phenyl alanine

REFERENCES

[1] Schmidt, K. F. *Angew. Chem.* 36 (1923) 511; *Ber. dt. chem. Ges.* 57 (1924) 704 [*Chem. Abstr.* 18 (1924) 2868]
[2] Wolff, H. *Org. React.* 3 (1946) 307
[3] Smith, P. A. S. *Org. React.* 3 (1946) 363; *Molec. Rearrangements* 1 (1963) 457
[4] Smith, P. A. S. *J. Am. chem. Soc.* 70 (1948) 320
[5] Badger, G. M., Howard, R. T. and Simons, A. *J. chem. Soc.* (1952) 2849
[6] Smith, P. A. S. and Antoniades, E. P. *Tetrahedron* 9 (1960) 210
[7] Vogel, A. I. *A Textbook of Practical Organic Chemistry*, 3rd edn., p. 919, London (Longmans) 1956
[8] McNamara, A. J. and Stothers, J. B. *Can. J. Chem.* 42 (1964) 2354
[9] Conley, R. T. *Chemy Ind.* (1958) 438; *J. org. Chem.* 23 (1958) 1330
[10] Conley, R. T. and Nowak, B. E. *Chemy Ind.* (1959) 1161; *J. org. Chem.* 26 (1961) 692

7
BOUVEAULT–BLANC REDUCTION

Nature of the reaction
Metallic sodium and ethanol are used to reduce esters to alcohols

$$RCOOC_2H_5 + 4Na + 4C_2H_5OH \longrightarrow RCH_2OH + 4C_2H_5ONa + C_2H_5OH$$

Historical development
   This procedure for forming primary alcohols was introduced by Bouveault and Blanc[1] in 1903 and elaborated in subsequent papers[2]. They found that, while the method was satisfactory with aliphatic esters, it did not work with aromatic esters such as ethyl benzoate, although ethyl phenyl acetate was reduced

$$Ph\text{-}CH_2CO_2C_2H_5 \xrightarrow{Na/C_2H_5OH} Ph\text{-}CH_2CH_2OH$$

   Yields from the preparation vary between 30 and 80 per cent; various modifications have been applied to enable higher yields to be obtained. The advent of lithium aluminium hydride and related reagents has diminished the preparative value of the Bouveault–Blanc reduction, but it is still frequently employed.

Mechanism
   The reduction is not simply a case of the reaction of nascent hydrogen with the ester. The work of Hansley[3] suggests that it proceeds *via* a sodium ketal

$$R\text{-}\underset{O}{\overset{\parallel}{C}}\text{-}OC_2H_5 + 2Na \longrightarrow R\text{-}\underset{Na}{\overset{ONa}{\underset{|}{\overset{|}{C}}}}\text{-}OC_2H_5 \xrightarrow{C_2H_5OH} R\text{-}\underset{H}{\overset{ONa}{\underset{|}{\overset{|}{C}}}}\text{-}OC_2H_5 + C_2H_5ONa$$

$$\text{Sodium ketal} \qquad \text{Sodium hemiacetal}$$

$$\longrightarrow R\overset{O}{\overset{\parallel}{C}}H + C_2H_5ONa \xrightarrow{2Na} R\text{-}\underset{Na}{\overset{ONa}{\underset{|}{\overset{|}{C}H}}} \xrightarrow{C_2H_5OH} RCH_2ONa + C_2H_5ONa$$

$$\xrightarrow{\text{Hydrolysis}} RCH_2OH + NaOH$$

This mechanism is supported by the fact that no absorption of hydrogen occurs and that aldehydes and ketones are readily reduced by sodium in ethanol.

General reaction conditions

The ester is dissolved in about 5–8 times its own volume of specially dried ethanol. An excess of sodium is gradually added at a rate that will maintain a steady reaction. When addition is complete, the mixture is gently heated until all the sodium has reacted. Water is added to hydrolyse the alkoxide, and the products isolated by extraction and fractional distillation.

Adkins and Gillespie[4] recorded 50 per cent yields of oleyl alcohol with this type of procedure

$$CH_3(CH_2)_7CH=CH(CH_2)_7CO_2C_2H_5 \xrightarrow{Na/C_2H_5OH} CH_3(CH_2)_7CH=CH(CH_2)_7CH_2OH$$

Modifications

Ford and Marvel[5] obtained several long-chain alcohols, with yields above 70 per cent, by adding the ester and alcohol to sodium in toluene

$$C_{11}H_{23}CO_2C_2H_5 \xrightarrow{Na/C_2H_5OH} C_{11}H_{23}CH_2OH$$

Ethyl laurate → Lauryl alcohol

Palfrey and Anglaret[6] used a 25 per cent excess of sodium in n-butanol under an inert atmosphere to reduce butyl oleate to oleyl alcohol and obtained a 95 per cent yield.

Application

Ruzicka *et al.*[7] used the Bouveault–Blanc procedure to reduce methyl abietate to abietinol as a step in determining the position of the acid grouping in abietic acid

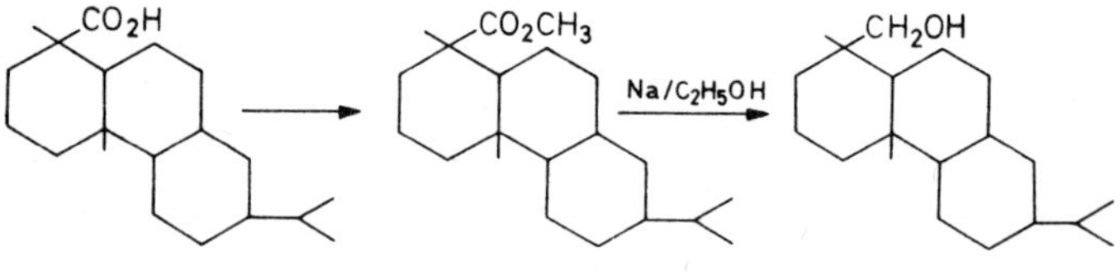

Abietic acid → Methyl abietate $\xrightarrow{Na/C_2H_5OH}$ Abietinol

REFERENCES

[1] Bouveault, L. and Blanc, G. *C. r. hebd. Séanc. Acad. Sci., Paris* 136 (1903) 1676 [*J. chem. Soc.* (A) 84 (1903) [1] 597]
[2] Bouveault, L. and Blanc, G. *C. r. hebd. Séanc. Acad. Sci., Paris* 137 (1903) 60, 328 [*J. chem. Soc.* (A) 84 (1903) [1] 673, 730]; *Bull. Soc. chim. Fr.* 31 (1904) [3] 1213 [*J. chem. Soc.* (A) 88 (1905) [1] 13]
[3] Hansley, V. L. *Ind. Engng Chem.* 39 (1947) 55
[4] Adkins, H. and Gillespie, R. H. *Org. Synth., Coll. Vol.* 3 (1955) 671
[5] Ford, S. G. and Marvel, C. S. *Org. Synth., Coll. Vol.* 2 (1943) 372
[6] Palfray, L. and Anglaret, P. *C. r. hebd. Séanc. Acad. Sci., Paris* 224 (1947) 404 [*Chem. Abstr.* 41 (1947) 3052]
[7] Ruzicka, L. and Meyer, J. *Helv. chim. Acta* 5 (1922) 581 [*Chem. Abstr.* 16 (1922) 3893]

8

CLEMMENSEN REDUCTION

Nature of the reaction

The reduction of carbonyl groups of aldehydes and ketones to methylene groups is accomplished by the action of hydrochloric acid and amalgamated zinc

$$\underset{R'}{\overset{R}{>}}C=O + 4(H) \xrightarrow[HCl]{Zn/Hg} \underset{R'}{\overset{R}{>}}CH_2 + H_2O$$

Historical development

Clemmensen's[1] reduction procedure was first announced at the 8th International Congress of Applied Chemistry of 1912 in Detroit. The reduction of acetophenone to ethyl benzene and of heptaldehyde to n-heptane were amongst the earliest examples of the procedure

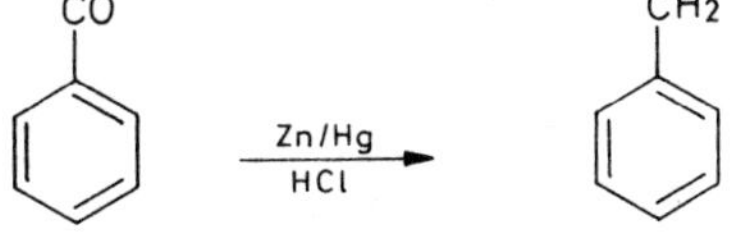

$CH_3(CH_2)_5CHO \xrightarrow[HCl]{Zn/Hg} CH_3(CH_2)_5CH_3$

The method has frequently been applied since its introduction; a review by Martin[2] covers the literature to 1942. Yields are generally between 50 and 80 per cent, although lower ones are not uncommon.

Mechanism

The full nature of the mechanism is still uncertain, although several attempts at elucidation have been made.

The route proposed by Brewster[3] was used as the basis for work reported by Nakabayashi[4] in which the following mechanism is suggested:

$$RR'CO + Zn + Cl^- \rightleftharpoons \underset{Zn^+ \cdots Cl^-}{R-\overset{O^-}{\underset{|}{C}}-R'} \xrightleftharpoons{H_3O^+} \underset{Zn^+\cdots Cl^-}{R-\overset{OH}{\underset{|}{C}}-R'} \xrightarrow{H_3O^+} \underset{Zn-Cl}{R-\overset{+}{\underset{|}{C}}-R'}$$

$$\xrightarrow{Zn} \underset{\underset{+Zn^+Cl^-}{Zn^-}}{R-\overset{\cdot}{\underset{|}{C}}-R'} \longleftrightarrow \underset{Zn^-}{R-\overset{+}{\underset{|}{C}}-R'} \xrightarrow{H_3O^+} \underset{Zn^+}{R-\overset{}{\underset{|}{CH}}-R'} \longrightarrow$$

$$R-\overset{-}{CH}-R' + Zn^{++} \xrightarrow{H_3O^+} R-CH_2-R'$$

Much of the variability of the reaction, however, is due to the dependence upon the solubilities of reactants and products. Some results of work in this field have been given by Risinger and Thompson[5] who investigated the reductions of alkyl aryl ketones under Clemmensen conditions with increasing proportions of acetic acid. The yield of hydrocarbon decreased with increase in acetic acid, whilst that of pinacols and pinacolones showed a corresponding increase. It appears that formation of the hydrocarbon is favoured if the concentration of ketone in the acidic medium is limited.

General reaction conditions

Among the well-established procedures for carrying out the Clemmensen reduction, both homogeneous aqueous[6] and heterogeneous organic systems[7] have been used.

Amalgamated zinc is prepared by shaking mossy zinc with mercuric chloride, water and hydrochloric acid. After decanting off the supernatant liquid, the ketone and hydrochloric acid are added, together with ethanol or toluene if necessary. The mixture is refluxed for an extended period—from 3 to 30 h—with stirring. On cooling, the product is isolated from the toluene layer, where used, or extracted from the aqueous layer by an appropriate solvent.

CLEMMENSEN REDUCTION

Schwarz and Hering[8] used ethanol to obtain complete solution for the reduction of vanillin to creosol. Yields of 60–67 per cent were obtained

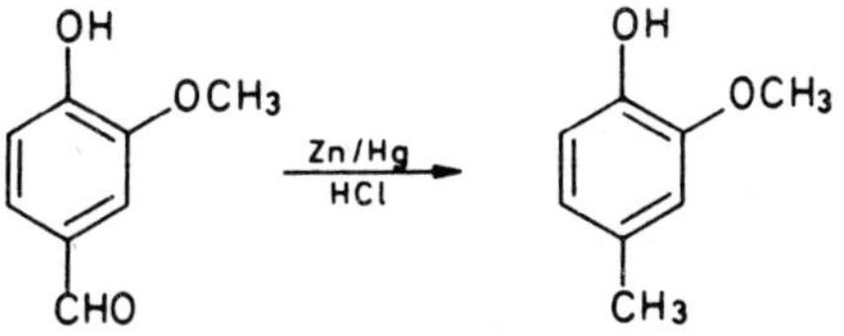

Applications

Reduction of carbonyl groups in steroids and terpenes is frequently achieved by the Clemmensen procedure. Windaus[9] reduced cholestanone to cholestane during the course of studies on the structure of cholesterol

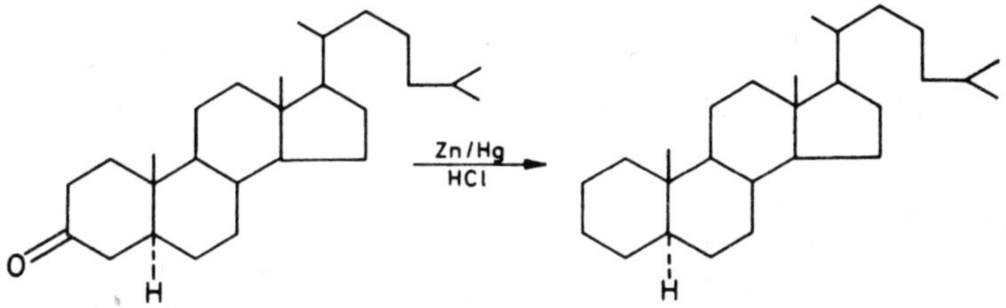

Similarly, Ruzicka's[10] studies on the structure of civetone involved a Clemmensen reduction of the carbonyl group

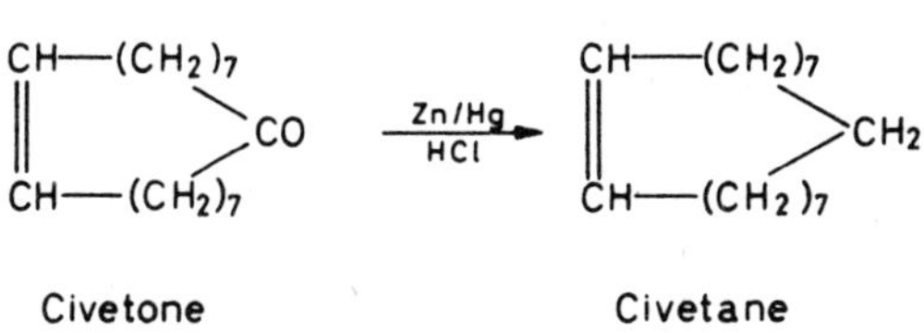

Civetone   Civetane

REFERENCES

[1] Clemmensen, E. *Proc. 8th int. Congr. appl. Chem.* 6 (1912) 68; *Ber. dt. chem. Ges.* 46 (1913) 1838; 47 (1914) 681 [*Chem. Abstr.* 6 (1912) 2919; 8 (1914) 1112, 1588]
[2] Martin, E. L. *Org. React.* 1 (1942) 155
[3] Brewster, J. H. *J. Am. chem. Soc.* 76 (1954) 6364
[4] Nakabayashi, T. *J. Am. chem. Soc.* 82 (1960) 3900, 3909
[5] Risinger, G. E. and Thompson, J. A. *J. appl. Chem.* 13 (1963) 346
[6] Read, R. R. and Wood, J. *Org. Synth., Coll. Vol.* 3 (1955) 444

[7] Martin, E. L. *Org. Synth., Coll. Vol.* 2 (1943) 499
[8] Schwarz, R. and Hering, H. *Org. Synth., Coll. Vol.* 4 (1963) 203
[9] Windaus, A. *Ber. dt. chem. Ges.* 53 (1920) 488 [*Chem. Abstr.* 14 (1920) 3083]
[10] Ruzicka, L., Schinz, H. and Seidel, C. F. *Helv. chim. Acta* 10 (1927) 695 [*Chem. Abstr.* 22 (1928) 581]

9

ROSENMUND (ROSENMUND–SAYTZEFF) REDUCTION

Nature of reaction

Acid chlorides are reduced to aldehydes by hydrogen in the presence of a palladium on barium sulphate catalyst

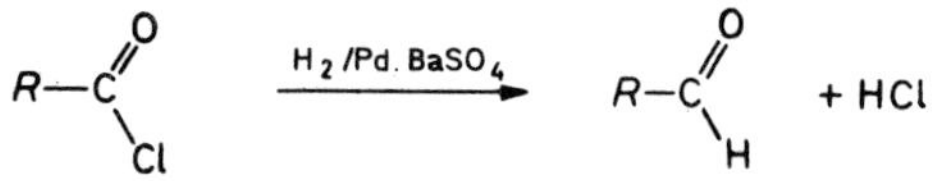

$$R-COCl \xrightarrow{H_2/Pd \cdot BaSO_4} R-CHO + HCl$$

Historical development

The first reduction of this type was carried out by Saytzeff[1] in 1872 in his preparation of benzaldehyde from benzoyl chloride

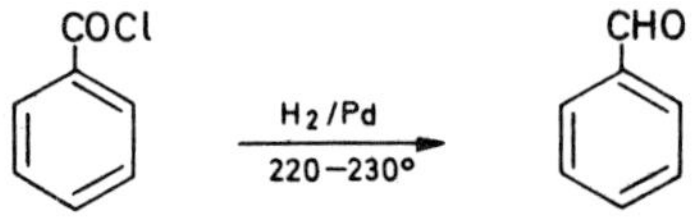

$$C_6H_5COCl \xrightarrow[220-230°]{H_2/Pd} C_6H_5CHO$$

The reaction was not exploited further until Rosenmund[2] carried out his extensive studies on catalytic reductions. In the early stages of the work controversies arose concerning the lack of reproducibility of reductions of acid chlorides to aldehydes. This was settled when it was found[3] that the catalyst has to be partially poisoned, usually by a thio compound, to prevent over-activation leading to further reduction products.

The method is of wide application to all organic systems and has the main advantage that other normally susceptible groups are not

ROSENMUND (ROSENMUND–SAYTZEFF) REDUCTION)

affected. The reaction, which usually gives high yields of aldehydes, 60–90 per cent, has been reviewed by Mosettig and Mozingo[4].

Mechanism

Little work has been carried out upon the catalytic mechanism. Affrossman and Thomson[5] concluded that the hydrogenation is consecutive rather than simultaneous, on the basis of their investigation into selective poisoning of the catalyst. Its function is to prevent later steps in the consecutive reaction from occurring and giving alcohols and hydrocarbons as products.

Stoddart and Kemball[6] found that, in the case of acetone, the carbonyl group is attached to the catalyst in the following manner

$$>\!C\!=\!O \xrightarrow{\text{Catalyst (M)}} >\!\!\underset{\underset{M}{|}}{C}\!\!-\!\!\underset{\underset{M}{|}}{O}$$

However, Affrossman and Thomson[5] suggest that, in the case of acid chlorides, replacement of chlorine occurs due to active hydrogen from the surface of the catalyst, without absorption of the acid chloride on the surface taking place. This may be represented as

$$H_2 \xrightarrow{\text{Catalyst (M)}} -\underset{\underset{M}{|}}{\overset{\overset{H}{|}}{\,}}-\underset{\underset{M}{|}}{\overset{\overset{H}{|}}{\,}}-$$

$$R-\overset{O}{\underset{\|}{C}}-Cl + H(M) \longrightarrow R-\overset{O}{\underset{\|}{C}}{}^{*} + HCl \xrightarrow{H(M)} R-\overset{O}{\underset{\|}{C}}H$$

Gianpaolo and Agnes[7], on the basis of other oxidation–reduction investigations, have suggested that the Rosenmund reduction involves formation of an intermediate between the acid chloride and catalyst; there is at present little evidence in support of this

$$RCOCl + Pd \longrightarrow RCO - Pd - Cl$$

General reaction conditions

The catalyst is made by heating a suspension of palladium chloride with dilute hydrochloric acid to give a clear red solution. A suspension of barium sulphate is added and the mixture neutralized by the addition of sodium hydroxide. The precipitated catalyst is filtered and dried.

Reactions are carried out by adding the acid chloride to a stirred mixture of the solvent, usually xylene, the catalyst and a small quantity of quinoline–sulphur poison. A steady stream of hydrogen is passed through the mixture which is heated to reflux temperature. The progress of the reduction may be followed by absorbing the evolved hydrogen chloride in water and titrating with sodium hydroxide.

On completion of the reaction, the catalyst is removed either by filtration or by centrifuging. The product is isolated either by forming a bisulphite addition compound which can be filtered or, if a solid, by steam-distilling the xylene and recrystallizing the residue[8].

Hershberg and Cason[9] have given a procedure for the preparation of β-naphthaldehyde which is of wide application:

$$\text{Naphthyl-COCl} \xrightarrow{H_2/Pd.BaSO_4} \text{Naphthyl-CHO} + HCl$$

Similar conditions were used for that of mesitaldehyde[10]

$$\text{Mesityl-COCl} \xrightarrow{H_2/Pd.BaSO_4} \text{Mesityl-CHO}$$

Applications

Gilman and co-workers[11] applied the reduction to a number of furoyl chlorides and obtained high yields (99 per cent) of the corresponding furaldehydes even when ester groups existed in the molecule

$$\text{Furyl(CO}_2\text{CH}_3\text{)-COCl} \xrightarrow{H_2/Pd.BaSO_4} \text{Furyl(CO}_2\text{CH}_3\text{)-CHO}$$

REFERENCES

[1] Saytzeff, M. *J. prakt. Chem.* 6 (1873) [2] 130
[2] Rosenmund, K. W. *Ber. dt. chem. Ges.* 51 (1918) 585; with Zetzsche, F. 55B (1922) 2774 [*Chem. Abstr.* 12 (1918) 2569; 16 (1922) 4115]
[3] Rosenmund, K. W., with Zetzsche, F. *Ber. dt. chem. Ges.* 54B (1921) 425; with Jordan, G. 58B (1925) 160 [*Chem. Abstr.* 15 (1921) 2435; 19 (1925) 1563]
[4] Mosettig, E. and Mozingo, R. *Org. React.* 4 (1948) 362
[5] Affrossman, S. and Thomson, S. J. *J. chem. Soc.* (1962) 2024
[6] Stoddart, C. T. H. and Kemball, C. *Proc. R. Soc.* A 241 (1957) 208
[7] Gianpaolo, C. and Agnes, G. *Chimica Ind. Agric. Biol.* 46 (1964) 548

[8] Ayres, D. C., Carpenter, B. G. and Denney, R. C. *J. chem. Soc.* (1965) 3578
[9] Hershberg, E. B. and Cason, J. *Org. Synth., Coll. Vol.* 3 (1955) 627
[10] Barnes, R. P. *Org. Synth., Coll. Vol.* 3 (1955) 551
[11] Gilman, H., Burtner, R. R. and Smith, E. W. *J. Am. chem. Soc.* 55 (1933) 403

10

WOLFF–KISHNER REDUCTION

Nature of the reaction
Carbonyl groups are reduced to methylenes by the alkaline decomposition of hydrazones or related derivatives of the carbonyl group

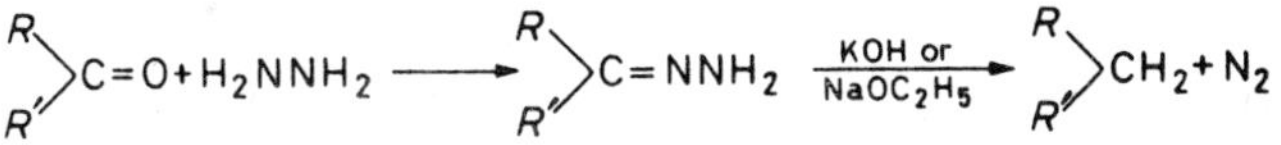

$$\underset{R'}{\overset{R}{>}}C=O + H_2NNH_2 \longrightarrow \underset{R'}{\overset{R}{>}}C=NNH_2 \xrightarrow{\text{KOH or NaOC}_2H_5} \underset{R'}{\overset{R}{>}}CH_2 + N_2$$

Historical development
Working independently, Wolff and Kishner developed their closely related techniques for reducing the carbonyl group to methylene.

Kishner[1] prepared hydrazones which were then added to hot potassium hydroxide in the presence of platinized porous plate to give the hydrocarbon

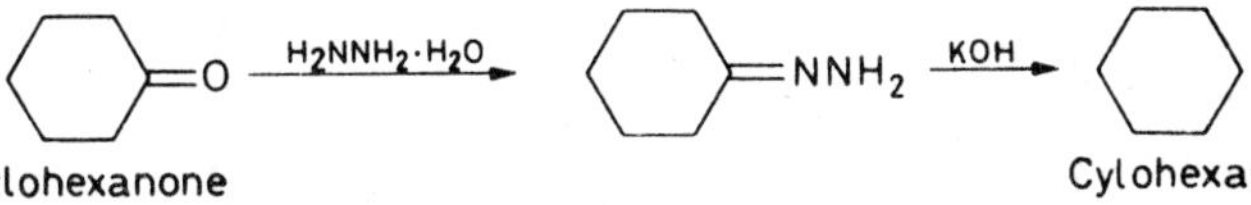

Cyclohexanone $\xrightarrow{H_2NNH_2 \cdot H_2O}$ (cyclohexylidene hydrazone) =NNH$_2$ $\xrightarrow{KOH}$ Cylohexane

Wolff's procedure[2] was to heat either hydrazones or semicarbazones with sodium ethoxide for prolonged periods above 150° in sealed tubes. By this means a wide variety of aldehydes and ketones, including camphor and furfural, were reduced

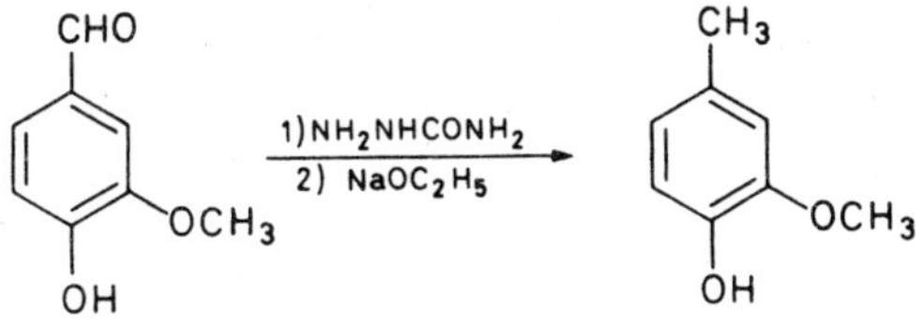

Vanillin $\xrightarrow[\text{2) NaOC}_2H_5]{\text{1) NH}_2\text{NHCONH}_2}$ (4-methyl-2-methoxyphenol)

31

Modifications have been introduced to enable reactions to be carried out under atmospheric conditions. The method is used extensively and often preferred to the Clemmensen reduction (No. 8), as it does not necessitate the use of acids. Mueller et al.[3] found that comparative studies of the Wolff–Kishner and Clemmensen reductions on medium-ring ketones showed the former to give more pure cycloalkanes in higher yields. Gas chromatography was used in the determination of comparative purities.

Yields from Wolff–Kishner reductions are generally high, above 60 per cent and frequently greater than 80 per cent. Low yields have often been improved by a small change in reaction conditions.

The literature to 1947 has been reviewed by Todd[4].

Mechanism

The results of Szmant and co-workers[5] indicated that the function of the base in the reaction is to abstract a proton from the hydrazone to form an anion, followed by hydrogen transfer

$$R_2C{=}NNH_2 + B \rightleftharpoons R_2C{=}N\bar{N}H + \overset{+}{B}H$$

$$R_2C{=}N\bar{N}H \longrightarrow R_2\bar{C}H + N_2$$

$$R_2\bar{C}H + \overset{+}{B}H \longrightarrow R_2CH_2 + B$$

Recent work[6] has led to the suggestion that solvent molecules are involved in the hydrogen transfer, to form a transition state which enables the N–H bond in the anion to be broken.

General reaction conditions

The earlier methods of Kishner and Wolff had the undesirable features of high reaction temperatures and prolonged heating. Although reductions under these conditions are often still employed, it is now more common to use modified conditions devised by Huang-Minlon[7].

This method obviates the necessity of isolating the hydrazone before treating with base. The carbonyl compound is refluxed in diethylene glycol with 85 per cent hydrazine hydrate and 2 or 3 equivalents of potassium hydroxide. After about 1 h water and excess hydrazine are distilled off until the temperature attains 190–200°. Refluxing is continued at this temperature for 3–4 h to enable the hydrazone to decompose. On cooling and acidifying the product is isolated by extraction with benzene or ether.

WOLFF-KISHNER REDUCTION

A 90 per cent yield of hendecanedoic acid has been obtained[8] by the Huang-Minlon modification:

$$HO_2C(CH_2)_4 \overset{O}{\underset{\|}{C}} (CH_2)_4 CO_2H \xrightarrow[KOH]{H_2NNH_2} HO_2C(CH_2)_9 COOH$$

A similar method[9], using triethanolamine as solvent in place of the glycol, has been employed to prepare docosanedioic acid, $HO_2C(CH_2)_{20}CO_2H$.

Applications

Kishner[10] found his method of reduction to be of value in the terpene field; thus β-ionone was successfully converted to the ionane *via* the hydrazone

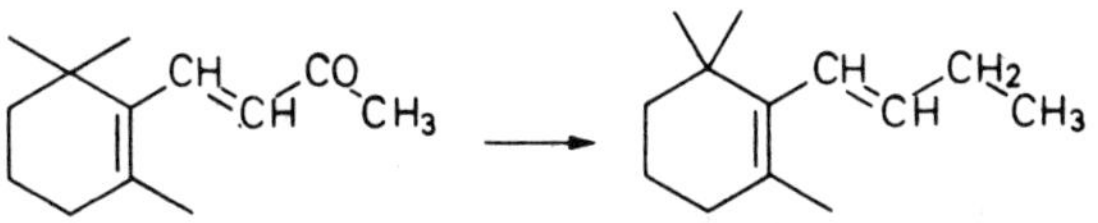

The Wolff-Kishner reduction has been of particular value in high-molecular-weight carbonyl compounds in the steroid field. Part of the proof of the structure of oestrone[11] depended upon the formation of 7-methoxy-1,2-cyclopentenophenanthrene after the carbonyl group had been reduced

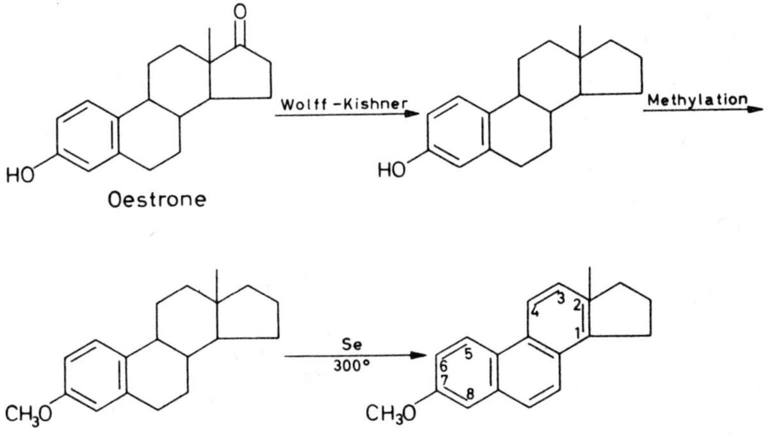

Oestrone

REFERENCES

[1] Kishner, N. *Zh. russk. fiz.-khim. Obshch.* 43 (1911) 582 [*Chem. Abstr.* 6 (1912) 347]
[2] Wolff, L. *Justus Liebigs Annln Chem.* 394 (1912) 86 [*Chem. Abstr.* 7 (1913) 790]
[3] Mueller, E. et al., *Z. Naturf.* 18B (1963) 5 [*Chem. Abstr.* 58 (1963) 11234]
[4] Todd, D. *Org. React.* 4 (1948) 378
[5] Szmant, H. H. et al., *J. Am. chem. Soc.* 74 (1952) 2724
[6] Szmant, H. H. and Harmuth, C. M. *J. Am. chem. Soc.* 86 (1964) 2909
[7] Huang-Minlon, *J. Am. chem. Soc.* 68 (1946) 2487
[8] Durham, L. J., McLeod, D. J. and Cason, J. *Org. Synth., Coll. Vol.* 4 (1963) 510
[9] Hünig, S., Lücke, E. and Brenninger, W. *Org. Synth.* 43 (1963) 34
[10] Kishner, N. *Zh. russk. fiz.-khim. Obshch.* 43 (1911) 1398 [*Chem. Abstr.* 6 (1912) 735]
[11] Cook, J. W. and Girard, A. *Nature, Lond.* 133 (1934) 377

11

CANNIZZARO REACTION

Nature of the reaction

The action of aqueous alkali on aldehydes possessing no α-hydrogen atoms leads to disproportionation between molecules and the formation of a mixture of acid and alcohol

$$2\ RCHO + NaOH \longrightarrow RCO_2Na + R\ CH_2OH$$

Historical development

Although Wöhler and Liebig[1] had shown that benzoic acid was formed when benzaldehyde was treated with aqueous alkali, it was several years before Cannizzaro[2] demonstrated that the reaction actually gave two products simultaneously

$$2\ C_6H_5CHO + NaOH \longrightarrow C_6H_5COONa + C_6H_5CH_2OH$$

Reactions may be carried out between a single molecular species or two dissimilar aldehydes. In the latter case, the process is referred to as a 'crossed' Cannizzaro reaction and may lead to a mixture of products.

CANNIZZARO REACTION

High yields are obtained, in most cases being above 70 per cent and often exceeding 85 per cent. Geissman[3] has listed many high yields in his review of the reaction.

Mechanism

The mechanism has been the subject of much investigation and is still being studied[4]. On the basis of earlier work, Hammett[5] proposed the following:

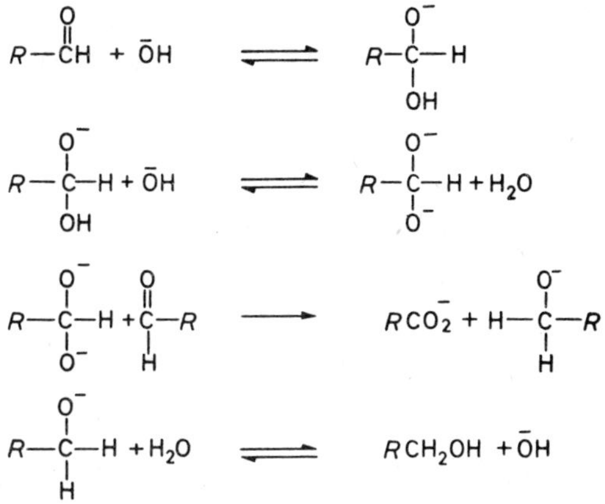

$$R-\overset{O}{\overset{\|}{C}}H + \bar{O}H \rightleftharpoons R-\overset{O^-}{\underset{OH}{\overset{|}{C}}}-H$$

$$R-\overset{O^-}{\underset{OH}{\overset{|}{C}}}-H + \bar{O}H \rightleftharpoons R-\overset{O^-}{\underset{O^-}{\overset{|}{C}}}-H + H_2O$$

$$R-\overset{O^-}{\underset{O^-}{\overset{|}{C}}}-H + \overset{O}{\overset{\|}{C}}-R \longrightarrow RCO_2^- + H-\overset{O^-}{\underset{H}{\overset{|}{C}}}-R$$

$$R-\overset{O^-}{\underset{H}{\overset{|}{C}}}-H + H_2O \rightleftharpoons RCH_2OH + \bar{O}H$$

It has been found, however, that weak bases, e.g. $Ca(OH)_2$, catalyse the reaction better than do strong ones, and Pfeil[6] has suggested that in such cases the metal hydroxide forms a complex (I) with two molecules of aldehyde

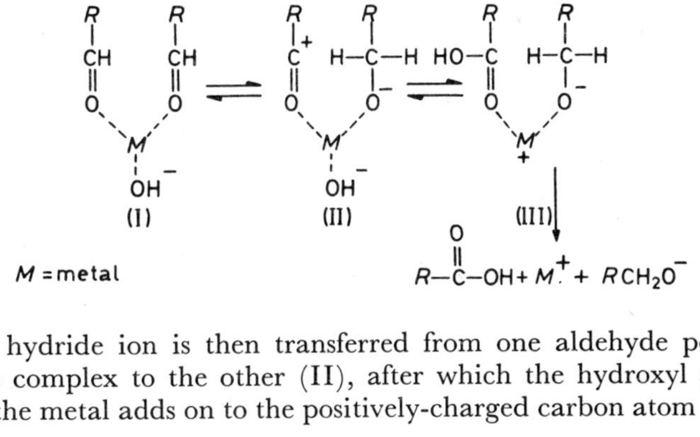

(I)   (II)   (III)

$$\longrightarrow R-\overset{O}{\overset{\|}{C}}-OH + \overset{+}{M} + RCH_2O^-$$

M = metal

An hydride ion is then transferred from one aldehyde portion of the complex to the other (II), after which the hydroxyl group from the metal adds on to the positively-charged carbon atom (III).

General reaction conditions

Simple Cannizzaro reactions are carried out by adding 50 per cent aqueous sodium hydroxide dropwise, with vigorous stirring, to the aldehyde. The mixture is heated on a steam bath, sometimes for as long as 12 h. After cooling, it is extracted with ether to remove the alcohol. Acidification of the aqueous alkaline solution then allows isolation of the carboxylic acid. A simple procedure of this type has been described in detail for the disproportionation of furfural[7] to 2-furoic acid and 2-furfuryl alcohol

$$2 \text{ (furfural-CHO)} \xrightarrow{\text{NaOH}} \text{(furfural-CO}_2\text{Na)} + \text{(furfural-CH}_2\text{OH)}$$

Modification

It has been reported[8] that the period of heating in Cannizzaro reactions can be cut down to less than 1 h by using metals (Ag) and metal oxides (CuO, PbO_2) as catalysts. This could have a profound effect upon industrial processes based upon this reaction and is currently being further investigated[9].

Applications

Crossed Cannizzaro reactions were little used until Davidson and Bogert[10] showed that aromatic alcohols could be made in high yields (85–90 per cent) by a disproportionation between formaldehyde and aromatic aldehydes

$$\text{ArCHO} + CH_2O + NaOH \longrightarrow \text{ArCH}_2OH + HCOONa$$

The reaction, which is carried out in methanol, has been described in detail for the preparation of 4-methyl benzylalcohol[11] from 4-methyl benzaldehyde and may be widely applied.

REFERENCES

[1] Wöhler, F. and Liebig, J. *Justus Liebigs Annln Chem.* 3 (1832) 249
[2] Cannizzaro, S. *Justus Liebigs Annln Chem.* 88 (1853) 129 [*J. chem. Soc.* (*A*) 7 (1854) 192]

[3] Geissman, T. A. *Org. React.* 2 (1944) 94
[4] Paucescu, S. D. *Studii Cerc. Chim.* 8 (1960) 623 [*Chem. Abstr.* 55 (1961) 26999]
[5] Hammett, L. *Physical Organic Chemistry*, p. 350, New York (McGraw-Hill) 1940
[6] Pfeil, E. *Ber. dt. chem. Ges.* 84 (1951) 229 [*Chem. Abstr.* 45 (1951) 7012]
[7] Wilson, W. C. *Org. Synth.*, *Coll. Vol.*, 2nd edn., 1 (1941) 276
[8] Isacescu, D. A. and Paucescu, S. D. *Studii Cerc. Chim.* 8 (1960) 579 [*Chem. Abstr.* 57 (1962) 3384]
[9] Lachowicz, D. R. and Gritter, R. J. *J. org. Chem.* 28 (1963) 106
[10] Davidson, D. and Bogert, M. T. *J. Am. chem. Soc.* 57 (1935) 905
[11] Davidson, D. and Weiss, M. *Org. Synth.*, *Coll. Vol.* 2 (1943) 590

12

TISHCHENKO REACTION

Nature of the reaction
The intermolecular condensation of aldehydes in the presence of aluminium alkoxides to give esters

$$2 RCHO \xrightarrow{Al(OC_2H_5)_3} RCOOCH_2R$$

With mixed aldehydes, crossed condensations occur, leading to a mixture of products

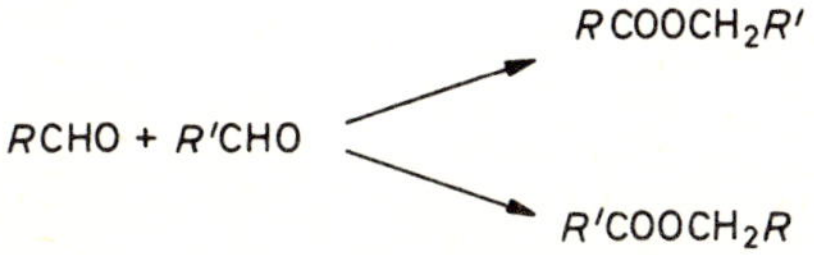

$$RCHO + R'CHO \longrightarrow \begin{array}{c} RCOOCH_2R' \\ R'COOCH_2R \end{array}$$

Historical development
  The reaction, which is similar to Cannizzaro's, was introduced by Tishchenko[1] in 1906 and differs from the former in that it is applicable to aldehydes possessing α-hydrogen atoms. The action of aluminium alkoxides in giving esters was first noticed with **paraformaldehyde** which gave methyl formate

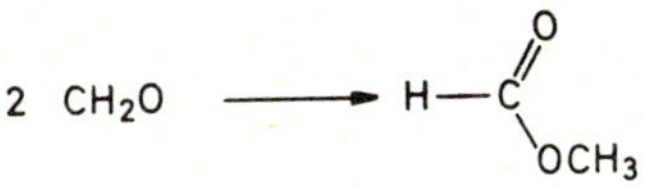

$$2\ CH_2O \longrightarrow H-\overset{O}{\underset{OCH_3}{C}}$$

The procedure was repeated with benzaldehyde to give benzyl benzoate and then extended to a range of aldehydes to demonstrate its wide applicability

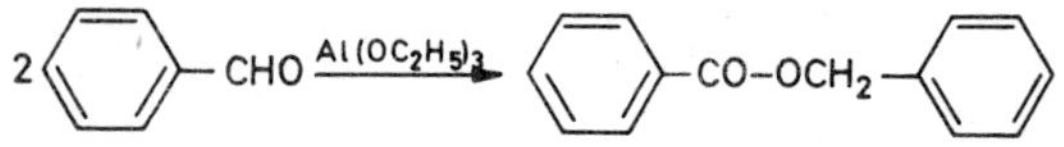

$$2\ \text{C}_6\text{H}_5\text{—CHO} \xrightarrow{\text{Al(OC}_2\text{H}_5)_3} \text{C}_6\text{H}_5\text{—CO—OCH}_2\text{—C}_6\text{H}_5$$

Wagner and Zook[2] have summarized the early work on this reaction. Although the method tends to give a number of products, it has been found that in any reaction the yields of the main product are usually above 50 per cent, and often much higher.

Mechanism

Only a limited amount of work has been applied to the study of the Tishchenko reaction.

On the basis of the reaction between n-butyraldehyde and 1,3-dichloroacetone, Lin and Day[3] suggested that in this case the ketone added to a carbonium ion formed in the aldehyde–metal oxide complex; the route is completed by an hydride transfer

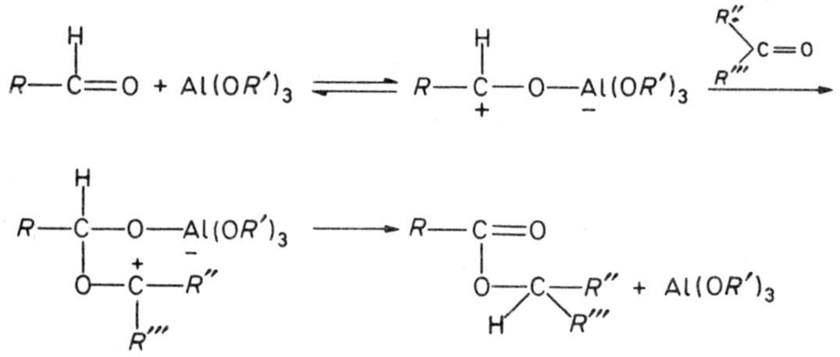

$$R\text{—CH=O} + \text{Al(OR')}_3 \rightleftharpoons R\text{—CH(+)—O(−)—Al(OR')}_3 \xrightarrow{R''R'''C=O}$$

$$\begin{array}{c} R\text{—CH(+)—O(−)—Al(OR')}_3 \\ | \\ O\text{—C(R'')(R''')} \end{array} \longrightarrow \begin{array}{c} R\text{—C=O} \\ | \\ O\text{—CH(R'')(R''')} \end{array} + \text{Al(OR')}_3$$

Results obtained with acetaldehyde by Ogata and co-workers[4] have indicated a second mechanism in which an alkoxide ion is transferred from the catalyst to the carbonyl carbon atom:

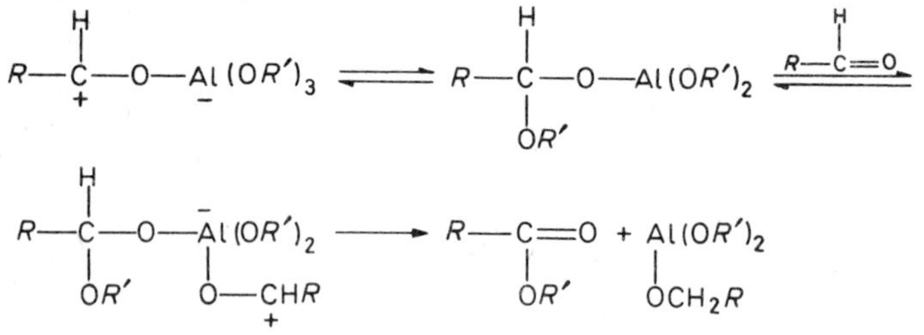

$$R\text{—CH(+)—O(−)—Al(OR')}_3 \rightleftharpoons R\text{—CH(OR')—O—Al(OR')}_2 \xrightleftharpoons{R\text{—CH=O}}$$

$$\begin{array}{c} R\text{—CH(OR')—O—Al(OR')}_2 \\ | \\ O\text{—CHR(+)} \end{array} \longrightarrow R\text{—C(OR')=O} + \text{Al(OR')}_2\text{OCH}_2R$$

TISHCHENKO REACTION

The first mechanism has met with general acceptance; however, further work is necessary in this field.

General reaction conditions

Tishchenko's early experiments were carried out by mixing the aldehydes with the catalyst; no solvent was employed. Child and Adkins[5] carried out an extensive study of the conditions necessary for high yields and verified Tishchenko's earlier work.

Alkoxides other than those of aluminium may be used and solvents are often added to the mixture to form an homogeneous system. Reactions are carried out under strictly anhydrous conditions, as the presence of moisture or free alcohols promotes the production of Cannizzaro products.

The aldehyde, or mixture of aldehydes, is added gradually to the catalyst (1–3 per cent of the total weight of aldehyde). Heating is unnecessary, as the reaction is exothermic and usually needs to be controlled by external cooling. The product is isolated by distillation.

Hawkins and co-workers[6] obtained a 90–95 per cent yield of 1,2,3,6-tetrahydrobenzyl-1,2,3,6-tetrahydrobenzoate by this procedure

$$2 \bigcirc\!\!-CHO \xrightarrow{Al[OCH(CH_3)_2]_3} \bigcirc\!\!-CH_2-O-CO-\bigcirc$$

Benzene has been used as a solvent for the reaction; its first application was by Tishchenko for the condensation of chloral and of bromal

$$2\ CCl_3CHO \xrightarrow{Al(OC_2H_5)_3} CCl_3CO_2CH_2CCl_3$$

Applications

The most interesting recent development in this work has been the use of dialdehydes, under Tishchenko conditions, to give polymers. Mitin and co-workers[7] used aluminium ethoxide in xylene to polymerize terephthalaldehyde into chains with molecular weights above 100,000. Further studies in this realm have been reported by Sweeney[8] who established that the polymer from terephthalaldehyde is a random copolymer of p-methylenebenzoate and p-xylyleneterephthalate units

$$2n\ OHC-\bigcirc\!\!-CHO \longrightarrow \left[-O-CH_2-\bigcirc\!\!-CH_2-O-CO-\bigcirc\!\!-CO- \right]_n$$

REFERENCES

[1] Tishchenko, V. E. *Zh. russk. fiz.-khim. Obshch.* 38 (1906) 355, 482 [*J. chem. Soc. (A)* 92 (1907) [1] 182, 282]
[2] Wagner, R. B. and Zook, H. D. *Synthetic Organic Chemistry*, p. 494, New York (Wiley) 1953
[3] Lin, I. and Day, A. R. *J. Am. chem. Soc.* 74 (1943) 5133
[4] Ogata, Y., Kawasaki, A. and Kishi, I. *Tetrahedron* 23 (1967) 825
[5] Child, W. C. and Adkins, H. *J. Am. chem. Soc.* 45 (1923) 3023
[6] Hawkins, E. G. E., Long, D. J. G. and Major, F. W. *J. chem. Soc.* (1955) 1462
[7] Mitin, Y. V. et al., *Polym. Sci. U.S.S.R.* 2 (1961) 423 [*Chem. Abstr.* 55 (1962) 7894]
[8] Sweeney, W. *J. appl. Polym. Sci.* 7 (1963) 1983

13

CLAISEN CONDENSATION

Nature of the reaction

Base-catalysed condensations between esters and compounds possessing active methylene groups are referred to collectively as Claisen condensations. More specifically, two molecules of an ester may be condensed to give a β-keto-ester (*A*), or an ester and a ketone to give a 1,3-diketone (*B*)

(A) $R-CO-OC_2H_5 + R'CH_2-CO-OC_2H_5 \xrightarrow{Na/C_2H_5OH} R-CO-\underset{\underset{R'}{|}}{C}HCO_2C_2H_5$

(B) $R-CO-OC_2H_5 + CH_3-\overset{O}{\overset{\|}{C}}-R' \xrightarrow{Na/C_2H_5OH} R-CO-CH_2-CO-R'$

Historical development

Condensation reactions involving the use of sodium ethoxide had been known for several years[1,2] before Claisen[3] commenced his investigations. Dibenzoylmethane was the first compound prepared by Claisen's method and led him to his extensive study of the process[4]

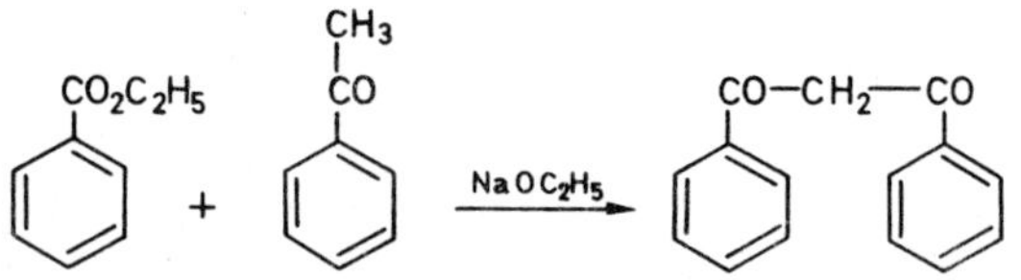

Ph–$CO_2C_2H_5$ + Ph–$COCH_3$ $\xrightarrow{NaOC_2H_5}$ Ph–CO–CH_2–CO–Ph

CLAISEN CONDENSATION

The best-known example of the reaction is the formation of ethyl acetoacetate (E.A.A.)

$$CH_3CO_2C_2H_5 + CH_3CO_2C_2H_5 \xrightarrow{NaOC_2H_5} CH_3COCH_2CO_2C_2H_5$$

Both the ester/ester[5] and ester/ketone[6] condensations have been reviewed by Hauser and his colleagues. Yields vary a great deal depending upon the nature of the reactants and the catalyst employed, but are frequently greater than 60 per cent and rarely below 30 per cent.

Mechanism

The accepted mechanism[7] proceeds by the formation of a carbanion by reaction of the ester with the base catalyst, followed by the carbanion adding at the carbonyl carbon atom of another molecule of the ester

$$CH_3CO_2C_2H_5 + B^- \rightleftharpoons BH + \bar{C}H_2CO_2C_2H_5$$

$$\underset{OC_2H_5}{\overset{O}{\underset{|}{CH_3\overset{\|}{C}}}} + \bar{C}H_2CO_2C_2H_5 \rightleftharpoons \underset{OC_2H_5}{\overset{O^-}{\underset{|}{CH_3\overset{|}{C}CH_2CO_2C_2H_5}}} \rightleftharpoons$$

$$CH_3\overset{O}{\overset{\|}{C}}CH_2CO_2C_2H_5 + C_2H_5O^- \rightleftharpoons CH_3CO\bar{C}HCO_2C_2H_5 + C_2H_5OH$$
$$(A)$$

$$\xrightarrow{H^+} CH_3COCH_2CO_2C_2H_5$$

The overall success of the reaction depends upon the formation of the relatively stable ion (A) which causes the various reversible steps to be displaced in the desired direction.

General reaction conditions

The base commonly employed in the Claisen condensation is the sodium alkoxide corresponding to the ester group. Other bases used include triphenylmethyl sodium and sodium hydride[8].

Sodium ethoxide is prepared by dissolving sodium, often as freshly prepared sodium powder, in absolute alcohol. The reactants are added to the mixture at a rate that ensures steady refluxing, which is prolonged for 1 h or more after addition is complete. After cooling to room temperature, the mixture is treated with dilute sulphuric acid and the required product isolated by separation or extraction followed by distillation or recrystallization.

Detailed experimental conditions have been given for the preparation of dibenzoylmethane[9] and of acetylacetone[10]

$$CH_3CO_2C_2H_5 + CH_3COCH_3 \xrightarrow{NaOC_2H_5} CH_3COCH_2COCH_3 \quad 45\%$$
Acetylacetone

Applications

Burdon and McLoughlin[11] have shown that the Claisen condensation can be used to give β-keto-esters containing fluorine atoms. Careful control of the vigorous reactions which occurred between trifluoroacetic and other esters enabled yields of up to 80 per cent to be obtained

$$CF_3CO_2C_2H_5 + CH_3CO_2C_2H_5 \longrightarrow CF_3COCH_2CO_2C_2H_5$$
Ethyl-γ,γ,γ - trifluoro-
acetoacetate

A Claisen condensation was employed by von Kostanecki[12] as the first step in his procedure for the synthesis of flavones. Substituted acetophenones were condensed with ethyl-*o*-methoxybenzoate to give an intermediate that was ring-closed by hydriodic acid.

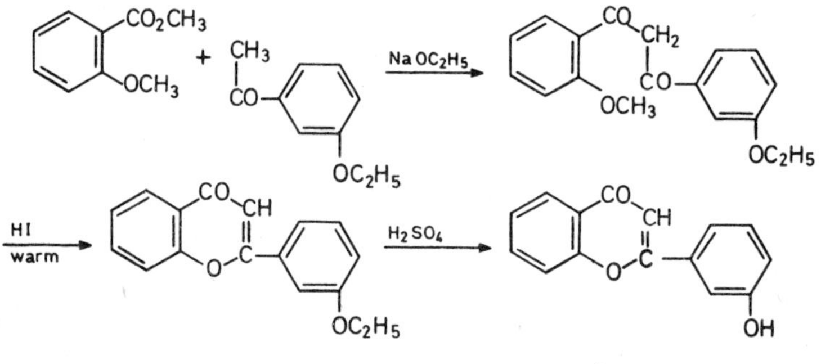

3'-Hydroxyflavone

REFERENCES

[1] Geuther, A. *Arch. Pharm., Berl.* 116 (1863) 97
[2] Hellon, R. and Oppenheim, A. *Ber. dt. chem. Ges.* 10 (1877) 699
[3] Claisen, L. *Ber. dt. chem. Ges.* 20 (1887) 655 [*J. chem. Soc.* (A) 52 (1887) 575]
[4] Beyer, C. and Claisen, L. *Ber. dt. chem. Ges.* 20 (1887) 2178; Claisen, L. and Stylos, N. *ibid.* 2188 [*J. chem. Soc.* (A) 52 (1887) 943, 917]
[5] Hauser, C. R. and Hudson, B. E. *Org. React.* 1 (1942) 266

[6] Hauser, C. R., Swamer, F. W. and Adams, J. T. *Org. React.* 8 (1954) 59
[7] Sykes, P. *A Guidebook to Mechanism in Organic Chemistry*, 2nd edn., p. 176, London (Longmans) 1965
[8] Swamer, F. W. and Hauser, C. R. *J. Am. chem. Soc.* 72 (1950) 1352
[9] Magnani, A. and McElvain, S. *Org. Synth., Coll. Vol.* 3 (1955) 251
[10] Adkins, H. and Rainey, J. L. *Org. Synth., Coll. Vol.* 3 (1955) 17
[11] Burdon, J. and McLoughlin, V. C. R. *Tetrahedron* 20 (1964) 2163
[12] Kostanecki, S. von and Tambor, J. *Ber. dt. chem. Ges.* 34 (1901) 1690 [*J. chem. Soc. (A)* 80 (1901) [1] 558]

14

MICHAEL CONDENSATION (ADDITION)

Nature of the reaction
Base-catalysed addition of compounds possessing active methylene groups to activated unsaturated systems

$$\text{C}_6\text{H}_5\text{CH}=\text{CHCO}_2\text{C}_2\text{H}_5 + \text{CH}_2(\text{CO}_2\text{C}_2\text{H}_5)_2 \xrightarrow{\text{NaOC}_2\text{H}_5} \text{C}_6\text{H}_5\text{CH}(\text{CH}(\text{CO}_2\text{C}_2\text{H}_5)_2)\text{CH}_2\text{CO}_2\text{C}_2\text{H}_5$$

Historical development
In 1887, Michael[1] reported that the condensation between ethyl sodiomalonate and ethyl cinnamate in alcohol produced an ester that was hydrolysed and decarboxylated to give phenylglutaric acid

$$\text{C}_6\text{H}_5\text{CH}(\text{CH}(\text{CO}_2\text{C}_2\text{H}_5)_2)\text{CH}_2\text{CO}_2\text{C}_2\text{H}_5 \xrightarrow{\text{KOH}} \text{C}_6\text{H}_5\text{CH}(\text{CH}_2\text{COOH})\text{CH}_2\text{COOH}$$

Although earlier work of Kommenos[2] and Claisen and Crismer[3] had already shown that aldehydes could condense with two molecules of diethylmalonate under these conditions, Michael's name is given to the reaction due to the extensive researches he made into it over a number of years[4]. The reaction is of very wide application and takes place between addends possessing the system

$$-\underset{|}{CH}-\underset{|}{C}=O$$

and acceptors having a conjugated system

$$>C=\underset{|}{C}-\underset{|}{C}=O$$

In reviews of the reaction[5] it has been pointed out that its scope now includes molecules where activation arises from groups such as nitriles and sulphones. The reaction of nitromethane with methyl vinyl ketone under these terms constitutes a Michael condensation[6]

$$CH_3-CO-CH=CH_2 + CH_3NO_2 \xrightarrow{NaOCH_3} CH_3COCH_2CH_2CH_2NO_2$$

Most yields in successful condensations are above 50 per cent and frequently > 60 per cent.

The reversibility of the Michael condensation, known for many years, has been the subject of extensive studies[7]. In general, condensations are favoured by low temperatures while elevated temperatures lead to the reverse reaction. Ingold[8] has discussed the theoretical basis of the conditions affecting equilibria in this and related reactions.

Mechanism

The necessity for base catalysis for condensations to occur suggests that the first step in the mechanism involves loss of a proton from an active methylene group[8]

$$CH_2(CO_2C_2H_5)_2 \xrightarrow{Base} \bar{C}H(CO_2C_2H_5)_2$$

$$R-CH=CH-CN + \bar{C}H(CO_2C_2H_5)_2 \longrightarrow R-CH-\bar{C}H-CN$$
$$\underset{}{} \qquad \qquad \qquad \qquad \qquad \qquad \underset{CH(CO_2C_2H_5)_2}{|}$$

$$\xrightarrow{H_2O} R-\underset{\underset{CH(CO_2C_2H_5)_2}{|}}{CH}-CH_2-CN$$

This mechanism, along with those that apply under abnormal Michael conditions, has been discussed in detail by House[9]. It is emphasized that reversal of the Michael condensation can occur if a large quantity of base is employed at an elevated temperature for a prolonged period of time.

MICHAEL CONDENSATION (ADDITION)

Certain aspects of the general process are still under investigation. Finley and co-workers[10] studied the classical reaction between ethyl cinnamate and diethyl malonate with various proportions of sodium ethoxide. The results, checked by computer, supported the accepted mechanism.

General reaction conditions

Catalysts used in condensations have included the more commonly prepared metal alkoxides as well as ammonia, pyridine and piperidine. Reactions are carried out in alcoholic solution or, less frequently, dioxane or benzene.

Condensations are brought about by mixing the reactants together with the catalyst and solvent at room temperature. After standing for a length of time ranging from 30 min to several days, the solution is acidified and the product is then extracted if it has not precipitated from solution.

In some cases, the reaction mixture is heated to a moderate temperature to enable reaction to occur. The procedure used by Moffett[11] to prepare γ-methyl-γ-nitrovalerate also employs dioxane as solvent and benzyltrimethylammonium hydroxide (Triton B) as catalyst (80–86 per cent yield).

$$(CH_3)_2 CHNO_2 + CH_2=CHCO_2CH_3 \longrightarrow (CH_3)_2 \underset{NO_2}{C} CH_2CH_2 CO_2 CH_3$$

Horning and Finelli[12] have established a procedure for condensing ethyl phenyl cyanoacetate with acrylonitrile in t-butanol; a temperature of 45° or less was maintained during the reaction

$$\underset{CO_2C_2H_5}{\underset{|}{Ph-CH}}-CN + CH_2=CHCN \xrightarrow{KOCH_3} \underset{CO_2C_2H_5}{\underset{|}{Ph-\underset{|}{C}-CH_2CH_2CN}}-CN$$

α-Phenyl-α-carbethoxy-glutaronitrile

Yields in most additions can be radically altered by variations in concentrations and conditions.

Applications

Several synthetic routes to alkaloids, terpenes and steroids involve

Michael additions. An example of this is Wohl's[13] synthesis of cincholoiponic acid

(±) Cincholoiponic acid

Intramolecular Michael condensations are well known; in these cases the condensation readily occurs in most instances if the difunctional compound is heated in alcoholic sodium ethoxide. By this means Koelsch[14] carried out several cyclizations in his work on intermediates for the synthesis of morphine

REFERENCES

[1] Michael, A. *J. prakt. Chem.* 35 (1887) 349 [*J. chem. Soc.* (*A*) 52 (1887) 672]
[2] Kommenos, T. *Justus Liebigs Annln Chem.* 218 (1883) 145 [*J. chem. Soc.* (*A*) 46 (1884) 422]
[3] Claisen, L. and Crismer, L. *Justus Liebigs Annln Chem.* 218 (1883) 129 [*J. chem. Soc.* (*A*) 46 (1884) 444]
[4] Michael, A. *Am. chem. J.* 9 (1887) 112; *Ber. dt. chem. Ges.* 33 (1900) 3731 [*J. chem. Soc.* (*A*) 80 (1901) [1] 123]
[5] Bergmann, E. D., Ginsburg, D. and Pappo, R. *Org. React.* 10 (1959) 179; Nielsen, A. T. and Houlihan, W. J. 16 (1968) 47
[6] Shechter, H., Ley, D. E. and Zeldin, L. *J. Am. chem. Soc.* 74 (1952) 3664
[7] Grob, C. A. and Baumann, W. *Helv. chim. Acta* 38 (1955) 594 [*Chem. Abstr.* 50 (1956) 9304]
[8] Ingold, C. K. *Structure and Mechanisms in Organic Chemistry*, p. 691, Ithaca, N.Y. (Cornell Univ. Press) 1953
[9] House, H. O. *Modern Synthetic Reactions*, p. 204, New York (Benjamin) 1965
[10] Finley, K. T. et al., *Can. J. Chem.* 45 (1967) 571
[11] Moffett, R. B. *Org. Synth., Coll. Vol.* 4 (1963) 652
[12] Horning, E. C. and Finelli, A. F. *Org. Synth., Coll. Vol.* 4 (1963) 776
[13] Wohl, A. and Losanitsch, M. S. *Ber. dt. chem. Ges.* 40 (1907) 4698 [*Chem. Abstr.* 2 (1908) 839]
[14] Koelsch, C. F. *J. Am. chem. Soc.* 67 (1945) 569

15

CLAISEN–SCHMIDT CONDENSATION (CLAISEN REACTION)

Nature of the reaction
The condensation between an aromatic and an aliphatic aldehyde (or ketone) in the presence of dilute alkali produces α,β-unsaturated aldehydes or ketones in which the double bond is in conjugation with the aromatic system

$$ArCHO + RCH_2CHO \xrightarrow{NaOH} ArCH=C(R)CHO + H_2O$$

$$C_6H_5CHO + CH_3CH_2CHO \xrightarrow{NaOH} C_6H_5CH=C(CH_3)CHO + H_2O$$

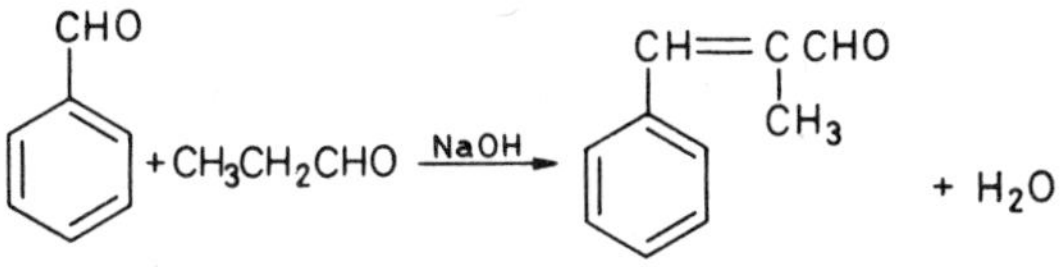

Historical development
The reaction is a modification of the aldol condensation introduced by Wurtz[1] in 1872

$$CH_3-CH=O + CH_3-CH=O \longrightarrow CH_3-CH(OH)-CH_2-CHO$$
$$\text{Aldol}$$

Schmidt[2] established that the use of aqueous sodium hydroxide to bring about the condensation causes the loss of a molecule of water and formation of a double bond. This led Claisen and his colleagues[3] to carry out an extensive study of the reaction which clarified some of the previously ambiguous results

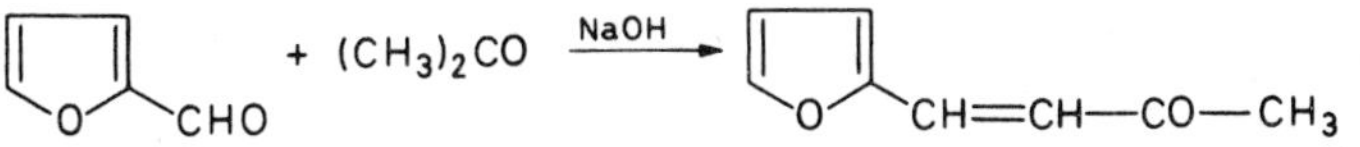

$$\text{(furfural)} + (CH_3)_2CO \xrightarrow{NaOH} \text{(furyl)}-CH=CH-CO-CH_3$$

Yields from the reaction are usually >60 per cent and frequently attain 80 per cent.

Mechanism

The Claisen–Schmidt condensation is often considered, from a mechanistic point of view, in conjunction with related reactions such as the Knoevenagel condensation[4] (No. 16). It is generally represented[5] as being initiated by attack of an hydroxyl group on the aliphatic carbonyl compound, forming a carbanion

$$HO^- + HCHRCHO \rightleftharpoons H_2O + {}^-CHRCHO$$

$$Ar\overset{O}{\overset{\|}{C}}H + {}^-CHRCHO \rightleftharpoons Ar\overset{O^-}{\overset{|}{C}}HCHRCHO \underset{H_2O}{\rightleftharpoons} Ar\overset{OH}{\overset{|}{C}}H-\overset{-}{C}HRCHO + \overset{-}{O}H$$

$$\rightleftharpoons Ar-\overset{OH}{\overset{|}{C}}H-\overset{-}{C}RCHO + H_2O \longrightarrow ArCH=CRCHO + OH^-$$

General reaction conditions

Numerous examples of the condensation appear in the standard preparative publications[6, 7].

The reaction is often carried out by mixing the two carbonyl compounds, with stirring, in water. A 10 per cent aqueous solution of sodium hydroxide is added to the mixture at a rate such that the temperature is maintained below 30°, external cooling being frequently required. Stirring is continued for several hours after addition is complete. The mixture is made just acid with hydrochloric acid, causing the product to separate out.

Using conditions similar to these, Conrad and Dolliver[8] employed appropriate molar proportions to form 1,5-diphenyl-pentan-3-one from benzaldehyde and acetone

$$2\,C_6H_5\text{-CHO} + CH_3COCH_3 \xrightarrow{NaOH} C_6H_5\text{-CH=CH-CO-CH=CH-}C_6H_5 + 2H_2O$$

Applications

Amongst the useful compounds which may be prepared by the Claisen–Schmidt condensation are the antibacterial chalcones. The parent substance, chalcone, is the product of a condensation between benzaldehyde and acetophenone[9]

$$C_6H_5\text{-CHO} + CH_3CO\text{-}C_6H_5 \xrightarrow{NaOH} C_6H_5\text{-CH=CH-CO-}C_6H_5 + H_2O$$

CLAISEN–SCHMIDT CONDENSATION (CLAISEN REACTION)

The formation of chalcones has been studied in great detail by Davey and co-workers[10]. Several different approaches for preparations were studied and optimum conditions established. Treatment of acetophenone and terephthalaldehyde in methanol with methanolic sodium hydroxide gave 96 per cent yields of 'dichalcone'.

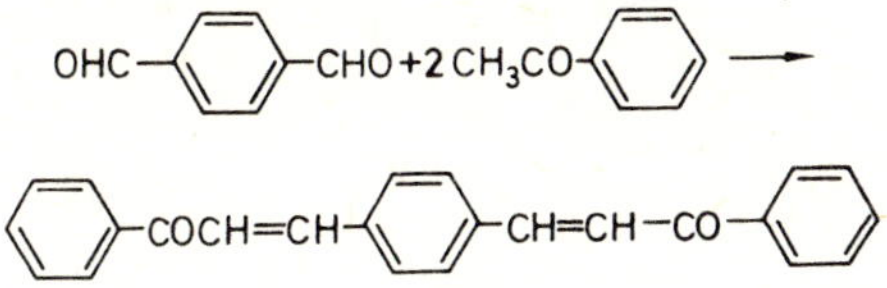

$$OHC-\underset{}{\bigcirc}-CHO + 2\,CH_3CO-\underset{}{\bigcirc} \longrightarrow$$

$$\underset{}{\bigcirc}-COCH=CH-\underset{}{\bigcirc}-CH=CH-CO-\underset{}{\bigcirc}$$

REFERENCES

[1] Wurtz, A. *C. r. hebd. Séanc. Acad. Sci., Paris* 74 (1872) 1361; 76 (1873) 1165 [*J. chem. Soc.* (*A*) 25 (1872) 808; 26 (1873) 876]; Nielsen, A. T. and Houlihan, W. J. *Org. React.* 16 (1968) 3
[2] Schmidt, J. G. *Ber. dt. chem. Ges.* 13 (1880) 2342; 14 (1881) 1459 [*J. chem. Soc.* (*A*) 40 (1881) 247, 889]
[3] Claisen, L., with Claparede, A. *Ber. dt. chem. Ges.* 14 (1881) 359; with Ponder, A. C. *Justus Liebigs Annln Chem.* 223 (1884) 137 [*J. chem. Soc.* (*A*) 40 (1881) 422; 46 (1884) 1166]
[4] House, H. O. *Modern Synthetic Reactions*, p. 218, New York (Benjamin) 1965
[5] Finar, I. L. *Organic Chemistry*, 4th edn., 1, 158, London (Longmans) 1964
[6] Drake, N. L. and Allen, P. *Org. Synth., Coll. Vol.* 1 (1932) 69
[7] Hill, G. A. and Bramann, G. M. *Org. Synth., Coll. Vol.* 1 (1932) 74
[8] Conrad, C. R. and Dolliver, M. A. *Org. Synth., Coll. Vol.* 2 (1943) 167
[9] Kohler, E. P. and Chadwell, H. M. *Org. Synth., Coll. Vol.* 1 (1932) 71
[10] Davey, W., with Guilt, J. R. *J. chem. Soc.* (1957) 1008; with Tivey, D. J. (1958) 1230; with Mass, D. H. (1963) 4386

16
KNOEVENAGEL CONDENSATION

Nature of the reaction

Condensation of aldehydes with compounds possessing active methylene groups in the presence of organic bases give α,β-unsaturated acids or esters

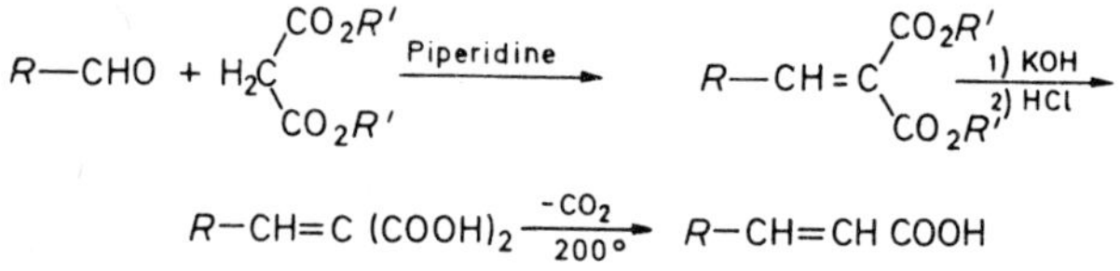

$$R-CHO + H_2C\begin{matrix}CO_2R'\\CO_2R'\end{matrix} \xrightarrow{\text{Piperidine}} R-CH=C\begin{matrix}CO_2R'\\CO_2R'\end{matrix} \xrightarrow[\text{2) HCl}]{\text{1) KOH}}$$

$$R-CH=C(COOH)_2 \xrightarrow[200°]{-CO_2} R-CH=CH\ COOH$$

Historical development

Knoevenagel[1] originally reported this reaction in connection with the condensation between formaldehyde and diethyl malonate and showed that a bis-condensation product was formed from which glutaric acid could be obtained

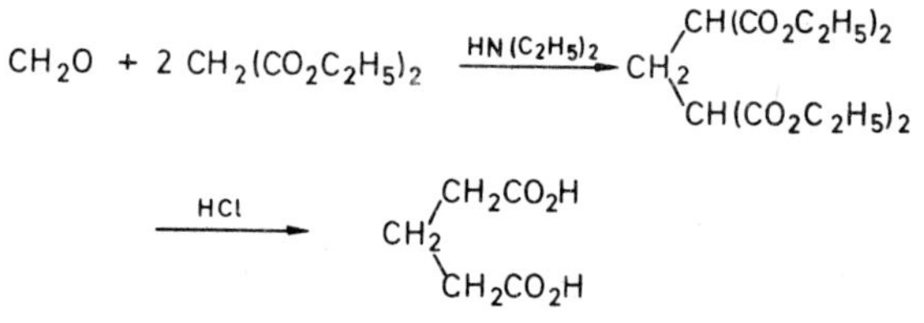

$$CH_2O + 2\ CH_2(CO_2C_2H_5)_2 \xrightarrow{HN(C_2H_5)_2} CH_2\begin{matrix}CH(CO_2C_2H_5)_2\\CH(CO_2C_2H_5)_2\end{matrix}$$

$$\xrightarrow{HCl} CH_2\begin{matrix}CH_2CO_2H\\CH_2CO_2H\end{matrix}$$

Further investigation[2] showed that, by altering the conditions, mono-condensation products could be produced, as with benzaldehyde and ethyl acetoacetate (95 per cent yield)

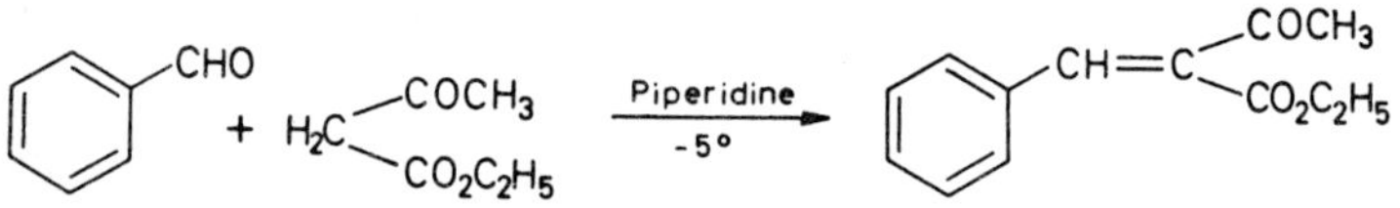

$$\text{PhCHO} + H_2C\begin{matrix}COCH_3\\CO_2C_2H_5\end{matrix} \xrightarrow[-5°]{\text{Piperidine}} \text{PhCH}=C\begin{matrix}COCH_3\\CO_2C_2H_5\end{matrix}$$

It was soon demonstrated that both aromatic and aliphatic aldehydes would undergo reaction in the presence of ammonia or primary or secondary amines. The use of pyridine as solvent and catalyst was introduced by Doebner[3] to enable a condensation

KNOEVENAGEL CONDENSATION

between crotonaldehyde and malonic acid to be effected; the refluxing temperature of the reaction automatically led to decarboxylation

$$CH_3CH=CHCHO + CH_2(CO_2H)_2 \xrightarrow[100°]{Pyridine} CH_3CH=CHCH=CHCOOH + CO_2 + H_2O$$

The Knoevenagel condensation has been reviewed[4] in conjunction with the Perkin reaction and several related processes. Recently, a detailed survey has been carried out by Jones[5].

Yields are usually high, 60–80 per cent, and may be even higher if the Doebner modification is employed.

Mechanism

The reaction has been conventionally represented as involving proton abstraction due to the base[6]

$$CH_2(CO_2C_2H_5)_2 + B \longrightarrow \bar{C}H(CO_2C_2H_5)_2 + B\overset{+}{H}$$

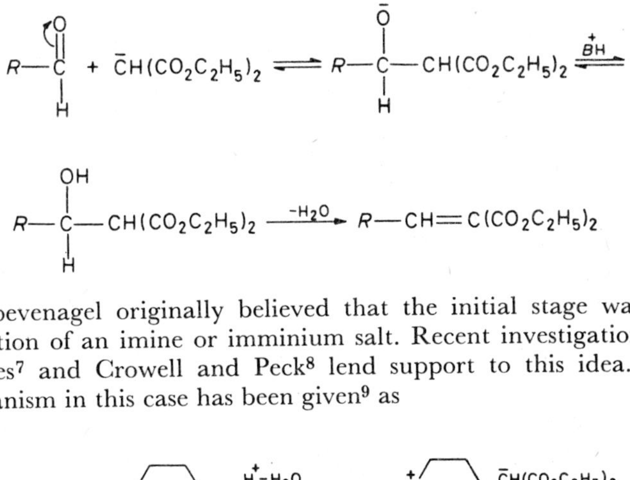

$$R-\underset{H}{\overset{O}{\underset{\|}{C}}} + \bar{C}H(CO_2C_2H_5)_2 \rightleftharpoons R-\underset{H}{\overset{\bar{O}}{\underset{|}{C}}}-CH(CO_2C_2H_5)_2 \xrightarrow{B\overset{+}{H}}$$

$$R-\underset{H}{\overset{OH}{\underset{|}{C}}}-CH(CO_2C_2H_5)_2 \xrightarrow{-H_2O} R-CH=C(CO_2C_2H_5)_2$$

Knoevenagel originally believed that the initial stage was the formation of an imine or imminium salt. Recent investigations by Charles[7] and Crowell and Peck[8] lend support to this idea. The mechanism in this case has been given[9] as

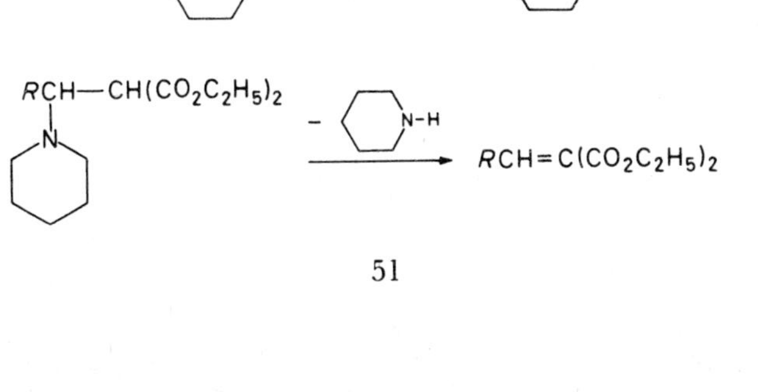

$$RCHO + H-N\bigcirc \xrightleftharpoons[]{H^+,-H_2O} RCH=\overset{+}{N}\bigcirc \xrightarrow{\bar{C}H(CO_2C_2H_5)_2}$$

$$RCH-CH(CO_2C_2H_5)_2 \atop N\bigcirc \xrightarrow{-\bigcirc N-H} RCH=C(CO_2C_2H_5)_2$$

NAMED ORGANIC REACTIONS

General reaction conditions

Most condensations are carried out under conditions of the Doebner modification. The aldehyde and a slight excess of malonic acid are dissolved in dry pyridine and a small quantity of piperidine added. The mixture is refluxed for as long as 3 h, until evolution of carbon dioxide is complete. The cooled solution is then poured into cold 10 per cent hydrochloric acid, from which the product may be filtered off and recrystallized.

A typical preparation of this form is that of sorbic acid from crotonaldehyde[10]

$$CH_3CH=CHCHO + H_2C(COOH)_2 \xrightarrow{Pyridine} CH_3CH=CHCH=CHCOOH + CO_2 + H_2O$$

Vogel[11] has given preparative details for a number of condensations under standard Knoevenagel conditions.

Modifications

To enable condensations of methylcyanoacetate to occur with ketones, Cope and co-workers[12] found ammonium and amine salts of organic acids to be better catalysts than free bases. By carrying out reactions in dry solvents and azeotroping off the water formed, yields were substantially increased.

Details of this procedure, with ammonium acetate catalyst in acetic acid–benzene solution, for the preparation of ethyl(1-ethylpropylidene)cyanoacetate have been given[13]

$$(C_2H_5)_2CO + \underset{CN}{CH_2-CO_2C_2H_5} \xrightarrow[CH_3COOH]{CH_3CO_2NH_4} (C_2H_5)_2C=\underset{CN}{C-CO_2C_2H_5} + H_2O$$

The mixture is heated at 160° for several hours until well after all water produced has been collected in a Dean and Stark trap. After removal of solvent, the product is isolated by forming a bisulphite addition product.

Patterson[14] employed ammonium acetate in pyridine to obtain a 75 per cent yield of 3-(2-furyl)acrylonitrile from furfural

$$\underset{CHO}{\text{furfuryl}} + \underset{CN}{CH_2-CO_2H} \xrightarrow[C_5H_5N]{CH_3CO_2NH_4} \underset{CH=CHCN}{\text{furyl}} + H_2O + CO_2$$

Applications

An interesting development has been the use of the Knoevenagel condensation to increase the chain length of sugars[15]. L-glycero-

KNOEVENAGEL CONDENSATION

L-galacto-heptose was synthesized from L-arabinose (protected by isopropylidene groups) by malonic acid in pyridine with piperidine

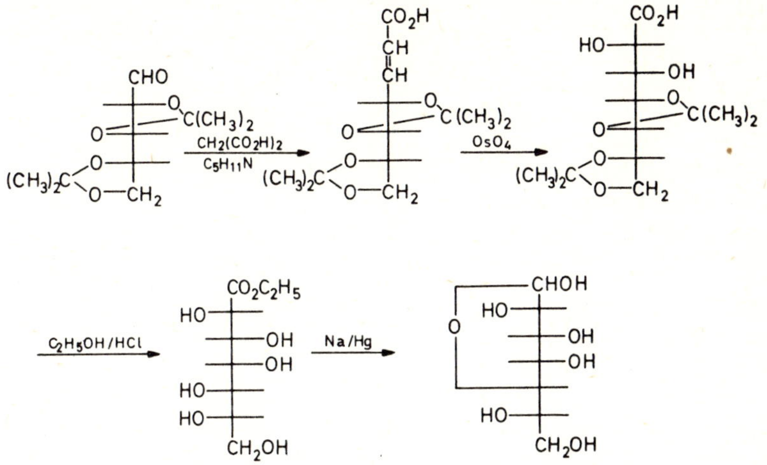

REFERENCES

[1] Knoevenagel, E. *Ber. dt. chem. Ges.* 27 (1894) 2345 [*J. chem. Soc. (A)* 66 (1894) [1] 570]
[2] Knoevenagel, E. *Ber. dt. chem. Ges.* 29 (1896) 172; 31 (1898) 730, 2596 [*J. chem. Soc. (A)* 70 (1896) [1] 232; 74 (1898) [1] 406; 76 (1899) [1] 144]
[3] Doebner, O. *Ber. dt. chem. Ges.* 33 (1900) 2140 [*J. chem. Soc. (A)* 78 (1900) [1] 536]
[4] Johnson, J. R. *Org. React.* 1 (1942) 233
[5] Jones, G. *Org. React.* 15 (1967) 204
[6] Finar, I. L. *Organic Chemistry*, 4th edn., 1, 280, London (Longmans) 1964
[7] Charles, G. *Bull. Soc. chim. Fr.* (1963) 1559, 1576 [*Chem. Abstr.* 60 (1964) 471,472]
[8] Crowell, T. I. and Peck, D. W. *J. Am. chem. Soc.* 75 (1953) 1075
[9] House, H. O. *Modern Synthetic Reactions*, p. 227, New York (Benjamin) 1965
[10] Allen, C. F. H. and Allen, J. van *Org. Synth., Coll. Vol.* 3 (1955) 783
[11] Vogel, A. I. *A Textbook of Practical Organic Chemistry*, 3rd edn., p. 711, London (Longmans) 1965
[12] Cope, A. C. *J. Am. chem. Soc.* 59 (1937) 2327; *et al.* 63 (1941) 3452
[13] Cope, A. C. and Hancock, E. M. *Org. Synth., Coll. Vol.* 3 (1955) 399
[14] Patterson, J. M. *Org. Synth.* 40 (1960) 47
[15] Kochetkov, N. K. and Dmitriev, B. A. *Chemy Ind.* (1963) 115

17

PERKIN REACTION

Nature of the reaction

Aromatic aldehydes condense with acid anhydrides possessing two hydrogen atoms on an α-carbon atom, in the presence of base to give β-aryl acrylic acids

$$ArCHO + (CH_3CO)_2O \xrightarrow{Base} ArCH{=}CHCOOH + CH_3COOH$$

Historical development

Perkin's[1] first condensation was of sodium salicylaldehyde with acetic anhydride and led to the formation of coumarin (the structure of which had not then been established)

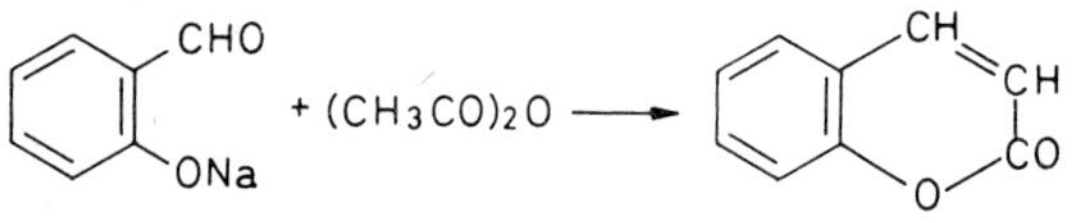

Further investigation resulted in the development of a standard procedure for preparing cinnamic acid and related compounds. With acid anhydrides other than acetic, the condensation still occurs on the α-carbon atom, giving rise to an α,β-disubstituted acrylic acid

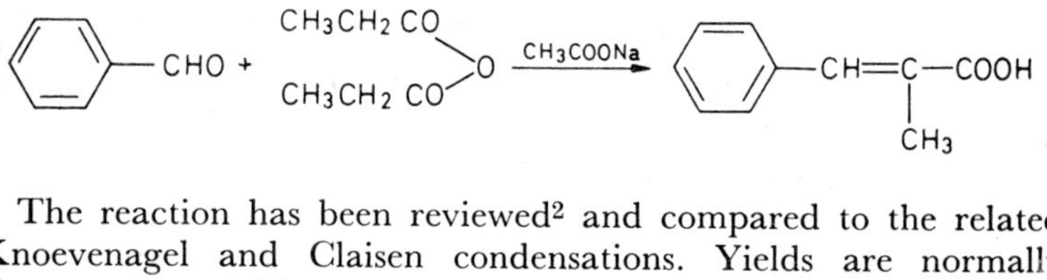

The reaction has been reviewed[2] and compared to the related Knoevenagel and Claisen condensations. Yields are normally between 55 and 80 per cent.

Mechanism

Much of the early work on the mechanism was contradictory or misinterpreted[3]. Breslow and Hauser[4] established that the sodium salt of the acid serves to abstract a proton from the α-methylene group of the anhydride. It is, therefore, the latter that undergoes the condensation rather than the acid radical. Since other catalysts, such as sodium carbonate and triethylamine, may be used in place of the sodium salt, the point has been well established.

The currently accepted mechanism[5] may be considered to be

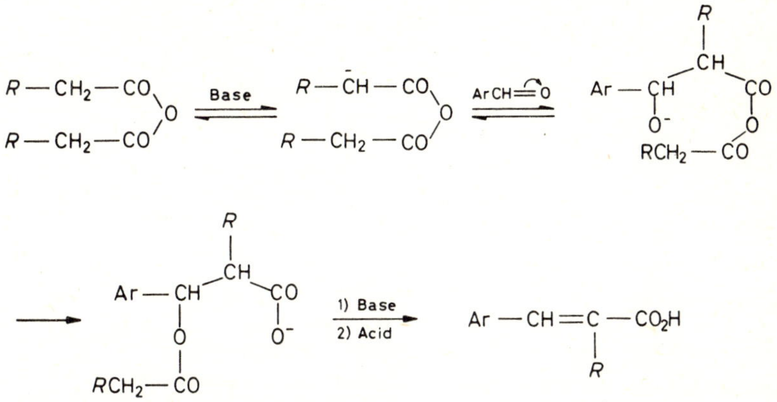

The reaction can theoretically give both *cis-* and *trans-*isomers in every condensation. In practice, the carboxyl and aromatic groups show a marked preference for the *trans* orientation.

General reaction conditions

The most straightforward preparations are carried out by mixing an aldehyde, acid anhydride, and base together, then heating between 170° and 200° for several hours. The hot mixture is poured into iced water and acidified to precipitate the product. A detailed preparation of this type for furylacrylic acid, given by Johnson[6], is of wide application

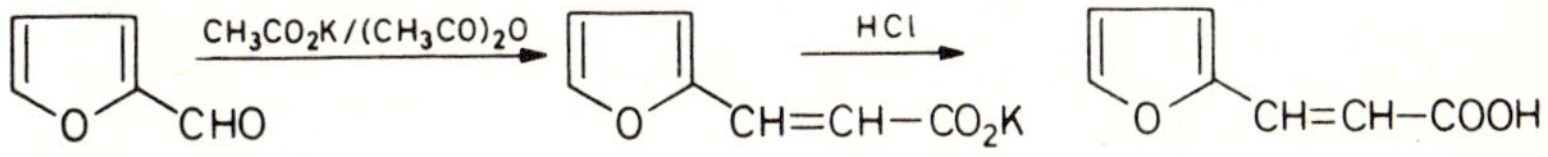

Modifications

It has been found that condensations may be carried out between aldehydes and carboxylic acids possessing α-hydrogen atoms if these are reacted in the presence of acetic anhydride and a base, the anhydride being sufficient to enable the mixture to be refluxed.

De Tar[7] used this procedure in the condensation of *o*-nitrobenzaldehyde with phenylacetic acid, with triethylamine as catalyst

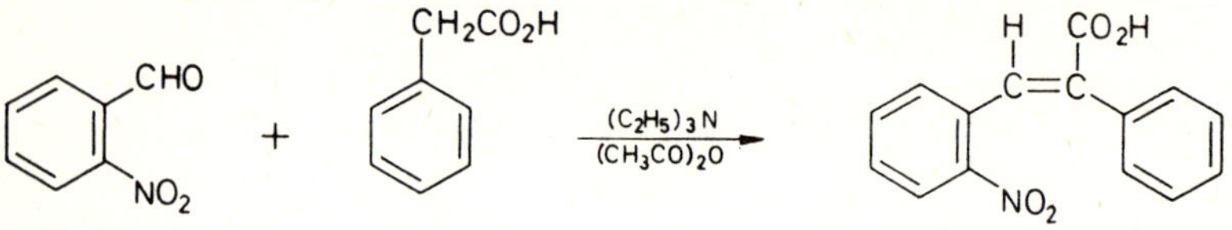

o-Nitro-α-phenyl cinnamic acid

Applications

The largest single application of the Perkin reaction has been as the first step in the 4-stage Pschorr synthesis of substituted phenanthrenes[8] in which substituted nitrobenzaldehydes are condensed with substituted phenylacetic acids

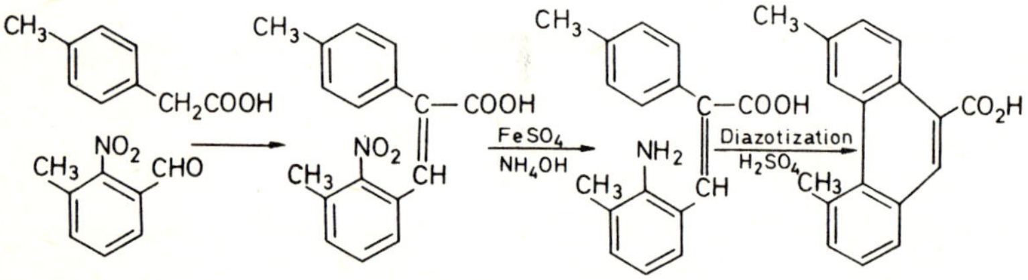

Ketcham[9] has used the modified method as basis for a students' project in the study of the *cis*- and *trans*-isomers formed by condensing *para*-substituted aldehydes with *para*-substituted phenylacetic acids in the presence of acetic anhydride and pyridine.

REFERENCES

[1] Perkin, W. H. *J. chem. Soc.* 21 (1868) 53, 181; 31 (1877) 388
[2] Johnson, J. R. *Org. React.* 1 (1942) 210
[3] Watson, H. B. *Rep. Prog. Chem.* 36 (1939) 210
[4] Breslow, D. S. and Hauser, C. R. *J. Am. chem. Soc.* 61 (1939) 786
[5] House, H. O. *Modern Synthetic Reactions*, p. 236, New York (Benjamin) 1965
[6] Johnson, J. R. *Org. Synth., Coll. Vol.* 3 (1955) 426
[7] De Tar, D. F. *Org. Synth., Coll. Vol.* 4 (1963) 730
[8] Leake, P. H. *Chem. Rev.* 56 (1956) 27
[9] Ketcham, R. *J. chem. Educ.* 41 (1964) 565

18

STOBBE CONDENSATION

Nature of the reaction

In the presence of basic catalysts, succinic esters condense with aldehydes or ketones to give alkylidenesuccinic esters

$$R\text{—CHO} + \begin{array}{c} CH_2CO_2R' \\ | \\ CH_2CO_2R' \end{array} \xrightarrow{NaOR'} \begin{array}{c} R\text{—CH}=C\text{—}CO_2R' \\ | \\ CH_2\text{—}CO_2H \end{array}$$

Historical development

In 1893, Stobbe[1] reported that his attempt to carry out a Claisen condensation between acetone and diethyl succinate had not given the expected β-diketo ester. Instead, the product was teraconic acid, produced by condensation of the ketone carbonyl with a methylene group in the ester molecule

$$\begin{array}{c} CH_3 \\ \diagdown \\ \diagup C=O \\ CH_3 \end{array} + \begin{array}{c} CH_2CO_2C_2H_5 \\ | \\ CH_2CO_2C_2H_5 \end{array} \xrightarrow{NaOC_2H_5} \begin{array}{c} CH_3 \\ \diagdown \\ \diagup C=C\text{—}CO_2C_2H_5 \\ CH_3 | \\ CH_2\text{—}CO_2H \end{array}$$

The reaction was rapidly extended[2] to condensations between aromatic aldehydes and diethyl succinate, leading to the formation of phenylitaconic acids

$$\text{Ph—CHO} + \begin{array}{c} CH_2CO_2C_2H_5 \\ | \\ CH_2CO_2C_2H_5 \end{array} \xrightarrow{NaOC_2H_5} \begin{array}{c} \text{Ph—CH}=C\text{—}CO_2C_2H_5 \\ | \\ CH_2\text{—}CO_2H \end{array}$$

During Stobbe's investigations[3] it was found that the main product from condensation is the half-ester, which has to be hydrolysed to give the free dicarboxylic acid. Reactions can also lead to the formation of disubstituted compounds, and these may be made to predominate by using excess carbonyl compound and base.

Although sodium ethoxide is most commonly employed as the base, potassium t-butoxide has often been used, and sodium hydride has also been found[4] of value in giving high yields.

Johnson and Daub[5] have given an extensive survey of the reaction, including details of condensations with substituted succinic esters. Yields vary greatly and, when low, can often be improved by use of a different base or reaction temperature; 50–90 per cent have been reported for both monoalkyl and arylidenesuccinic acids.

Mechanism

Stobbe believed the reaction to involve the intermediate formation of a paraconic ester, and chemical evidence supporting this has been reported[6]. A satisfactory mechanism including this intermediate and accounting for the loss of one of the ester groups during the process has been represented[7] as

$$\underset{R'}{\overset{R}{>}}C=O + \begin{array}{c}{}^-CHCO_2C_2H_5\\|\\CH_2CO_2C_2H_5\end{array} \longrightarrow \begin{array}{c}R_2C\!-\!CH\!-\!CO_2C_2H_5\\|\quad\quad|\\^-O\quad CH_2\\\diagdown C \diagup\\O^{\diagup\diagdown}OC_2H_5\end{array}$$

$$\longrightarrow \begin{array}{c}\overset{H}{|}\\R_2C\!-\!\overset{|}{C}\!-\!CO_2C_2H_5\\|\quad\;\,|\\O\quad CH_2\\\diagdown C\diagup\\\|\\O\end{array} \longrightarrow \begin{array}{c}R_2C\!=\!C\!-\!CO_2C_2H_5\\|\\CH_2CO_2^-\end{array}$$

Paraconic ester

General reaction conditions

Early reactions were carried out by maintaining a mixture of sodium ethoxide, diethyl succinate and the carbonyl compound in ether at low temperatures (below 10°) for several days. The mixture was then poured on to ice, acidified and the monoester extracted with extra ether.

Now it is more common to add the carbonyl compound and the diester in t-butanol to a refluxing solution of potassium t-butoxide in the same solvent. The mixture is refluxed for 1–2 h, cooled and acidified. Isolation of the product is achieved by concentrating the solution and extracting with ether.

Johnson and Schneider[8] employed a potassium t-butoxide procedure in their condensation of benzophenone with diethyl succinate (92–94 per cent yield)

$$(C_6H_5)_2CO + \begin{array}{c}CH_2CO_2C_2H_5\\|\\CH_2CO_2C_2H_5\end{array} \xrightarrow{KOC(CH_3)_3} \begin{array}{c}(C_6H_5)_2C\!=\!C\;CO_2C_2H_5\\|\\CH_2CO_2K\end{array}$$

$$\xrightarrow{HCl} \begin{array}{c}(C_6H_5)_2C\!=\!C\;CO_2C_2H_5\\|\\CH_2CO_2H\end{array}$$

STOBBE CONDENSATION

The monoester obtained in this type of reaction is normally an oil or semicrystalline mass. For characterization it is often necessary to hydrolyse the ester to the dicarboxylic acid or to form the acid anhydride by treating with acetyl chloride[9].

Applications

The method has been employed to give a number of diarylidene compounds. Ayres and co-workers[10] prepared unsymmetrical compounds of this type by using two separate condensation steps. The first monoester was re-esterified to a diester prior to the second condensation

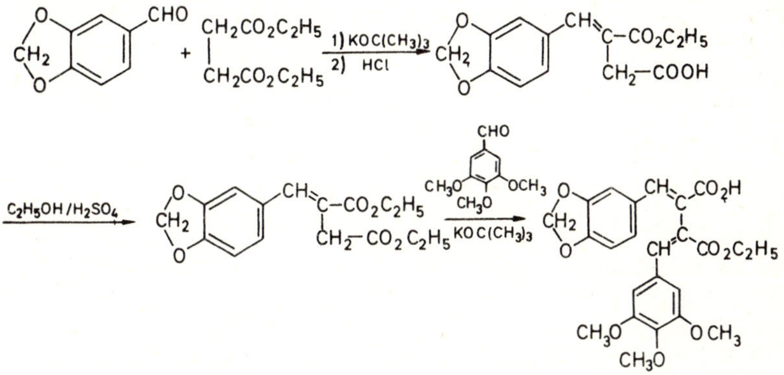

One great advantage of the Stobbe condensation is that the succinyl group can be used to form a ring system. Johnson and Stromberg[11] used it in this manner to form 14,15-dehydroequilenin methyl ether as an intermediate in the synthesis of the hormone equilenin

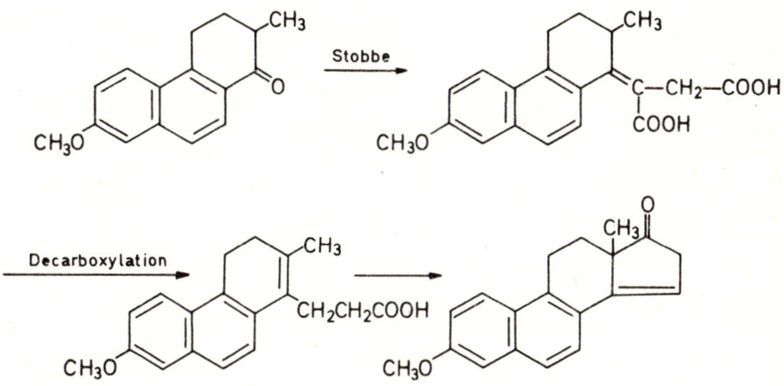

NAMED ORGANIC REACTIONS

Martin and his collaborators[12] have reported a series of Stobbe condensations between ketones and the monophosphonate analogue of diethyl succinate (diethyl β-carbethoxyethylphosphonate)

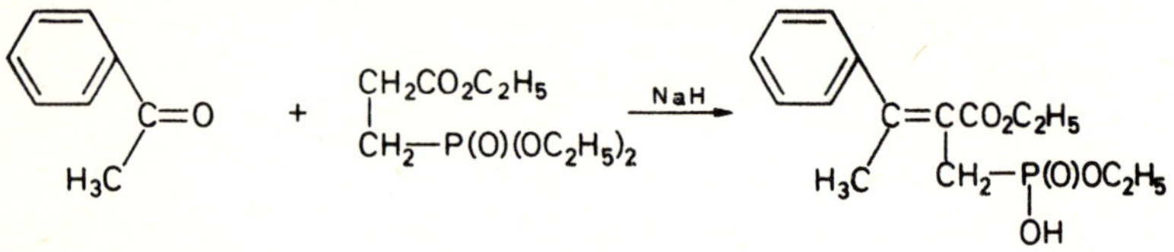

$$\underset{H_3C}{\overset{C_6H_5}{\diagdown}}C=O \;+\; \underset{CH_2-P(O)(OC_2H_5)_2}{\overset{CH_2CO_2C_2H_5}{|}} \xrightarrow{NaH} \underset{H_3C}{\overset{C_6H_5}{\diagdown}}C=\underset{CH_2-\underset{OH}{\overset{|}{P(O)OC_2H_5}}}{\overset{CCO_2C_2H_5}{|}}$$

All products were exclusively the β,γ-unsaturated phosphonates, no condensations on the α-methylene being detected.

REFERENCES

[1] Stobbe, H. *Ber. dt. chem. Ges.* 26 (1893) 2312 [*J. chem. Soc.* (A) 66 (1894) [1] 15

[2] Stobbe, H. and Kloeppel E. *Ber. dt. chem. Soc.* 27 (1894) 2405 [*J. chem. Soc.* (A) 66 (1894) [1] 594]

[3] Stobbe, H., Ljungren, G. and Freyberg, J. *Ber. dt. chem. Ges.* 59B (1926) 265 [*Chem. Abstr.* 20 (1926) 1796]

[4] Daub, G. H. and Johnson, W. S. *J. Am. chem. Soc.* 72 (1950) 501

[5] Johnson, W. S. and Daub, G. H. *Org. React.* 6 (1951) 1

[6] Johnson, W. S., McCloskey, A. L. and Dunnigan, D. A. *J. Am. chem. Soc.* 72 (1950) 514

[7] House, H. O. *Modern Synthetic Reactions*, p. 238, New York (Benjamin) 1965

[8] Johnson, W. S. and Schneider, W. P. *Org. Synth., Coll. Vol.* 4 (1963) 132

[9] Baddar, F. G., El-Assal, L. S. and Gindy, M. *J. chem. Soc.* (1948) 1270

[10] Ayres, D. C., Carpenter, B. G. and Denney, R. C. *J. chem. Soc.* (1965) 3578

[11] Johnson, W. S. and Stromberg, V. L. *J. Am. chem. Soc.* 72 (1950) 505

[12] Martin, D. J., Gordon, M. and Griffin, C. E. *Tetrahedron* 23 (1967) 1831

19

WITTIG REACTION

Nature of the reaction
Alkenes are formed by reacting carbonyl compounds with methylenetriphenylphosphorane under mild, alkaline conditions

$$\underset{R'}{\overset{R}{>}}C=O \;+\; (C_6H_5)_3P=CHR'' \longrightarrow \underset{R'}{\overset{R}{>}}C=CHR'' \;+\; (C_6H_5)_3P\rightarrow O$$

Historical development
   The Wittig reaction has had a profound effect upon organic chemistry during recent years. The procedure has developed from the reaction of methylenetriphenylphosphorane with benzophenone, first described by Wittig and Geissler[1] in 1953

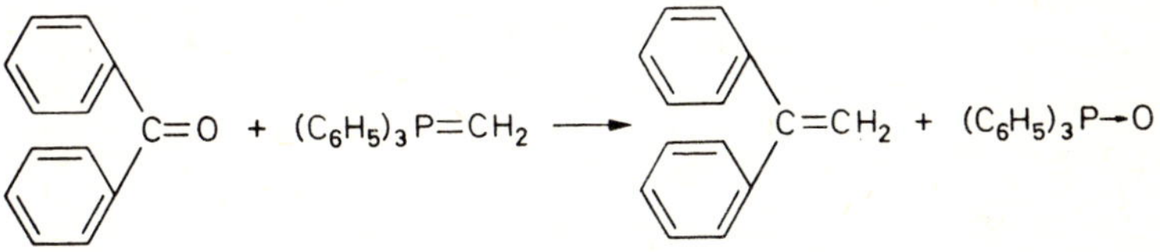

$$\underset{\phi}{\overset{\phi}{>}}C=O \;+\; (C_6H_5)_3P=CH_2 \longrightarrow \underset{\phi}{\overset{\phi}{>}}C=CH_2 \;+\; (C_6H_5)_3P\rightarrow O$$

   Within a short time, Wittig and his collaborators[2] developed the method into a well-established route for preparing alkenes. The alkylidene phosphoranes, known as Wittig reagents, are prepared[3] by treating alkyl triphenylphosphonium halides with methyl or phenyl lithium

$$R-\underset{\underset{H}{|}}{C}HX \xrightarrow{(C_6H_5)_3P} \left[R-\underset{\underset{H}{|}}{C}H-\overset{+}{P}(C_6H_5)_3 \right] X^- \xrightarrow{CH_3Li} R-\underset{\underset{}{|}}{\overset{R'}{C}}=P(C_6H_5)_3$$

   Even with complex carbonyl molecules, alkenes are formed in good yields, usually above 60 per cent and frequently almost quantitative. The main value of the reaction, to give a specific product with a high degree of purity, has led to the process being extensively studied. Numerous reviews have appeared[4], that of Maercker[5] covering the literature to 1963.

Mechanism

Some aspects of the Wittig reaction are still being investigated, but the nature of the process seems well established[6].

The phosphorane (also known as an ylide) adds to the polarized carbonyl group to produce an intermediate, called a betaine, which decomposes to give the alkene and the phosphine oxide

$$\begin{array}{c}R\\R'\end{array}\!\!>\!C=\bar{O} \qquad \begin{array}{c}R\\R'\end{array}\!\!>\!C-\bar{O} \qquad \begin{array}{c}R\\R'\end{array}\!\!>\!C$$
$$\begin{array}{c}R''\\R'''\end{array}\!\!>\!\bar{C}-\overset{+}{P}(C_6H_5)_3 \rightleftharpoons \begin{array}{c}R''\\R'''\end{array}\!\!>\!C-\overset{+}{P}(C_6H_5)_3 \longrightarrow \begin{array}{c}R''\\R'''\end{array}\!\!>\!C \quad + \quad \overset{O}{\underset{\downarrow}{P}}(C_6H_5)_3$$

Betaine

Such a mechanism can give rise to a mixture of *cis-* and *trans-*isomers. Current investigations on the mechanism are mainly concerned with the factors influencing the ratio of isomers formed[7]. Jones and Trippett[8] have studied the formation of *cis-* and *trans-*stilbene and obtained results consistent with reversible betaine formation.

General reaction conditions

The complete procedure involves three stages:
(*a*) formation of phosphonium salts;
(*b*) preparation of the phosphorane;
(*c*) reaction of the phosphorane with a carbonyl compound.

Phosphonium salts are obtained by reacting alkyl halides with triphenyl phosphine. In most cases this is achieved by warming the reagents together in a solvent. Alternatively, the two compounds may be fused together to give the required salt.

The phosphorane is prepared by suspending the dry phosphonium salt in tetrahydrofuran and adding a molecular amount of phenyl lithium in the same solvent. The appearance of an orange-red coloration is an indication of ylide formation. The ylide is not usually isolated but employed *in situ* for elaboration to the alkene. A carbonyl compound is stirred into the solution, forming the betaine which is then decomposed to the alkene by warming the mixture.

Many well established procedures are fully documented in the standard literature. Methylenecyclohexane has been obtained[9], with 40 per cent yields, from methyl bromide and cyclohexanone

WITTIG REACTION

$(C_6H_5)_3P + CH_3Br \longrightarrow (C_6H_5)_3P^+CH_3Br^- \xrightarrow{H_4H_9Li} (C_6H_5)_3P=CH_2$

[cyclohexanone] $\longrightarrow$ [cyclohexylidene]$=CH_2 + (C_6H_5)_3P\rightarrow O$

A procedure employing potassium t-butoxide in place of the more common lithium alkyl, has been given by Speziale and co-workers[10]

$(C_6H_5)_3P + HCCl_3 \xrightarrow{(CH_3)_3COK} (C_6H_5)_3P=CCl_2$

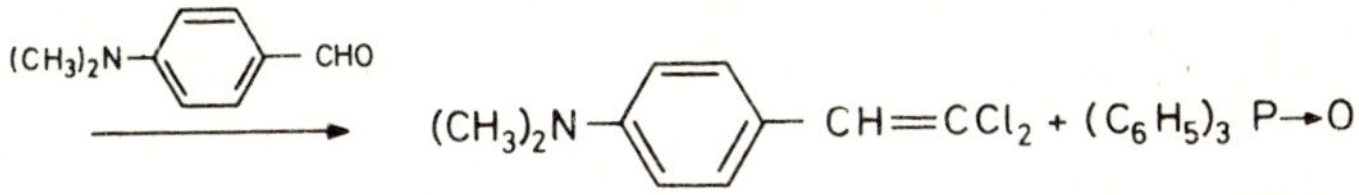

$(CH_3)_2N-\text{[C}_6\text{H}_4\text{]}-CHO \longrightarrow (CH_3)_2N-\text{[C}_6\text{H}_4\text{]}-CH=CCl_2 + (C_6H_5)_3P\rightarrow O$

β,β -Dichloro -p- dimethylaminostyrene

Applications

The Wittig reaction has been used in the synthesis of a number of natural products. Isler *et al.*[11] have reported the preparation of various carotenoids by procedures similar to that shown for lycopene

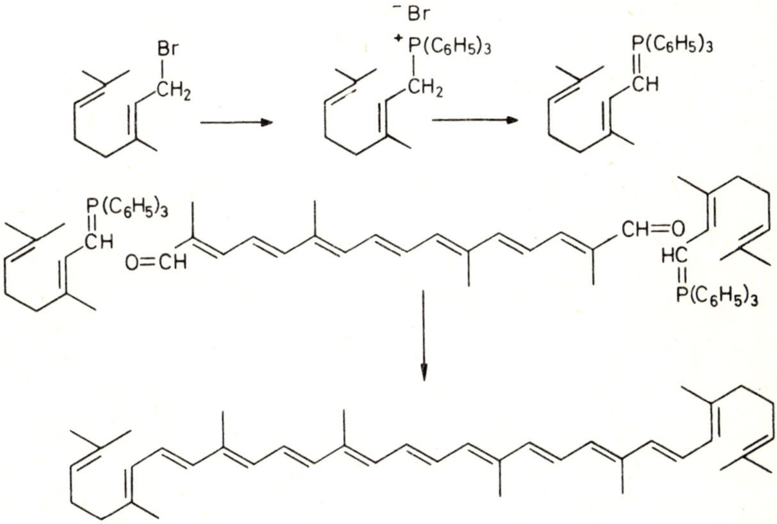

63

NAMED ORGANIC REACTIONS

The method has been applied to steroids to replace carbonyl groups by methylene groups[12], as in the formation of 3-methylene-$\triangle^4$-cholestene from $\triangle^4$-cholesten-3-one. In these reactions the ylide employed is triphenylphosphine-methylene obtained from methyltriphenylphosphonium bromide.

REFERENCES

[1] Wittig, G. and Geissler, G. *Justus Liebigs Annln Chem.* 580 (1953) 44 [*Chem. Abstr.* 48 (1954) 7566]

[2] Wittig, G., with Schöllkopf, U. *Ber. dt. chem. Ges.* 87 (1954) 1318; with Haag, A. 96 (1963) 1535 [*Chem. Abstr.* 49 (1955) 13926; 59 (1963) 6436]

[3] Wittig, G. and Rieber, M. *Justus Liebigs Annln Chem.* 562 (1949) 187 [*Chem. Abstr.* 44 (1950) 562]

[4] Schöllkopf, U. *Newer Methods of Preparative Organic Chemistry* (ed. Foerst, W.), 3, 111, New York (Academic Press) 1964; Trippett, S. *Advances in Organic Chemistry. Methods and Results* (ed. Raphael, R. A.), 1, 83, New York (Interscience) 1960

[5] Maercker, A. *Org. React.* 14 (1965) 270

[6] House, H. O. *Modern Synthetic Reactions*, p. 245, New York (Benjamin) 1965

[7] Ayres, D. C. *Carbanions in Synthesis*, p. 115, London (Oldbourne Press) 1966

[8] Jones, M. E. and Trippett, S. *J. chem. Soc.* C (1966) 1090

[9] Wittig, G. and Schöllkopf, U. *Org. Synth.* 40 (1960) 66

[10] Speziale, A. J., Ratts, K. W. and Bissing, D. E. *Org. Synth.* 45 (1965) 33

[11] Isler, O. *et al.*, *Helv. chim. Acta* 39 (1956) 463 [*Chem. Abstr.* 50 (1956) 15471]

[12] Sondheimer, F. and Mechonlam, R. *J. Am. chem. Soc.* 79 (1957) 5029

20

CLAISEN REARRANGEMENT

Nature of the reaction

The action of heat upon enol or phenol allyl ethers leads to an intramolecular migration of the allyl group

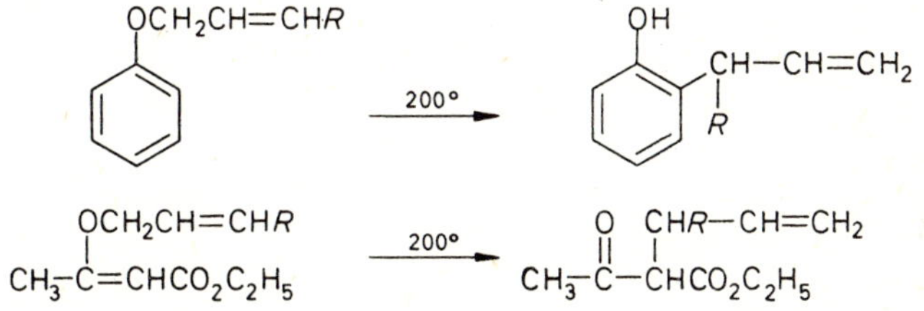

$$\underset{}{\text{C}_6\text{H}_5\text{OCH}_2\text{CH}=\text{CHR}} \xrightarrow{200°} \underset{R}{\text{o-HOC}_6\text{H}_4\text{CH(R)CH}=\text{CH}_2}$$

$$\text{CH}_3\text{-C(OCH}_2\text{CH}=\text{CHR})=\text{CHCO}_2\text{C}_2\text{H}_5 \xrightarrow{200°} \text{CH}_3\text{-CO-CH(CHR-CH}=\text{CH}_2)\text{CO}_2\text{C}_2\text{H}_5$$

Historical development

In 1912, Claisen[1] reported that the allyl ether of 2-naphthol could be rearranged to 1-allyl-2-naphthol by heating at 210°. The reaction, occurring without the use of catalysts, was found to be of wide application.

Tarbell[2], who has reviewed it, pointed out that in general the rearrangement is possible when there is an arrangement of atoms of the type

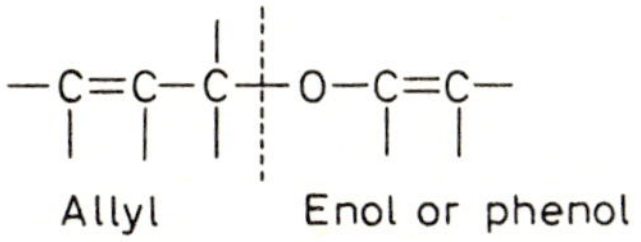

$$\underset{\text{Allyl}}{-\text{C}=\text{C}-\text{C}-}\underset{\text{Enol or phenol}}{-\text{O}-\text{C}=\text{C}-}$$

In aromatic systems where the positions *ortho* to the ether are occupied, migration to the *para* position occurs.

Yields from the rearrangement are normally above 50 per cent and often >80 per cent.

Mechanism

The intramolecular nature of the Claisen rearrangement was demonstrated by Hurd and Schmerling[3]. Other workers[4] have contributed to the knowledge of substituent and solvent effects;

White et al.[5] showed that reactivity is increased if electron-donating groups are attached to the allyl group and the aromatic system.

The *ortho* rearrangement is represented as

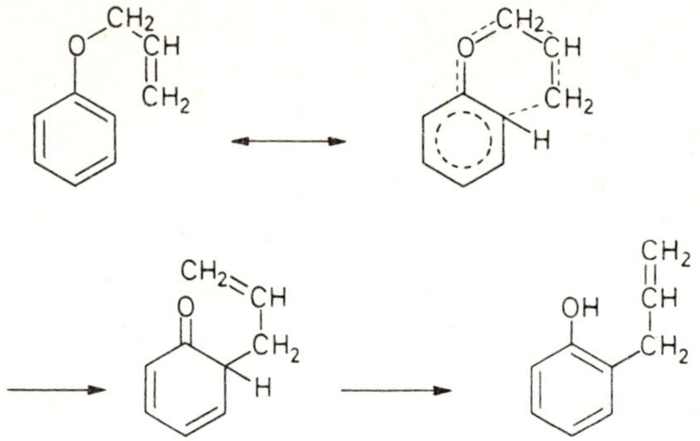

In the case of *para* rearrangements, Hurd and Pollack[6] have proposed a mechanism proceeding by a route analogous to the previous but undergoing a second rearrangement to give an intermediate quinone structure that enolizes to the *para*-phenol

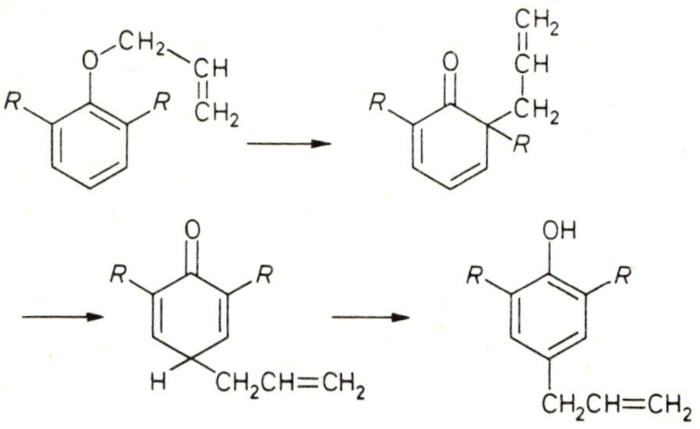

General reaction conditions

Allyl ethers are prepared by direct reaction between a phenol and an allyl halide in the presence of potassium carbonate and acetone. After filtering and removing the solvent, the crude ether is refluxed above 200° for several hours either by itself or in a high-boiling solvent such as dimethyl aniline or paraffin oil. The rearranged product is extracted by base and isolated by acidifying the aqueous solution.

CLAISEN REARRANGEMENT

o-Eugenol has been obtained with 80–90 per cent yields by the Claisen rearrangement of guiacol allyl ether[7]

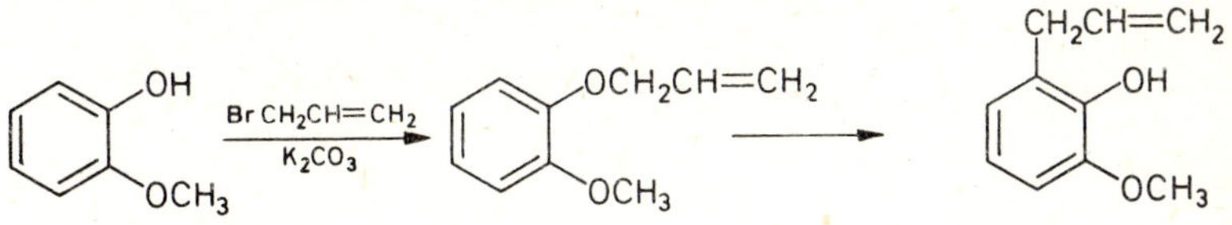

Recent developments

A great deal of interest is now being shown in the stereochemical features of the rearrangement. Marvell and co-workers[8] have studied the *cis*- and *trans*-isomers of α,γ-dimethylallyl ethers and found that the major product in each case was the *trans*-α,γ-dimethyl allyl phenol. Huestis and Andrews[9] have shown that for such isomeric ethers the *trans* normally rearranges more rapidly than does the *cis*. The results obtained support the accepted rearrangement mechanism.

Related to the Claisen rearrangement is the photochemical one of diaryl ethers observed by Kelly and Pinhey[10]. This process is still under investigation but probably proceeds by a free-radical mechanism

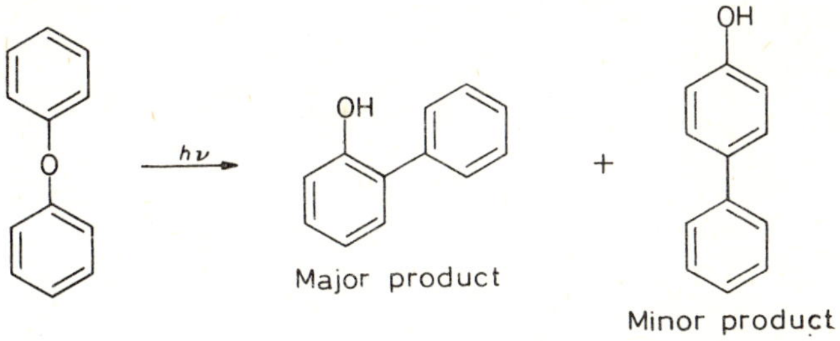

Major product   +   Minor product

REFERENCES

[1] Claisen, L. *Ber. dt. chem. Ges.* 45 (1912) 3157; with Tietze, E. 58 (1925) 275; 59 (1926) 2344 [*Chem. Abstr.* 7 (1913) 1016; 19 (1925) 1565; 21 (1927) 396]
[2] Tarbell, D. S. *Chem. Rev.* 27 (1940) 495; *Org. React.* 2 (1944) 1
[3] Hurd, C. D. and Schmerling, L. *J. Am. chem. Soc.* 59 (1937) 107
[4] Kalberer, F. and Schmid, H. *Helv. chim. Acta* 40 (1957) 13 [*Chem. Abstr.* 52 (1958) 299]; Jefferson, A. and Scheinmann, F. *Q. Rev. chem. Soc.* 22 (1968) 391
[5] White, W. N. *et al.*, *J. Am. chem. Soc.* 80 (1958) 3271; with Slater, C. D. and Fife, W. K. *J. org. Chem.* 26 (1961) 627

[6] Hurd, C. D. and Pollack, M. A. *J. org. Chem.* 3 (1939) 550
[7] Allen, C. F. H. and Gates, J. W. *Org. Synth., Coll. Vol.* 3 (1955) 418
[8] Marvell, E. N., Stephenson, J. L. and Ong, J. *J. Am. chem. Soc.* 87 (1965) 1267
[9] Huestis, L. D. and Andrews, L. J. *J. Am. chem. Soc.* 83 (1961) 1963
[10] Kelly, D. P. and Pinhey, J. T. *Tetrahedron Lett.* 46 (1964) 3427; with Rigby, J. D. G. 48 (1966) 5953

21

FRIES REARRANGEMENT

Nature of the reaction

Phenyl esters rearrange in the presence of metal halide catalysts to form *o*- or *p*-hydroxyketones

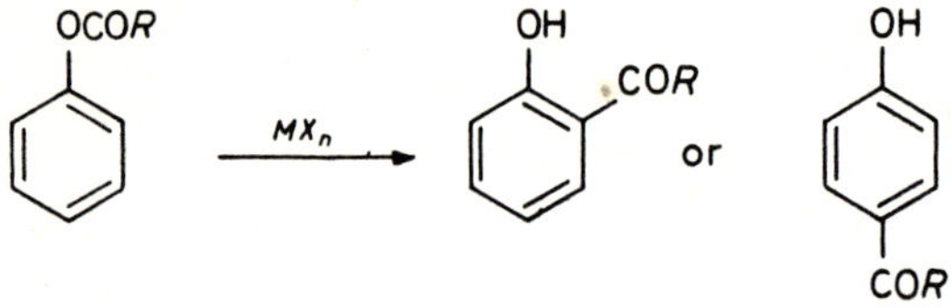

$$\text{PhOCOR} \xrightarrow{MX_n} \text{o-HOC}_6\text{H}_4\text{COR} \text{ or } \text{p-HOC}_6\text{H}_4\text{COR}$$

Historical development

The rearrangement was discovered by Fries and Finck[1] during their work on derivatives of cumaranone. When *p*-cresylchloroacetate was heated at 140° with aluminium chloride, it rearranged with a 90 per cent yield to 5-methyl-2-hydroxy-ω-chloroacetophenone

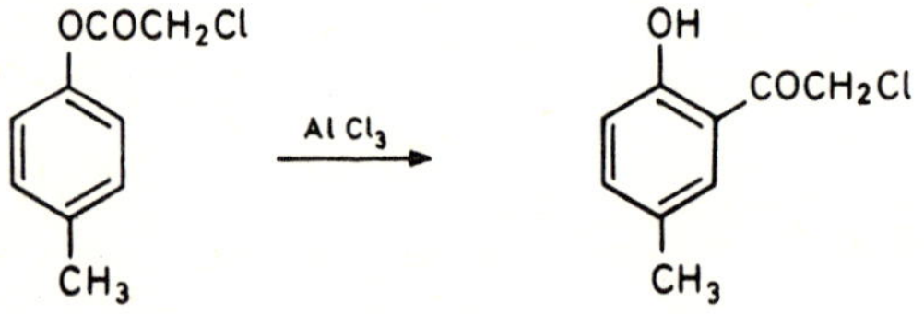

$$\text{4-CH}_3\text{-C}_6\text{H}_4\text{-OCOCH}_2\text{Cl} \xrightarrow{AlCl_3} \text{5-CH}_3\text{-2-HO-C}_6\text{H}_3\text{-COCH}_2\text{Cl}$$

FRIES REARRANGEMENT

Fries and his co-workers[2] subsequently demonstrated the generality of the reaction, including its applicability to naphthalene systems

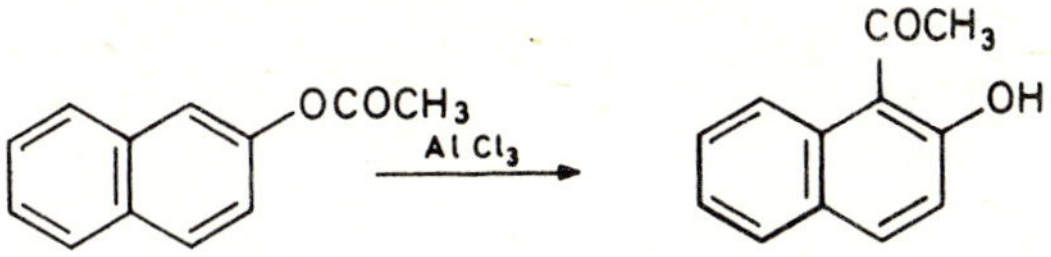

[reaction scheme: 2-naphthyl acetate + AlCl₃ → 1-acetyl-2-naphthol]

The rearrangement occurs more readily when the group R is aliphatic than when it is aromatic. Early work on the development of the rearrangement, reviewed by Blatt[3] in 1942, dealt with the variabilities of temperature, solvents and concentrations. Rearrangement to the *ortho* or *para* position depends upon both steric and temperature factors. When both positions are unsubstituted, a mixture of the two isomers will often result; if one is occupied, rearrangement to the vacant position occurs. At temperatures below 100° the *para* isomer is normally formed preferentially, but above this temperature the *ortho* isomer predominates. The nature of the metallic halide also affects the proportion of isomers. Total yields from the reaction are normally high, between 70 and 90 per cent, but in some cases separation of isomers is a major problem.

Mechanism

Investigators of the reaction have reported both inter- and intramolecular processes[4]. The intermolecular mechanism is believed to occur by electrophilic attack of the acyl cation formed by the action of the catalyst

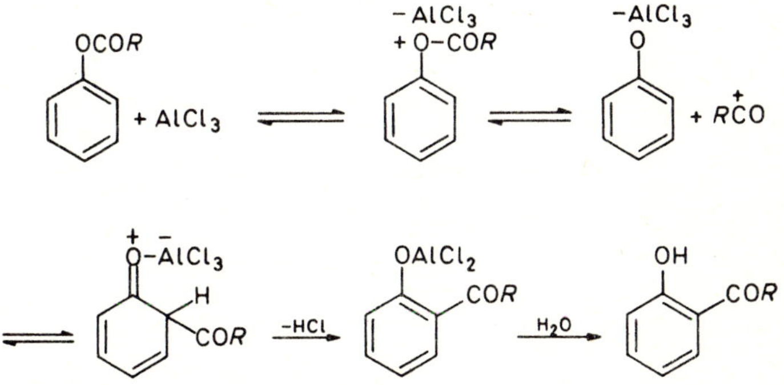

[mechanism scheme showing: ArOCOR + AlCl₃ ⇌ Ar-O⁺(COR)-AlCl₃⁻ ⇌ ArO-AlCl₃⁻ + RCO⁺ → cyclohexadienone intermediate → (−HCl) → o-OAlCl₂-aryl-COR → (H₂O) → o-hydroxyaryl ketone]

Ogata and Tabuchi[5] used tracer techniques to study the rearrangement of phenyl acetate to *o*- and *p*-hydroxyacetophenones and proposed the intramolecular mechanism for this reaction involving a π complex

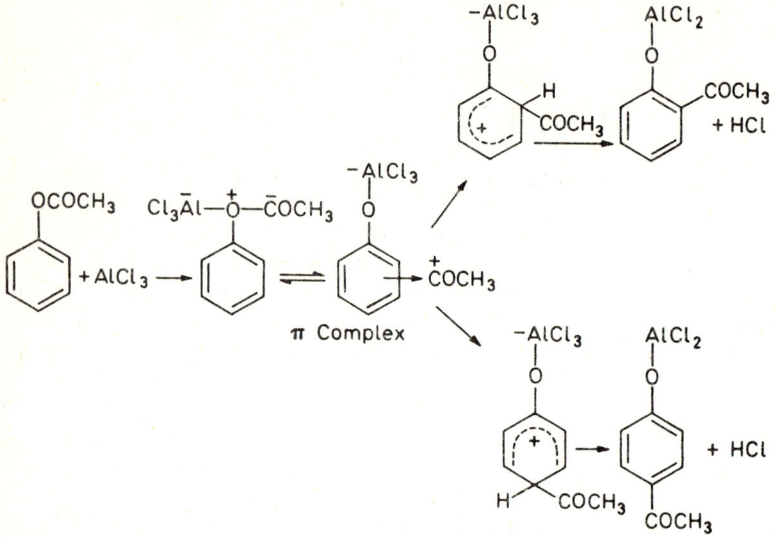

General reaction conditions

The Fries rearrangement is normally a straightforward procedure. Low-temperature reactions are usually carried out in carbon disulphide, high-temperature ones in nitrobenzene.

At least 1 mol of catalyst is stirred into the solvent and the ester added gradually at a rate that ensures steady refluxing. When addition is complete, the mixture is refluxed for several hours before being cooled and carefully treated with dilute hydrochloric acid. The product is isolated by filtration or extraction.

Applications

Miller and Hartung[6] used an aluminium chloride–carbon disulphide low-temperature approach to obtain a mixture of *o*- and *p*-hydroxypropiophenone from phenyl propionate, with a total yield of 90 per cent. Separation of isomers in this case presented no great difficulty, since the *p*-isomer readily crystallized out and the *o*-isomer was obtained from the mother liquors.

Russell and Frye[7], by comparison, used a large excess (3·5 mol) of aluminium chloride and no solvent to prepare 4-methyl-7-hydroxy-8-acetyl coumarin in 75 per cent yields from 4-methyl-7-acetoxycoumarin

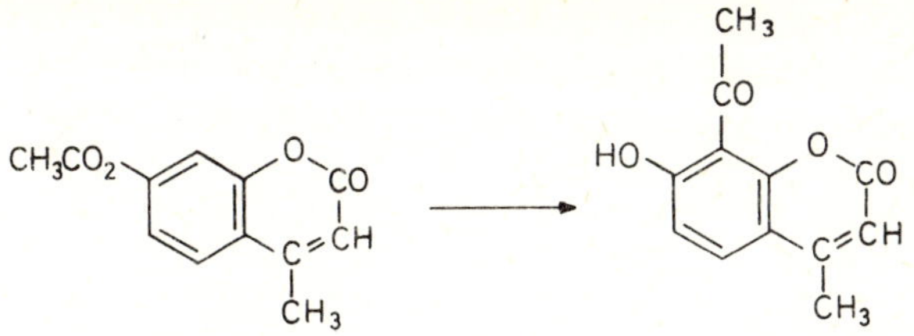

FRIES REARRANGEMENT

Price and Israelstam[8] have reported that Fries rearrangements can be carried out in media employing a cation-exchange resin at 100°. Esters prepared from phenols and carboxylic acids, or acid anhydrides, underwent spontaneous rearrangement in the presence of the heated resin

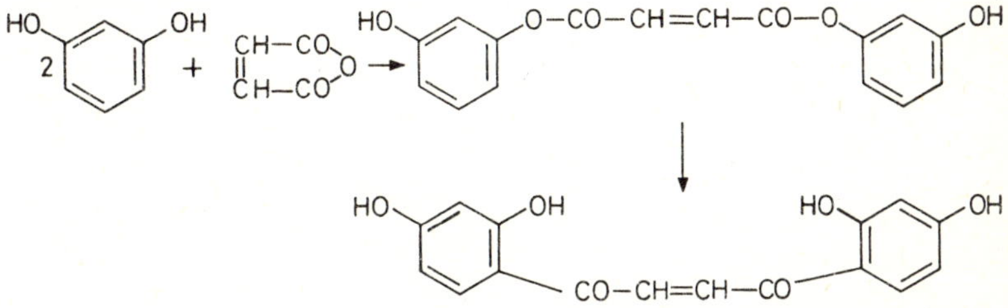

Physiologically active diphenyl sulphones were obtained by Thoi and Long[9] by rearranging substituted phenylbenzene sulphonates

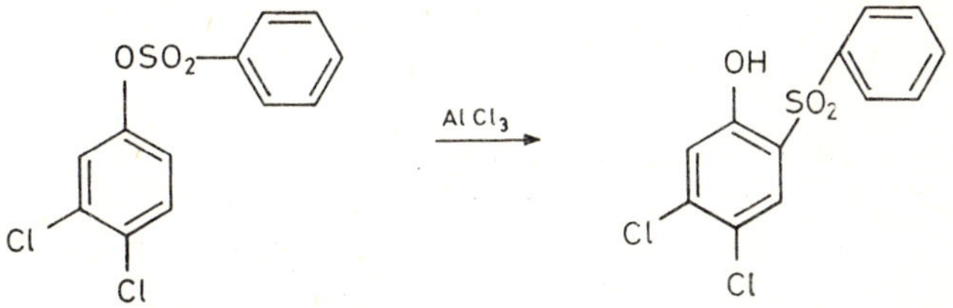

REFERENCES

[1] Fries, K. and Finck, G. *Ber. dt. chem. Ges.* 41 (1908) 4271 [*Chem. Abstr.* 3 (1909) 647]
[2] Fries, K. *Ber. dt. chem. Ges.* 54 (1921) 709; with Schimmelschmidt, K. 58 (1925) 2835 [*Chem. Abstr.* 15 (1921) 2865; 20 (1926) 1616]
[3] Blatt, A. H. *Org. React.* 1 (1942) 342; *Chem. Rev.* 27 (1940) 413
[4] Cullinane, N. M., Woolhouse, R. A. and Edwards, B. F. R. *J. chem. Soc.* (1961) 3842; Tarbell, D. S. and Fanta, P. E. *J. Am. chem. Soc.* 65 (1943) 2169
[5] Ogata, Y. and Tabuchi, H. *Tetrahedron* 20 (1964) 1661

[6] Miller, E. and Hartung, W. *Org. Synth., Coll. Vol.* 2 (1943) 543
[7] Russell, A. and Frye, J. R. *Org. Synth., Coll. Vol.* 3 (1955) 282
[8] Price, P. and Israelstam, S. S. *J. org. Chem.* 29 (1964) 2800
[9] Thoi, L. V. and Long, C. T. *Annls Fac. Sci. Univ. Saigon* (1962) 73 [*Chem. Abstr.* 62 (1965) 2730]

22

CRUM-BROWN–WALKER REACTION

Nature of the reaction

Potassium salts of monoalkyl esters of dibasic carboxylic acids are electrolysed in aqueous solution to give long-chain diesters of ω,ω'-dicarboxylic acids

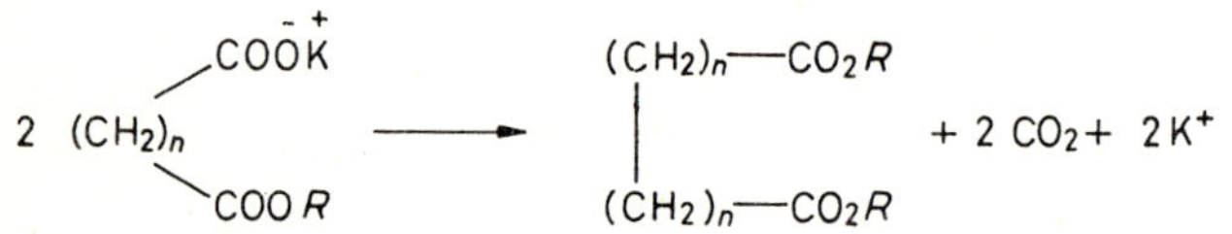

$$2 \ (CH_2)_n \begin{matrix} COO^-K^+ \\ \\ COOR \end{matrix} \longrightarrow \begin{matrix} (CH_2)_n-CO_2R \\ | \\ (CH_2)_n-CO_2R \end{matrix} + 2 CO_2 + 2 K^+$$

Historical development

Kolbe's synthesis[1] of paraffins by the electrolysis of carboxylic acids (No. 23) was the basis on which Crum-Brown and Walker[2] developed their procedure for the formation of long-chain dicarboxylic acids. They demonstrated the general applicability of their method by preparing all the even homologues of the dicarboxylic acids as far as octadecanedioic acid, $HO_2C(CH_2)_{16}CO_2H$

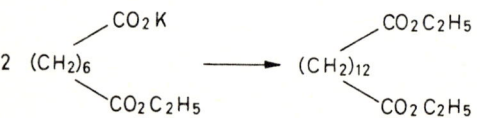

Potassium ethyl suberate   Diethyltetradecanedioate

Yields are usually above 50 per cent and in some cases attain 90 per cent.

CRUM-BROWN–WALKER REACTION

Mechanism

Weedon[3] has reviewed the electrolytic reactions which take place with carboxylic acids and considers the Crum-Brown–Walker reaction to be a special type of the Kolbe synthesis.

Hickling and Westwood[4] investigated the effect of variabilities such as acidity, temperature and current density in the process and concluded that hydrogen peroxide is the primary product of electrolysis at the anode. With ethyl malonate, the reaction occurring is believed to be

$$2\ \mathrm{CH_2}\begin{smallmatrix}CO_2C_2H_5\\ \\COO^-\end{smallmatrix} + H_2O_2 \longrightarrow \mathrm{CH_2}\begin{smallmatrix}CO_2C_2H_5\\ \\CH_2\\CO_2C_2H_5\end{smallmatrix} + 2\ CO_2 + 2\ \overline{O}H$$

Waters[5], however, considers the reaction to involve the formation of free radicals, so that it should be expressed as

$$\begin{smallmatrix}CO_2^-\\|\\CH_2COOC_2H_5\end{smallmatrix} \xrightarrow{-e} \begin{smallmatrix}O^{\cdot}\\\diagdown\\C\\\diagup\\O\\|\\CH_2COOC_2H_5\end{smallmatrix} \longrightarrow CO_2 + \dot{C}H_2{-}CO_2C_2H_5$$

$$2\ \dot{C}H_2{-}CO_2C_2H_5 \longrightarrow \begin{smallmatrix}CH_2CO_2C_2H_5\\|\\CH_2CO_2C_2H_5\end{smallmatrix}$$

General reaction conditions

The electrolysis is carried out in a large beaker, the temperature during the process being controlled between 0° and 50° by a cold-water cooling coil. The potassium ethyl dicarboxylate is electrolysed between platinum electrodes with a current of 3–10 A and a potential of 30–60 V. The desired product floats to the surface if an aqueous solution is employed. Other workers have successfully employed methanolic[6] and aqueous alcoholic solutions[7].

The detailed procedure for the preparation of diethyloctadecanedioate given by Swann and co-workers[8] is of general application.

REFERENCES

[1] Kolbe, H. *Justus Liebigs Annln Chem.* 69 (1849) 257 [*J. chem. Soc.* 2 (1849) 157]

[2] Crum-Brown, A and Walker, J. *Proc. R. Soc. Edinb.* 17 (1890) 54; *Trans. R. Soc. Edinb.* 18 (1891) 95; *Justus Liebigs Annln Chem.* 261 (1891) 107; 274 (1893) 41 [*J. chem. Soc.* (A) 58 (1890) 583; 60 (1891) 1192, 1193; 64 (1893) 394]

[3] Weedon, B. C. L. *Q. Rev.* **6** (1952) 383
[4] Hickling, A. and Westwood, J. V. *J. chem. Soc.* (1938) 1039
[5] Waters, W. A. *The Chemistry of Free Radicals*, p. 142, Oxford (Clarendon Press) 1950
[6] Bennett, G. M. and Gudgeon, H. *J. chem. Soc.* (1938) 1679
[7] Fairweather, D. A. *Proc. R. Soc. Edinb.* **45** (1925) 283 [*Chem. Abstr.* **20** (1926) 47]
[8] Swann, S., Oehler, R. and Pinkney, P. S. *Org. Synth., Coll. Vol.* **3** (1955) 401

23
KOLBE REACTION

Nature of the reaction

Hydrocarbons are formed by electrolysis of the alkali salts of aliphatic acids

$$2\ RCO_2K + 2H_2O \xrightarrow{\text{Electrolysis}} \underset{\text{At anode}}{R-R + 2CO_2} + \underset{\text{At cathode}}{2KOH + H_2}$$

Historical development

This electrolysis procedure devised by Kolbe[1] occurred with potassium acetate. The original work was precise and reported in considerable detail. The proportions of all products, including by-products, were listed.

$$2\ CH_3COO^- \longrightarrow CH_3-CH_3 + 2\ CO_2$$

Wurtz[2] extended the reaction to mixtures of fatty acids which gave both symmetrical and crossed coupling

$$\begin{array}{ll} RCO_2H & a)\ R-R \\ + & \longrightarrow b)\ R'-R' \\ R'CO_2H & c)\ R-R' \end{array}$$

Since then it has been extensively studied and the subject

reviewed[3]. Side reactions in the process include the formation of olefins, short-chain paraffins and alcohols

$$C_nH_{2n+1}COOH \longrightarrow \begin{array}{l} a) \; C_nH_{2n} \\ b) \; C_nH_{2n+2} \\ c) \; C_nH_{2n+1}OH \end{array}$$

The choice of electrolysis conditions can be extremely important, but under precisely determined conditions the Kolbe electrolysis procedure gives excellent yields, between 60 and 90 per cent.

Mechanism

Crum-Brown and Walker[4] were among the early investigators of the mechanism and concluded that carboxylate ions, RCO_2^-, are discharged at the anode. Eventually, on the basis of work by Walker[5] and Clusius[6], evidence for a radical mechanism was collated

$$RCO_2^- \longrightarrow RCO_2^{\bullet} + e$$
$$RCO_2^{\bullet} \longrightarrow R^{\bullet} + CO_2$$
$$2R^{\bullet} \longrightarrow R-R \qquad \text{Main product}$$
$$RCO_2^{\bullet} + R^{\bullet} \longrightarrow RCO_2R \qquad \text{By-product}$$
$$R^{\bullet} + R^{\bullet} \longrightarrow RH + \text{olefin} \qquad \text{By-products}$$

This is generally accepted, in preference to other mechanisms which assume the intermediate formation of peroxides.

The work of Eberson[7] shows that a second process can occur at the anode, giving rise to carbonium ions

$$R^{\bullet} \longrightarrow R^+ + e$$

The results obtained show that this happens when the ionization potential of the radical is less than 8 eV and can account for non-standard results in many electrolyses.

A detailed treatment of the various mechanistic theories has been given by Vijh and Conway[8] in their comprehensive review of the electrode kinetic aspects of the Kolbe reaction.

General reaction conditions

Electrolysis is carried out in a tall beaker with external cooling. Platinum electrodes placed up to a couple of centimetres apart are used and a high current density employed (particularly necessary for reactions in aqueous solutions) for periods greater than theoretically necessary[9].

Concentrated solutions of the sodium or potassium salts are employed with a small quantity of free carboxylic acid and electrolysis is carried out until the solution is alkaline. The hydrocarbon formed can then be separated from the electrolyte. In cases where salts of high-molecular-weight acids are insoluble in water, methanol may be successfully used instead.

The preparation of 2,7-dimethyl-2,7-dinitro-octane from 4-methyl-4-nitrovaleric acid has been described in detail by Sharkey and Langkammerer[10]. Using a current of 6–8 A at 50–60 V for 8 h and methanolic solution, a 55 per cent yield was obtained

$$O_2N-\underset{CH_3}{\underset{|}{\overset{CH_3}{\overset{|}{C}}}}-CH_2CH_2CO_2^- \longrightarrow O_2N\underset{CH_3}{\underset{|}{\overset{CH_3}{\overset{|}{C}}}}-CH_2CH_2^{\bullet} + CO_2 + e$$

$$2\ O_2N\underset{CH_3}{\underset{|}{\overset{CH_3}{\overset{|}{C}}}}-CH_2CH_2^{\bullet} \longrightarrow O_2N\underset{CH_3}{\underset{|}{\overset{CH_3}{\overset{|}{C}}}}-(CH_2)_4-\underset{CH_3}{\underset{|}{\overset{CH_3}{\overset{|}{C}}}}-NO_2$$

Similar conditions were employed by Swann and Garrison[11] to prepare dimethyl octadecanedioate from methyl sebacate

$$CH_3OCO(CH_2)_8CO_2^- \longrightarrow CH_3OCO(CH_2)_8^{\bullet} + CO_2 + e$$

$$2\ CH_3OCO(CH_2)_8^{\bullet} \longrightarrow CH_3OCO(CH_2)_{16}COOCH_3$$

Applications

Woolford[12] has used the Kolbe reaction to prepare long-chain ω,ω'-dibromohydrocarbons which can only be obtained with considerable difficulties by other methods. By using precise conditions, 55–70 per cent yields were recorded for chains possessing 10–22 carbon atoms

$$2\ Br(CH_2)_n COOH \longrightarrow Br(CH_2)_{2n}Br \qquad n = 5-11$$

REFERENCES

[1] Kolbe, H. *Justus Liebigs Annln Chem.* 69 (1849) 257 [Transl. *J. chem. Soc.* 2 (1849) 157]
[2] Wurtz, A. *Annls Chim. Phys.* 44 (1855) [3] 275
[3] Weedon, B. C. L. *Q. Rev.* 6 (1952) 380; *Adv. org. Chem.* 1 (1960) 1
[4] Crum-Brown, A. and Walker, J. *Justus Liebigs Annln Chem.* 261 (1891) 107; *Trans. R. Soc. Edinb.* 36 (1891) 291

[5] Shukla, S. N. and Walker, O. J. *Trans. Faraday Soc.* 27 (1931) 722; Fairweather, D. A. and Walker, O. J. *J. chem. Soc.* (1926) 3111
[6] Hölemann, P. and Clusius, K. *Z. phys. Chem.* 35B (1937) 361; Clusius, K. and Schanzer, W. 192A (1943) 237 [*Chem. Abstr.* 31 (1937) 4213; 38 (1944) 2275]
[7] Eberson, L. *Acta chem. scand.* 17 (1963) 2004; with Nyberg, K. 18 (1964) 1567 [*Chem. Abstr.* 60 (1964) 5074; 62 (1965) 2680]
[8] Vijh, A. K. and Conway, B. F. *Chem. Rev.* 67 (1967) 623
[9] Swann, S. *Technique of Organic Chemistry* (ed. Weissberger, A.), 2, 196, New York (Interscience) 1948
[10] Sharkey, W. H. and Langkammerer, C. M. *Org. Synth.* 41 (1961) 24
[11] Swann, S. and Garrison, W. E. *Org. Synth.* 41 (1961) 33
[12] Woolford, R. G. *Can. J. Chem.* 40 (1962) 1846

24

HELL–VOLHARD–ZELINSKY HALOGENATION

Nature of the reaction

The formation of α-halogeno acids by the action of halogen and phosphorus (or phosphorus halides) upon carboxylic acids

$$2P + 6\,RCH_2CO_2H + 3\,Br_2 \longrightarrow 6\,RCH_2COBr + 2\,H_3PO_3$$

$$RCH_2COBr + Br_2 \longrightarrow RCHCOBr + HBr$$
$$|$$
$$Br$$

$$RCHCOBr + H_2O \longrightarrow RCHCO_2H + HBr$$
$$||$$
$$BrBr$$

Historical development

The original procedure, reported by Hell[1], involved using sealed glass tubes. Volhard[2] modified the technique to enable reactions to be carried out under normal reflux conditions, at the same time as Zelinsky[3] used the approach to prepare large quantities of ethyl-α-bromopropionate with a 65 per cent yield

$$CH_3CH_2CO_2H \xrightarrow{P/Br_2} CH_3CH_2COBr \xrightarrow{Br_2} CH_3CHCOBr$$
$$\qquad\qquad\qquad\qquad\qquad\qquad\qquad\qquad\quad |$$
$$\qquad\qquad\qquad\qquad\qquad\qquad\qquad\qquad\quad Br$$

$$\xrightarrow{C_2H_5OH} CH_3CHCO_2C_2H_5$$
$$\qquad\qquad\quad |$$
$$\qquad\qquad\quad Br$$

The acid halide is formed owing to the presence of phosphorus trihalides on adding bromine to the phosphorus mixed with the carboxylic acid. In many cases, phosphorus is no longer employed; the carboxylic acid is treated instead with catalytic amounts of phosphorus trihalide. The method usually gives yields above 60 per cent.

Mechanism

It is apparent that formation of the acid halide leads to activation of the α-hydrogen atoms in the molecule. Hine[4] suggests that the bromination in the Hell–Volhard–Zelinsky method probably proceeds by a mechanism similar to the acid-catalysed halogenation of ketones

$$R-CH_2-C{\overset{X}{\underset{O}{\diagdown}}} \longrightarrow R-CH=C{\overset{X}{\underset{OH}{\diagdown}}} \xrightarrow{X_2} R-CH-C{\overset{X}{\underset{O}{\diagdown}}} + H^+ + X^-$$
$$\qquad\qquad\qquad\qquad\qquad\qquad\qquad\qquad\quad |$$
$$\qquad\qquad\qquad\qquad\qquad\qquad\qquad\qquad\quad X$$

As is suggested by this mechanism, the second α-hydrogen atom may similarly be replaced by halogen.

General reaction conditions

The carboxylic acid is mixed with phosphorus or phosphorus trihalide in a flask fitted with a reflux condenser. Bromine is then added gradually with stirring during about 1 h. When addition is complete, the mixture is heated at 90°–100° for as long as 20 h. Excess bromine and hydrogen bromide may be distilled off, and phosphorous acid usually separates out. The product may be purified by distillation, or isolated by hydrolysis to the acid or by esterification.

Bromine and phosphorus have been used to convert, with an 80 per cent yield, isobutyric acid to α-bromoisobutyryl bromide[5]

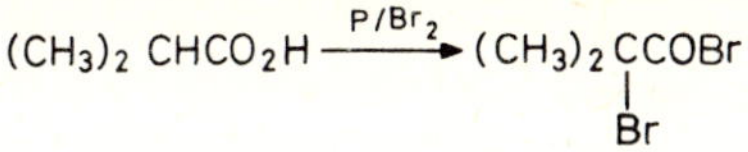

HELL–VOLHARD–ZELINSKY HALOGENATION

$$(CH_3)_2CHCO_2H \xrightarrow{P/Br_2} (CH_3)_2CCOBr$$
$$\phantom{(CH_3)_2CHCO_2H \xrightarrow{P/Br_2} (CH_3)_2C}|$$
$$\phantom{(CH_3)_2CHCO_2H \xrightarrow{P/Br_2} (CH_3)_2C}Br$$

Application

Marvel[6] has used the phosphorus trichloride–bromine approach to form α-bromo acids as intermediates in the preparation of amino acids

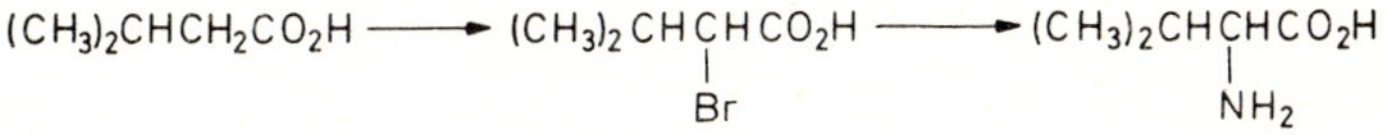

$$(CH_3)_2CHCH_2CO_2H \longrightarrow (CH_3)_2CHCHCO_2H \longrightarrow (CH_3)_2CHCHCO_2H$$
$$| |$$
$$Br NH_2$$

α-Bromoisovaleric acid   (±) Valine

REFERENCES

[1] Hell, C. *Ber. dt. chem. Ges.* 14 (1881) 891 [*J. chem. Soc.* (A) 40 (1881) 711]
[2] Volhard, J. *Justus Liebigs Annln Chem.* 242 (1887) 141 [*J. chem. Soc.* (A) 54 (1888) 129]
[3] Zelinsky, N. *Ber. dt. chem. Ges.* 20 (1887) 2026 [*J. chem. Soc.* (A) 52 (1887) 912]
[4] Hine, J. *Physical Organic Chemistry*, 2nd edn., p. 462, New York (McGraw-Hill) 1962
[5] Smith, C. W. and Norton, D. G. *Org. Synth., Coll. Vol.* 4 (1963) 348
[6] Marvel, C. S. *Org. Synth., Coll. Vol.* 3 (1955) 523, 848

25

WOHL–ZIEGLER REACTION

Nature of the reaction

N-bromo amides are used for selective bromination at carbon centres adjacent to unsaturated carbon–carbon systems without attacking the double bond; N-bromosuccinimide is the general reagent for this purpose

$$-CH=CH-CH_2- + BrN\begin{matrix}CO-\\CO-\end{matrix} \longrightarrow -CH=CH-CH- + HN\begin{matrix}CO-\\CO-\end{matrix}$$
$$\phantom{-CH=CH-CH_2- + BrN\begin{matrix}CO-\\CO-\end{matrix} \longrightarrow -CH=CH-CH}_{Br}$$

Historical development

Wohl found[1] that tetramethylethylene could be brominated on one of the methyl groups by using N-bromoacetamide; the reaction could also be applied to a number of other unsaturated compounds without the double bonds being affected

$$\begin{matrix}CH_3\\CH_3\end{matrix}C=C\begin{matrix}CH_3\\CH_3\end{matrix} + CH_3CONHBr \longrightarrow \begin{matrix}CH_3\\CH_3\end{matrix}C=C\begin{matrix}CH_2Br\\CH_3\end{matrix} + CH_3CONH_2$$

The procedure was little used until Ziegler and co-workers[2] started searching for a reagent to achieve halogenation at carbon atoms adjacent to olefinic double bonds. It was found that N-bromosuccinimide (NBS) was far superior to other halogenoimides and amides investigated, and this reagent is now normally employed for these brominations. N-bromophthalimide (NBP) has found limited application.

Cyclohexene was one of the first substances treated in this manner, giving 3-bromo-1-cyclohexene with a yield of 51 per cent

$$\bigcirc + \begin{matrix}CH_2-CO\\CH_2-CO\end{matrix}NBr \longrightarrow \bigcirc-Br + \begin{matrix}CH_2-CO\\CH_2-CO\end{matrix}NH$$

The reaction has been applied to a wide range of organic compounds. In some cases a mixture of products results owing to the free-radical nature of the process. The yield of α-bromoalkene varies but is normally above 40 per cent and frequently between 60 and 70 per cent.

A large number of successful brominations have been listed by Djerassi[3], while more recent aspects have been discussed by Horner and Winkelmann[4].

Mechanism

Soon after Ziegler's work was reported, Bloomfield[5] proposed a free-radical mechanism involving the steps: (*a*) initiation, (*b*) propagation, (*c*) termination. Dauben and McCoy[6] obtained results indicating that the pertinent steps are

Initiation:

$$\begin{array}{c}CH_2-CO\\ |\\ CH_2-CO\end{array}\!\!\!\!\!\!\!\!>\!\!NBr \longrightarrow \begin{array}{c}CH_2-CO\\ |\\ CH_2-CO\end{array}\!\!\!\!\!\!\!\!>\!\!N^{\cdot} + Br^{\cdot}$$

Propagation:

$$\begin{array}{c}CH_2-CO\\ |\\ CH_2-CO\end{array}\!\!\!\!\!\!\!\!>\!\!N^{\cdot} + -\overset{|}{C}=\overset{|}{C}-\overset{|}{\underset{H}{C}}- \longrightarrow \begin{array}{c}CH_2-CO\\ |\\ CH_2-CO\end{array}\!\!\!\!\!\!\!\!>\!\!NH\\ + -\overset{|}{C}=\overset{|}{C}-\overset{|}{\underset{\cdot}{C}}-\end{array}$$

Termination:

$$-\overset{|}{C}=\overset{|}{C}-\overset{|}{\underset{\cdot}{C}}- + Br^{\cdot} \longrightarrow -\overset{|}{C}=\overset{|}{C}-\overset{|}{\underset{Br}{C}}-$$

It has been suggested[7], however, that abstraction of hydrogen from the allyl carbon atom is due to a bromine rather than an imide radical. The reaction may, therefore, be more involved than is indicated by the representation given.

General reaction conditions

N-bromosuccinimide (or *N*-bromophthalimide) is added to a solution of the alkene in carbon tetrachloride or benzene. The mixture is refluxed until the insoluble *N*-bromoimide at the bottom of the vessel has all reacted and the surface of the solvent is covered by the floating imide. After cooling and filtering, the product is obtained by evaporation of the solvent. The reaction may be accelerated by free radical activators such as dibenzoyl peroxide, or retarded by inhibitors including quinones and iodine. Activation by ultra-violet light has also been used.

3-Methylthiophene has been brominated in the presence of

benzoyl peroxide to give 79 per cent yields of 3-bromomethylthiophene[8]

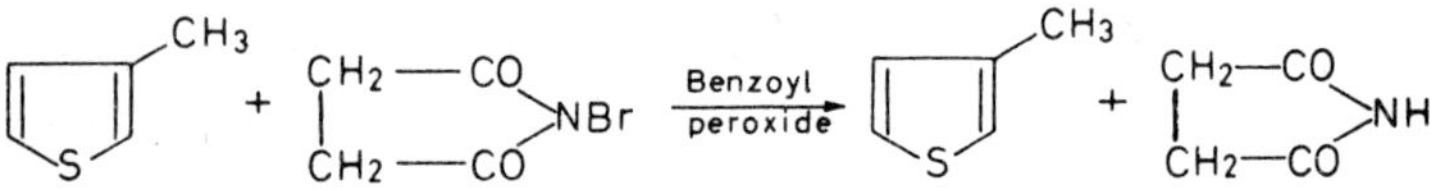

o-Nitrobenzyl bromide has been prepared[9] from *o*-nitrotoluene in yields of about 70 per cent without any form of activation.

Applications

The use of NBS in t-butyl alcohol to brominate indoles can give rise to either oxindoles or 3-bromo-oxindoles, depending upon whether a 1 or 2 molar ratio of the reagent is employed[10]

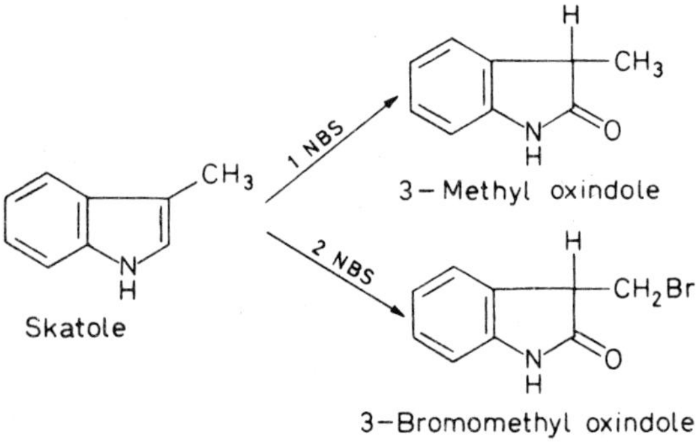

Schmid and Karrer[11] found that both NBS and NBP could be used to brominate cyclic ketones, such as cyclohexanone, in positions α to the carbonyl group. This has been employed[12] to prepare 2- and 4-bromo-3-ketosteroids from the corresponding saturated 3-ketosteroids

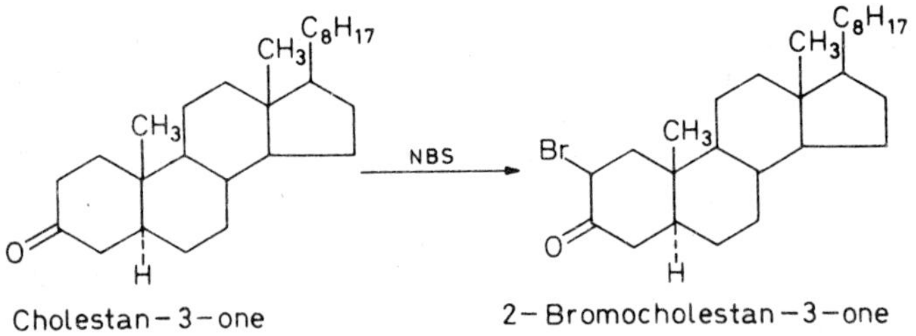

REFERENCES

[1] Wohl, A. *Ber. dt. chem. Ges.* 52B (1919) 51; with Jaschinowski, K. 54B (1921) 476 [*Chem. Abstr.* 13 (1919) 1588; 15 (1921) 2411]
[2] Ziegler, K. et al., *Justus Liebigs Annln Chem.* 551 (1942) 1, 80 [*Chem. Abstr.* 37 (1943) 5376, 5032]
[3] Djerassi, C. *Chem. Rev.* 43 (1948) 271
[4] Horner, L. and Winkelmann, E. H. *Newer Methods of Preparative Organic Chemistry* (ed. Foerst, W.), 3, 151, New York (Academic Press) 1964
[5] Bloomfield, G. F. *J. chem. Soc.* (1944) 114
[6] Dauben, H. J. and McCoy, L. L. *J. Am. chem. Soc.* 81 (1959) 4863
[7] House, H. O. *Modern Synthetic Reactions*, p. 158, New York (Benjamin) 1965
[8] Campaigne, E. and Tullar, B. F. *Org. Synth., Coll. Vol.* 4 (1963) 921
[9] Kalir, A. *Org. Synth.* 46 (1966) 81
[10] Hinman, R. L. and Bauman, C. P. *J. org. Chem.* 29 (1964) 1206
[11] Schmid, H. and Karrer, P. *Helv. chim. Acta* 29 (1946) 576 [*Chem. Abstr.* 40 (1946) 4670]
[12] Djerassi, C. and Scholz, C. R. *Experientia* 3 (1947) 107

26

HINSBERG'S REACTION (SEPARATION OF AMINES)

Nature of the reaction

Sulphonyl halides are used to form sulphonamides, as a procedure for the separation of amines. The sulphonamides of primary amines are soluble in alkali, those of secondary amines are insoluble, and tertiary amines do not give sulphonamides

$$RNH_2 + ArSO_2Cl \longrightarrow RNHSO_2Ar \xrightarrow{NaOH} R\underset{Na}{-}N-SO_2Ar$$

$$R_2NH + ArSO_2Cl \longrightarrow R_2NSO_2Ar$$

Historical development

The method as originally proposed by Hinsberg[1] involved the use of benzene sulphonyl chloride which quickly led to some controversy over its efficacy[2]. Other related reagents were also employed; Rodd and Everatt[3] used chlorosulphonic acid, whereas Schreiber and Shriner[4] found o-nitrobenzenesulphonyl chloride to be satisfactory.

The reagent commonly employed in Hinsberg's procedure is now p-toluenesulphonyl chloride, as this gives fewer side reactions with the amines compared to most other organosulphonyl chlorides. Seaman and co-workers[5] carried out a comprehensive study of recovery of amines using Hinsberg's procedure. They showed that recovery of primary and secondary amines varied between 70 and 120 per cent whilst tertiary amines were obtained in 80–140 per cent recoveries. By modifying the procedure, amines could be separated with an error no greater than ±2 per cent.

Mechanism of solvation

The sulphonamides from primary amines are probably soluble in alkali solution, owing to formation of a potassium salt of the enol tautomer[6]

$$\underset{O}{\overset{O}{Ar\,S}}\underset{|}{\overset{H}{-N-R}} \rightleftharpoons \underset{O}{\overset{OH}{Ar\,S}}=N-R \xrightarrow{KOH} \underset{O}{\overset{\overset{-}{O}\overset{+}{K}}{Ar\,S}}=N-R$$

In the case of sulphonamides of secondary amines, there is no hydrogen atom to enable tautomerism to occur.

General reaction conditions

Under Hinsberg's original approach, the amine mixture was treated with the benzenesulphonyl chloride and sodium hydroxide, shaken for a few minutes and warmed. The unreacted tertiary amine may be steam-distilled off, the sulphonamide of the secondary amine is in the insoluble residue and that of the primary amine is obtained on acidifying the aqueous solution[7].

The modified procedure by Seaman *et al.*[5] is carried out with the amines dissolved in benzene. The reagent in pyridine solution is added and the mixture kept in an ice bath for 30 min while water is slowly added. After being treated with 5N sodium hydroxide, the mixture is steam-distilled, and separation proceeds as before.

Regeneration of the primary and secondary amines is accomplished by hydrolysis with hydrochloric acid for an extended period.

Application

Hickinbottom[8] found that p-toluene sulphonamides were useful for characterizing secondary amines possessing tertiary alkyl

HINSBERG'S REACTION (SEPARATION OF AMINES)

groups. Although no derivatives were isolated under normal Hinsberg conditions, the addition of pyridine to the reaction mixture enabled good yields to be obtained

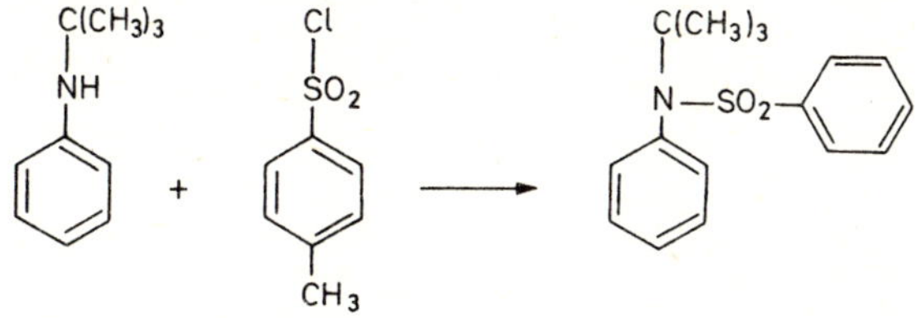

Coleman and co-workers[9] used Hinsberg's procedure with benzenesulphonyl chloride to separate 1-n-butylpyrrolidine from di-n-butylamine starting material.

REFERENCES

[1] Hinsberg, O. *Ber. dt. chem. Ges.* 23 (1890) 2962 [*J. chem. Soc.* (*A*) 60 (1891) 49]
[2] Solonina, W. *Zh. russk. fiz.-khim. Obshch.* 31 (1899) 640; Marckwald, W. *Ber. dt. chem. Ges.* 32 (1899) 3512; Hinsberg, O. and Kessler, J. 38 (1905) 906 [*J. chem. Soc.* (*A*) 78 (1900) [1] 147, 149; 88 (1905) [2] 338]
[3] Rodd, E. H. and Everatt, R. W. *Br. Pat.* 270, 930 (1926) [*Chem. Abstr.* 22 (1928) 1594]
[4] Schreiber, R. S. and Shriner, R. L. *J. Am. chem. Soc.* 56 (1934) 114 [*Chem. Abstr.* 28 (1934) 1674]
[5] Seaman, W. *et al.*, *J. Am. chem. Soc.* 67 (1945) 1571 [*Chem. Abstr.* 39 (1945) 5210]
[6] Finar, I. L. *Organic Chemistry*, 4th edn., p. 310, London (Longmans) 1963
[7] Vogel, A. I. *A Textbook of Practical Organic Chemistry*, 3rd edn., p. 650, London (Longmans) 1956
[8] Hickinbottom, W. J. *J. chem. Soc.* (1933) 946
[9] Coleman, G. H., Nichols, G. and Martens, T. F. *Org. Synth.*, *Coll. Vol.* 3 (1955) 161

27

SCHOTTEN–BAUMANN REACTION

Nature of the reaction

Substances possessing active hydrogen atoms are acylated by means of an acid chloride in the presence of dilute alkali, in particular by the action of benzoyl chloride

$$ROH + C_6H_5COCl + NaOH \longrightarrow C_6H_5COOR + NaCl + H_2O$$

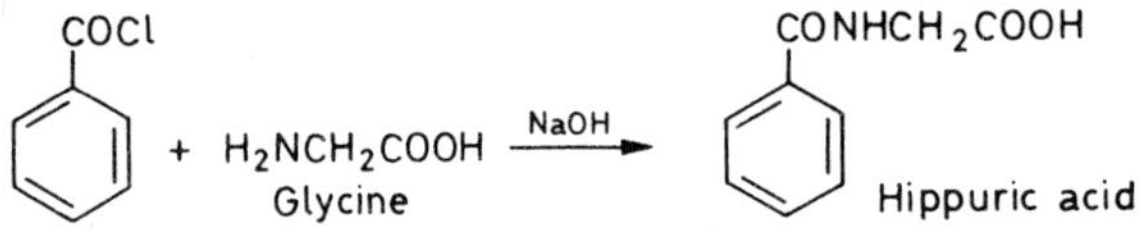

Ph-COCl + H$_2$NCH$_2$COOH $\xrightarrow{NaOH}$ Ph-CONHCH$_2$COOH (Hippuric acid)

(Glycine)

Historical development

In 1884, Schotten[1] used benzoyl chloride for the preparation of benzoyl piperidine. A short time later, Baumann[2] employed the same reagent in the benzoylation of hydroxy groups in glycerol

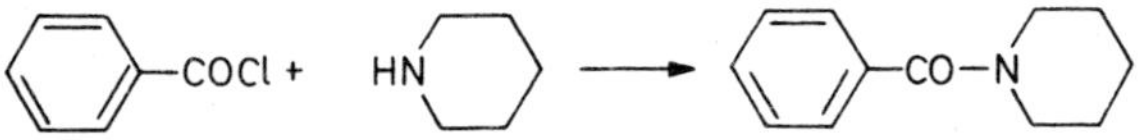

Ph-COCl + HN(piperidine) ⟶ Ph-CO-N(piperidine)

A detailed investigation of the value of benzoyl chloride as a reagent for active hydrogen atoms, carried out by Udransky and Baumann[3], firmly established its use for detecting and isolating compounds possessing amino and hydroxyl groups. Although the method is not restricted to benzoyl chloride[4], this reagent is commonly used, as the products formed are of low solubility in water and hydrolysis is generally slow.

Yields from the Schotten–Baumann reaction are normally high, 60–90 per cent, and often quantitative in the case of benzoylation. Hinsberg's method for the separation of amines (No. 26) is directly related to this reaction.

Mechanism

As the Schotten–Baumann reaction takes place in an heterogeneous system, the mechanism has been difficult to determine and is still obscure.

SCHOTTEN–BAUMANN REACTION

Alexander[5] pointed out that, owing to polarization, the effect of the carbonyl group in acyl halides hinders the release of the halogen, X^-, and encourages attack by a negatively charged ion on the carbonyl carbon atom. The overall reaction[6] has been represented in the case of amines as

$$R-\underset{\substack{\|\\O}}{C}-X \rightleftharpoons R-\underset{\substack{|\\+}}{\overset{O^-}{C}}-X \xrightarrow{R'NH_2} R-\underset{\substack{|\\H_2\overset{+}{N}-R'}}{\overset{O^-}{C}}-X \xrightarrow{-H^+} \rightleftharpoons$$

$$R-\underset{\substack{|\\HN-R'}}{\overset{O^-}{\underset{|}{C}}\overset{\frown}{-}X} \xrightarrow{-X^-} \rightleftharpoons R-\underset{\substack{|\\HN-R'}}{\overset{\|\\O}{C}}$$

The function of the base in the reaction is to reconvert any quaternary salt to the amine and to hydrolyse excess benzoyl chloride. With alcohols, a similar process takes place

$$R-\underset{\substack{\|\\O}}{C}-X + R'OH \rightleftharpoons R-\underset{\substack{|\\R'-\overset{+}{O}-H}}{\overset{O^-}{C}}-X \rightleftharpoons R-\underset{\substack{|\\R'-O}}{\overset{O^-}{C}}-X$$

$$\longrightarrow R-\underset{\substack{|\\R'-O}}{\overset{\|\\O}{C}} + X^-$$

General reaction conditions

The alcohol or amine is added to a 10 per cent aqueous sodium hydroxide solution, followed by a slight excess of the acid chloride. The mixture is either continuously shaken or vigorously stirred for about 30 min during which the product is precipitated from solution.

Benzoylation of compounds possessing several reactive groups requires heating of the mixture to 40°–50° to ensure complete reaction.

Vogel[7] has given several detailed procedures for Schotten–Baumann reactions with both amines and hydroxy compounds

$$\text{2-naphthol} \xrightarrow[\text{NaOH}]{C_6H_5COCl} \text{β-Naphthyl benzoate}$$

Modifications

Deninger[8] and Einhorn[9] used pyridine for many condensations in place of sodium hydroxide, as strong alkalis often have adverse effects. In this procedure the substance to be acylated is dissolved in pyridine and treated with a slight excess of acyl halide. The solution is kept at room temperature for up to 10 h and then added slowly to an excess of dilute sulphuric acid. The product usually separates out at this stage.

This type of modified procedure has been employed[10] to prepare o-benzoyloxyacetophenone as an intermediate in the formation of flavone

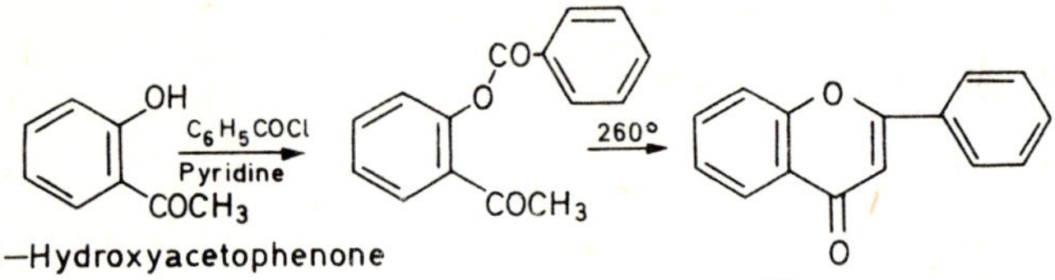

o—Hydroxyacetophenone → (intermediate) → Flavone

Applications

The Schotten–Baumann reaction has recently been shown to be of particular value in the formation of derivatives of polymers. Tsuda[11] has shown that by the conventional route with aqueous sodium hydroxide, and the addition of methyl ethyl ketone as an organic solvent, polyvinyl alcohol is readily alkylated by cinnamoyl chloride at 5°

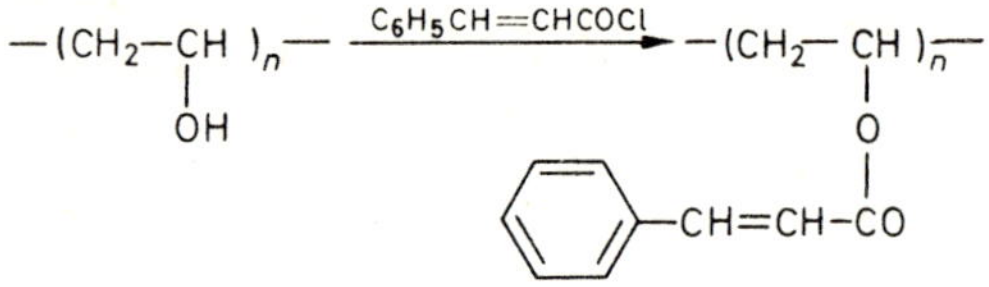

$$-(CH_2-CH)_n- \xrightarrow{C_6H_5CH=CHCOCl} -(CH_2-CH)_n-$$

REFERENCES

[1] Schotten, C. *Ber. dt. chem. Ges.* 17 (1884) 2544 [*J. chem. Soc.* (A) 48 (1885) 176]
[2] Baumann, E. *Ber. dt. chem. Ges.* 19 (1886) 3218 [*J. chem. Soc.* (A) 52 (1887) 228]
[3] Udransky, L. v. and Baumann, E. *Ber. dt. chem. Ges.* 21 (1888) 2744 [*J. chem. Soc.* (A) 54 (1888) 1296]
[4] Sonntag, N. O. V. *Chem. Rev.* 52 (1953) 272
[5] Alexander, E. R. *Principles of Ionic Organic Reactions*, p. 90, New York (Wiley) 1950
[6] Roberts, J. D. and Caserio, M. C. *Basic Principles of Organic Chemistry*, pp. 388, 664, New York (Benjamin) 1965
[7] Vogel, A. I. *A Textbook of Practical Organic Chemistry*, 3rd edn,. pp. 582, 780, London (Longmans) 1956

[8] Deninger, A. *Ber. dt. chem. Ges.* 28 (1895) 1322 [*J. chem. Soc. (A)* 68 (1895) [1] 461]
[9] Einhorn, A. and Holland, F. *Justus Liebigs Annln Chem.* 30 (1898) 95 [*J. chem. Soc. (A)* 74 (1898) [1] 577]
[10] Wheeler, T. S. *Org. Synth., Coll. Vol.* 4 (1963) 478
[11] Tsuda, M. *J. Polym. Sci.* 1B (1963) [5] 215; *Makromolec. Chem.* 72 (1964) 174

28

KILIANI–FISCHER SYNTHESIS

Nature of the reaction

This is a method of increasing the length of the carbon chain in a sugar by one carbon atom via the formation of a cyanohydrin, hydrolysis to the acid and reduction of the lactone from the acid

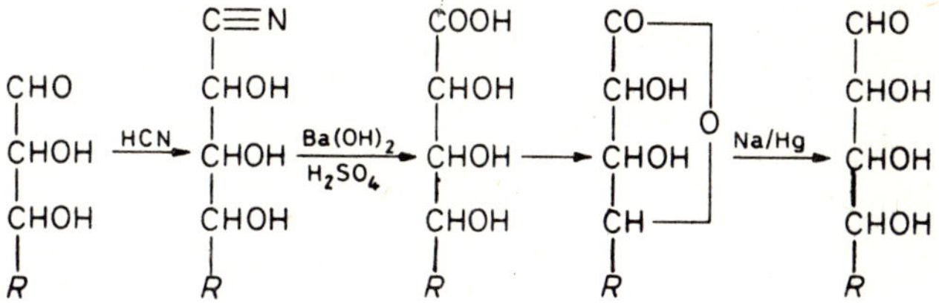

Historical development

The basis for the pioneer work of Kiliani, which led eventually to the conversion of sugars to their next higher homologues, was the preparation of mandelic acid from the cyanohydrin of benzaldehyde by Winckler[1]

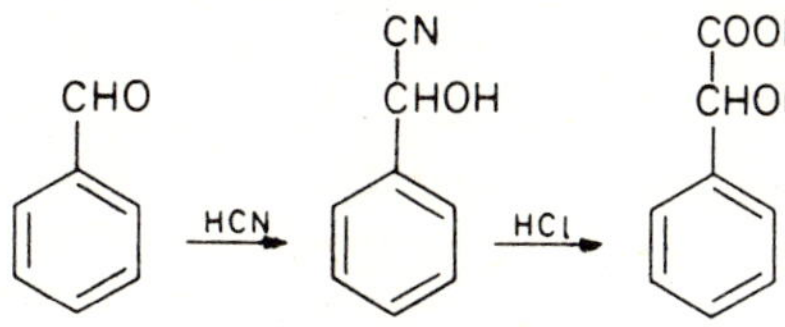

NAMED ORGANIC REACTIONS

Kiliani[2] applied this to reducing sugars to give polyhydroxy-carboxylic acids

$$
\begin{array}{cccc}
\text{CHO} & \text{CN} & \text{CONH}_2 & \text{COOH} \\
\text{H—C—OH} & \text{CHOH} & \text{CHOH} & \text{CHOH} \\
\text{HO—C—H} & \text{H—C—OH} & \text{H—C—OH} & \text{H—C—OH} \\
\text{HO—C—H} \xrightarrow{\text{HCN/H}_2\text{O}} & \text{HO—C—H} \longrightarrow & \text{HO—C—H} \xrightarrow[\text{H}^+]{\text{Ca(OH)}_2} & \text{HO—C—H} \\
\text{H—C—OH} & \text{HO—C—H} & \text{HO—C—H} & \text{HO—C—H} \\
\text{CH}_2\text{OH} & \text{H—C—OH} & \text{H—C—OH} & \text{H—C—OH} \\
 & \text{CH}_2\text{OH} & \text{CH}_2\text{OH} & \text{CH}_2\text{OH} \\
\text{D(+)-Galactose} & & & \text{Galactose carboxylic acid}
\end{array}
$$

He also found that, by treatment with sulphuric acid, the barium salts could be converted to lactones. Fischer[3] later used sodium amalgam to reduce the lactones to aldoses, thus completing the route

$$
\begin{array}{ccc}
\text{COOH} & \text{CO———} & \text{CHO} \\
\text{HO—C—H} & \text{HO—C—H} \Big| & \text{HO—C—H} \\
\text{HO—C—H} & \text{HO—C—H} \text{O} & \text{HO—C—H} \\
\text{H—C—OH} \longrightarrow & \text{H—C———} \longrightarrow & \text{H—C—OH} \\
\text{H—C—OH} & \text{H—C—OH} & \text{H—C—OH} \\
\text{CH}_2\text{OH} & \text{CH}_2\text{OH} & \text{CH}_2\text{OH} \\
 & & \text{D(+)-Mannose}
\end{array}
$$

The whole development of the Kiliani–Fischer synthesis has been summarized by Hudson[4].

In each case the addition of hydrocyanic acid to the aldose leads to the formation of two possible diasterioisomers, and a great deal of Fischer's work was involved with proving the relationship between the aldohexoses produced from aldopentoses.

The yields obtained during the various stages of the reaction are very dependent upon experimental conditions employed. It is possible to carry out an overall conversion with a high yield, although generally it is between 25 and 50 per cent.

KILIANI–FISCHER SYNTHESIS

General reaction conditions

Kiliani's original method employed the use of aqueous hydrogen cyanide which adds almost quantitatively to simple aldoses. The aldose was treated with 60 per cent aqueous hydrogen cyanide in a closed vessel at room temperature for several days. This was followed by heating at 40°–60° to expel any excess hydrogen cyanide. Barium hydroxide was added to the solution to form the barium salt of the acid. Addition of sulphuric acid and evaporation of the solution gave the lactones.

Hudson and co-workers[5] modified this procedure by mixing the aqueous solution of the aldose with sodium cyanide and calcium chloride at room temperature. The temperature in this case starts to rise almost immediately, and after a few hours' heating the solution expels most of the ammonia evolved. The reaction is complete within a day and the acid precipitated as the calcium salt by calcium hydroxide.

Varied conditions of pH and cyanide salts employed can be used to produce the two diasterioisomers in different proportions[6].

Fischer's sodium amalgam method for reducing the lactones to sugars was studied by Sperber and co-workers[7] to ascertain the optimum conditions. They found that the most important point was maintaining a pH of 3–4 during the course of the reaction. Reactions are usually carried out below 10°, as above this temperature the rate is too rapid to enable adequate pH control to be maintained. The lactone in aqueous solution is mixed with sodium amalgam and sulphuric acid added periodically to achieve the required pH conditions. Reduction is complete within about 15 min giving conversion yields for lactone to aldose of 60–75 per cent.

Modification

Wolfrom and Wood[8] found that sodium borohydride could be employed to reduce lactones in aqueous acid solution at pH 3–4; by adding it gradually at 0°, the rate of reaction was controlled, evolving hydrogen for about 1 h. Yields of 60–70 per cent are possible by this method.

Applications

The classical value of this synthesis was in Fischer's investigations into the configurational relationships between aldopentoses and aldohexoses.

Haworth[9] and Reichstein[10] used the Kiliani reaction in the synthesis of Vitamin C (L(+)-ascorbic acid)

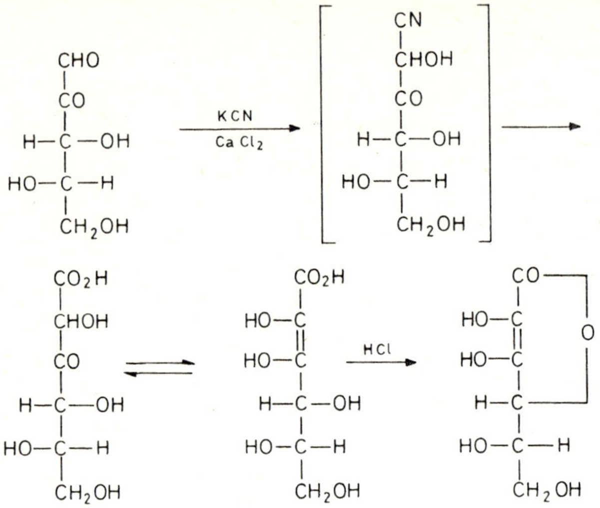

[Reaction scheme: CHO–CO–H−C−OH–HO−C−H–CH₂OH → (KCN/CaCl₂) → [CN–CHOH–CO–H−C−OH–HO−C−H–CH₂OH] →]

[Reaction scheme: CO₂H–CHOH–CO–H−C−OH–HO−C−H–CH₂OH ⇌ CO₂H–HO−C=C−OH(=)–H−C−OH–HO−C−H–CH₂OH → (HCl) → Vitamin C lactone structure]

Vitamin C

Wood and Fletcher[11] have employed the Kiliani–Fischer reaction to convert 2-deoxy-D-ribose to 3-deoxy-D-ribo-hexono-α-lactone (56 per cent yield) as an intermediate in the preparation of a number of other related compounds

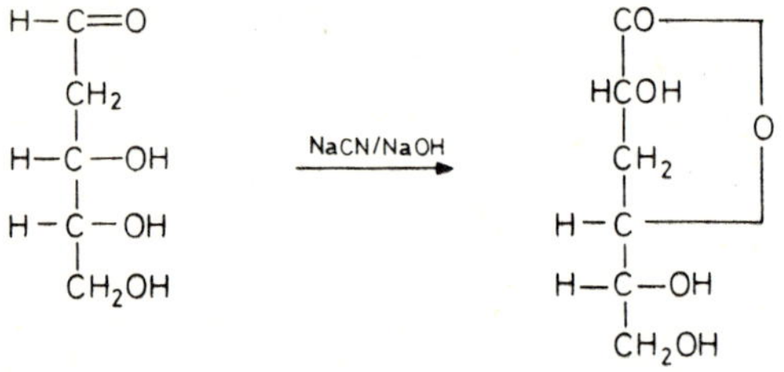

[Reaction scheme: H−C=O–CH₂–H−C−OH–H−C−OH–CH₂OH → (NaCN/NaOH) → CO–HCOH–CH₂–H−C−–H−C−OH–CH₂OH (lactone)]

3-Deoxy-D-ribo-hexono-
γ-lactone

REFERENCES

[1] Winckler, F. W. *Justus Liebigs Annln Chem.* 4 (1832) 242; 18 (1836) 310
[2] Kiliani, H. *Ber. dt. chem. Ges.* 18 (1885) 3066; 19 (1886) 3029; 21 (1888) 915; 22 (1889) 521; [*J. chem. Soc.* (A) 50 (1886) 219; 52 (1887) 229; 54 (1888) 581; 56 (1889) 589]
[3] Fischer, E. *Ber. dt. chem. Ges.* 22 (1889) 2204; with Stahel, R. 24 (1891) 528; with Fay, I. W. 28 (1895) 1975 [*J. chem. Soc.* (A) 56 (1889) 1149; 60 (1891) 667; 68 (1895) [1] 650]

4 Hudson, C. S. *Adv. Carbohyd. Chem.* 1 (1945) 1
5 Hudson, C. S., Hartley, O. and Purves, C. B. *J. Am. chem. Soc.* 56 (1934) 1248
6 Webber, J. M. *Adv. Carbohyd. Chem.* 17 (1962) 18
7 Sperber, N., Zangg, H. E. and Sandstrom, W. M. *J. Am. chem. Soc.* 69 (1947) 915
8 Wolfrom, M. L. and Wood, H. B. *J. Am. chem. Soc.* 73 (1951) 2933
9 Haworth, W. N. *et al.*, *J. chem. Soc.* (1933) 1419; (1934) 1192; with Hirst, E. L., *Helv. chim. Acta* 17 (1934) 520 [*Chem. Abstr.* 28 (1934) 4704]
10 Reichstein, T., Grussner, A. and Oppenauer, R. *Helv. chim. Acta* 16 (1933) 561, 1019; 17 (1934) 510 [*Chem. Abstr.* 27 (1933) 3977; 28 (1934) 570, 4704]
11 Wood, H. B. and Fletcher, H. G. *J. org. Chem.* 26 (1961) 1969

Note—Those interested in the historical development of Fischer's work on carbohydrates are referred to
Hudson, C. S., 'Emil Fischer's Discovery at the Configuration of Glucose', *J. chem. Educ.* 18 (1941) 353
Freudenberg, K. 'Emil Fischer and his Contribution to Carbohydrate Chemistry', *Adv. Carbohyd. Chem.* 21 (1966) 2

29

LOBRY DE BRUYN–ALBERDA VAN EKENSTEIN TRANSFORMATION

Nature of the reaction

Reducing sugars dissolved in dilute alkaline solutions are transformed into mixtures of aldoses and ketoses

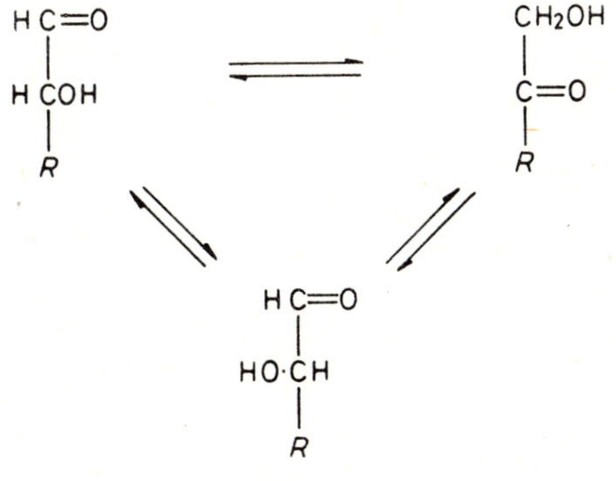

NAMED ORGANIC REACTIONS

Historical development

The extensive collaboration between Lobry de Bruyn and Alberda van Ekenstein on this transformation[1] began in 1895 and extended over many years[2]. It was first observed with an aqueous solution of glucose in which the optical activity diminished to zero in the presence of dilute potassium hydroxide. It was found that fructose and mannose could be isolated from the solution and that the three carbohydrates were interconvertable

$$\begin{array}{c} CHO \\ | \\ H-C-OH \\ | \\ HO-C-H \\ | \\ H-C-OH \\ | \\ H-C-OH \\ | \\ CH_2OH \end{array} \rightleftharpoons \begin{array}{c} CH_2OH \\ | \\ C=O \\ | \\ HO-C-H \\ | \\ H-C-OH \\ | \\ H-C-OH \\ | \\ CH_2OH \end{array} \rightleftharpoons \begin{array}{c} CHO \\ | \\ HO-C-H \\ | \\ H-C-OH \\ | \\ H-C-OH \\ | \\ H-C-OH \\ | \\ CH_2OH \end{array}$$

D-Glucose   D-Fructose   D-Mannose

The scope of the transformation has been shown to be extensive. It was quickly appreciated that it need not be limited to carbohydrates but could be applied to other compounds possessing a suitable α-hydroxy aldehyde ketone system.

A review of the transformation by Speck[3] includes tables of reactions effected by both base and enzymatic catalysis. Yields of ketoses from the transformation range from nearly 0 to 50 per cent.

Mechanism

Lobry de Bruyn and Alberda van Ekenstein believed the transformation to occur owing to interaction with water giving hemiacetals

$$\begin{array}{c} CHO \\ | \\ HCOH \\ | \\ R \end{array} \xrightleftharpoons[]{H_2O} \begin{array}{c} CHOH \\ \diagdown O \\ HC \\ | \\ R \end{array} \xrightleftharpoons[H_2O]{} \begin{array}{c} CH_2OH \\ | \\ CO \\ | \\ R \end{array} \xrightleftharpoons[]{H_2O} \begin{array}{c} CHOH \\ | \diagdown O \\ CH \\ | \\ R \end{array} \xrightleftharpoons[H_2O]{} \begin{array}{c} CHO \\ | \\ HOCH \\ | \\ R \end{array}$$

However, this was speculation and was soon displaced by Wohl and Neuberg's[4] idea of an enolization mechanism involving an enediol intermediate. Investigations carried out under deuterium and tritium exchange conditions are conflicting concerning the incorporation of the heavy atoms[5,6]. Although Topper[7] is satisfied that enediols are intermediates in the isomerization, this has not met with full agreement.

Under base-catalysed conditions the accepted mechanism[8] for the enolization route is

$$\begin{array}{c}H\\|\\C=O\\|\\H-C-OH\\|\\R\end{array} \xrightleftharpoons[H^+]{B^-} \left[\begin{array}{c}H\\|\\C=O\\|\\{}^-C-OH\\|\\R\end{array} \longleftarrow \begin{array}{c}H\\|\\C-O^-\\||\\C-OH\\|\\R\end{array}\right] \xrightleftharpoons[B^-]{H^+} \begin{array}{c}H\\|\\C=O\\|\\HO-C-H\\|\\R\end{array}$$

$$\updownarrow$$

$$\left[\begin{array}{c}H\\|\\{}^-C-OH\\|\\C=O\\|\\R\end{array} \longleftarrow \begin{array}{c}H\\|\\C-OH\\||\\C-O^-\\|\\R\end{array}\right] \xrightleftharpoons[B^-]{H^+} \begin{array}{c}H\\|\\H-C-OH\\|\\C=O\\|\\R\end{array}$$

General reaction conditions

Early transformations were achieved by adding a small quantity of aqueous alkali (about 5 per cent by vol. of a 1N solution) to an aqueous solution of the sugar which was left to stand for several days. It was later found that the transformation could be accelerated by working at elevated temperatures of 30°–70°, the proportions of products varying with the conditions employed.

Mendicino[9] has shown that yields of ketose from aldose can be increased by using dilute solutions of alkali and sugar and adding borate to the reaction mixture.

Isolation of products often presents difficulties. Where the procedure is used to prepare a ketose, it is necessary to destroy the residual aldose by hypobromite, followed by further separation on ion-exchange columns[10]. A number of transformations have also been carried out under enzymatic conditions.

Application

In work on the adrenal cortical hormones, Reichstein and von Euw[11] carried out a transformation of △[4]-pregnene-17,20-diol-3-

one-21-al in pyridine to give the active compound 'Substance *S*' (11-deoxy-17-hydroxy-corticosterone), with a 40 per cent yield

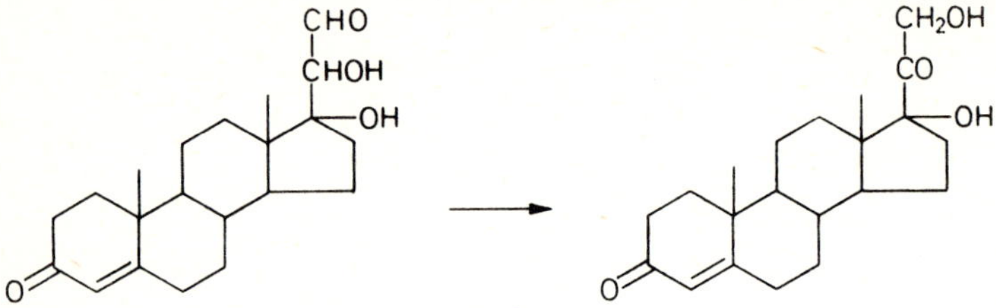

REFERENCES

[1] Lobry de Bruyn, C. A. *Recl Trav. chim. Pays-Bas Belg.* 14 (1895) 150; with Alberda van Ekenstein, W. *ibid.* 201 [*J. chem. Soc.* (*A*) 70 (1896) [1] 116]
[2] Lobry de Bruyn, C. A. and Alberda van Ekenstein, W. *Recl Trav. chim. Pays-Bas Belg.* 16 (1897) 262; 18 (1899) 147 [*J. chem. Soc.* (*A*) 74 (1898) [1] 225; 76 (1899) [1] 661]
[3] Speck, J. C. *Adv. Carbohyd. Chem.* 13 (1958) 63
[4] Wohl, A. and Neuberg, C. *Ber. dt. chem. Ges.* 33 (1900) 3095 [*J. chem. Soc.* (*A*) 80 (1901) [1] 12]
[5] Fredenhagen, H. and Bonhoeffer, K. F. *Z. phys. Chem.* 181A (1938) 392 [*Chem. Abstr.* 32 (1938) 4059]
[6] Topper, Y. J. and Stetten, D. *J. biol. Chem.* 189 (1951) 191
[7] Topper, Y. J. *J. biol. Chem.* 225 (1957) 419
[8] Hine, J. *Physical Organic Chemistry*, p. 271, New York (McGraw-Hill) 1962
[9] Mendicino, J. F. *J. Am. chem. Soc.* 82 (1960) 4975
[10] Pratt, J., Richtmyer, N. K. and Hudson, C. S. *J. Am. chem. Soc.* 74 (1952) 2210
[11] Reichstein, T. and von Euw, J. *Helv. chim. Acta* 23 (1940) 1258 [*Chem. Abstr.* 35 (1941) 2526]

30

RUFF DEGRADATION

Nature of the reaction

The carbon chain of an aldose sugar is degraded by one carbon atom by oxidizing the aldose sugar to an aldonic acid, followed by treatment with Fenton's reagent

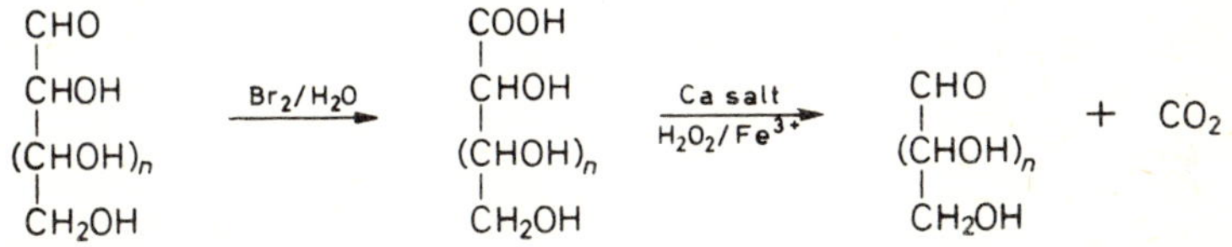

$$\begin{array}{c} CHO \\ | \\ CHOH \\ | \\ (CHOH)_n \\ | \\ CH_2OH \end{array} \xrightarrow{Br_2/H_2O} \begin{array}{c} COOH \\ | \\ CHOH \\ | \\ (CHOH)_n \\ | \\ CH_2OH \end{array} \xrightarrow[H_2O_2/Fe^{3+}]{Ca\ salt} \begin{array}{c} CHO \\ | \\ (CHOH)_n \\ | \\ CH_2OH \end{array} + CO_2$$

Historical development

The oxidizing action of hydrogen peroxide in the presence of ferrous (Fe^{2+}) ions was noticed in connection with tartaric acid by Fenton[1]. Ruff[2] applied the procedure to salts of aldonic acids and substituted ferric (Fe^{3+}) salts in place of ferrous. As the aldonic acids were obtained from the oxidation of aldoses by bromine water[3], this presented a complete route for the shortening of the carbohydrate chain. Overall yields from the degradation are low, 25–40 per cent, by the standard procedure but may often be improved by modified methods involving ion exchange.

Calcium gluconate is an example of this conversion, giving D-arabinose[4]

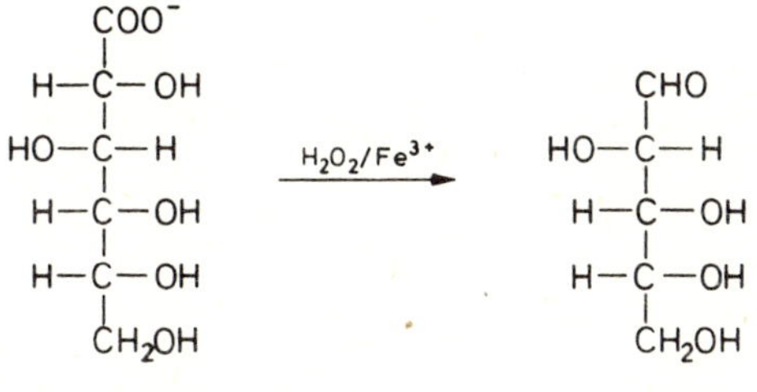

$$\begin{array}{c} COO^- \\ | \\ H-C-OH \\ | \\ HO-C-H \\ | \\ H-C-OH \\ | \\ H-C-OH \\ | \\ CH_2OH \end{array} \xrightarrow{H_2O_2/Fe^{3+}} \begin{array}{c} CHO \\ | \\ HO-C-H \\ | \\ H-C-OH \\ | \\ H-C-OH \\ | \\ CH_2OH \end{array}$$

Moody[5] has discussed this oxidation procedure in a review dealing with the action of hydrogen peroxide on carbohydrates.

Mechanism

Experiments have shown[6] that ferric salts in hydrogen peroxide lead to the formation of free radicals

$$Fe^{3+} + H_2O_2 \longrightarrow Fe^{2+} + HO_2^{\cdot} + H^+$$

$$HO_2^{\cdot} + H_2O_2 \longrightarrow H_2O + OH^{\cdot} + O_2$$

The free-radical nature of the reaction indicates that little selectivity is possible and explains why many yields from Ruff degradations are low.

It seems likely that the chain shortening arises from initial abstraction of hydrogen[7], followed by further attack due to hydrogen peroxide

$$\underset{R}{HO-\underset{|}{\overset{COO^-}{C}}-H} + {\cdot}OH \longrightarrow \underset{R}{HO-\underset{|}{\overset{COO^-}{C^{\cdot}}}} \xrightarrow{HO-OH} \underset{R}{HO-\underset{|}{\overset{COO^-}{C}}-OH} + {\cdot}OH$$

$$\longrightarrow \underset{R}{\underset{|}{\overset{COO^-}{C=O}}} \xrightarrow[-CO_2]{Heat} \underset{R}{\underset{|}{H-C=O}} + CO_2$$

General reaction conditions

The calcium salt of the aldonic acid is dissolved in boiling water to which ferric sulphate and barium acetate are added. The cooled solution is filtered and 30 per cent hydrogen peroxide added with stirring, which leads to a rise in temperature and evolution of gases. The temperature is controlled below 60° and reaction is complete when a purple coloration appears. The product is isolated by first concentrating the bulk of the aqueous solution by evaporation, then precipitating inorganic material by the addition of organic solvents. Filtration of the mixture and evaporation of the filtrate give the crude product, usually as a syrup.

Modification

With contemporary procedures it is now normal to pass the aqueous solution from the degradation through a series of ion-exchange columns to remove the inorganic ions prior to isolation of the sugar. This leads to increased yields and more easily purified products. Several preparations involving this ion-exchange step have been given in detail[8].

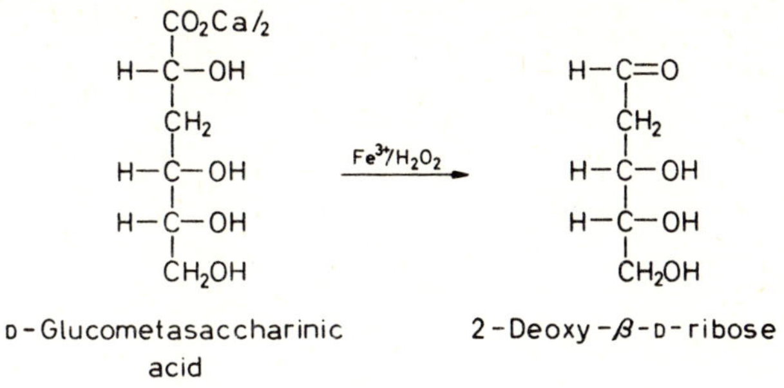

$$\begin{array}{c} CO_2Ca_{1/2} \\ H-C-OH \\ CH_2 \\ H-C-OH \\ H-C-OH \\ CH_2OH \end{array} \xrightarrow{Fe^{3+}/H_2O_2} \begin{array}{c} H-C=O \\ CH_2 \\ H-C-OH \\ H-C-OH \\ CH_2OH \end{array}$$

D-Glucometasaccharinic acid   2-Deoxy-β-D-ribose

Application

The Ruff degradation may be applied to systems other than the simple aldohexoses and aldopentoses. Gakhokidze[9] used the technique in his elimination of carbon atoms from rings in glycals.

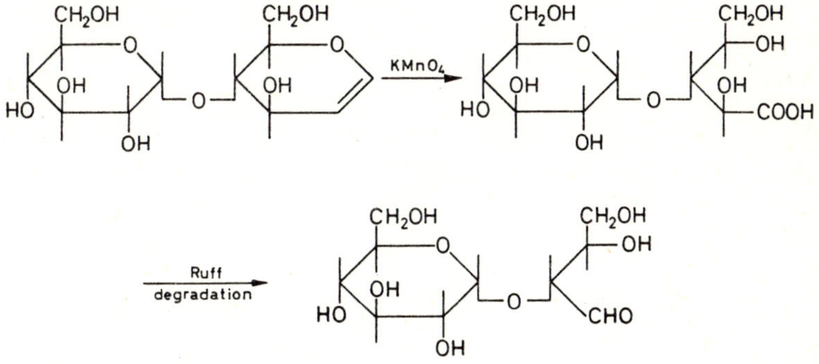

REFERENCES

[1] Fenton, H. J. H. *Proc. chem. Soc.* 9 (1893) 113; *J. chem. Soc.* 67 (1895) 774
[2] Ruff, O. *Ber. dt. chem. Ges.* 31 (1898) 1573; 32 (1899) 550; 34 (1901) 1362 [*J. chem. Soc.* (A) 74 (1898) [1] 516; 76 (1899) [1] 324; 80 (1901) [1] 449]
[3] Kiliani, H. *Ber. dt. chem. Ges.* 19 (1886) 3029 [*J. chem. Soc.* (A) 52 (1887) 229]
[4] Hockett, R. C. and Hudson, C. S. *J. Am. chem. Soc.* 56 (1934) 1632

[5] Moody, G. J. *Adv. Carbohyd. Chem.* 19 (1964) 149
[6] Haber, F. and Wilstätter, R. W. *Ber. dt. chem. Ges.* 64 (1931) 2644 [*Chem. Abstr.* 26 (1932) 3774]
[7] Waters, W. A. *Chemistry of Free Radicals*, p. 251, Oxford (Univ. Press) 1948
[8] *Meth. Carbohyd. Chem.* 1 (1962) 77, 79, 180
[9] Gakhokidze, A. M. *Zh. obshch. Khim.* 20 (1950) 289 [*Chem. Abstr.* 44 (1950) 6822]

31

WEERMAN DEGRADATION

Nature of the reaction

α-Hydroxy or α-methoxy amides are converted into aldehydes with one less carbon atom by treatment with aqueous sodium hypochlorite; the process is normally applied to the shortening of carbohydrate chains.

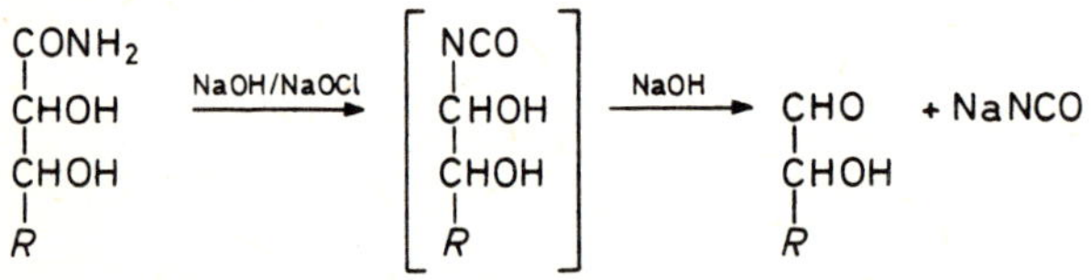

$$\begin{array}{c} CONH_2 \\ | \\ CHOH \\ | \\ CHOH \\ | \\ R \end{array} \xrightarrow{NaOH/NaOCl} \left[\begin{array}{c} NCO \\ | \\ CHOH \\ | \\ CHOH \\ | \\ R \end{array} \right] \xrightarrow{NaOH} \begin{array}{c} CHO \\ | \\ CHOH \\ | \\ R \end{array} + NaNCO$$

Historical development

In 1913, Weerman[1] applied the Hofmann reaction to unsaturated acid amides and obtained aldehydes. Cinnamide gave phenylacetaldehyde

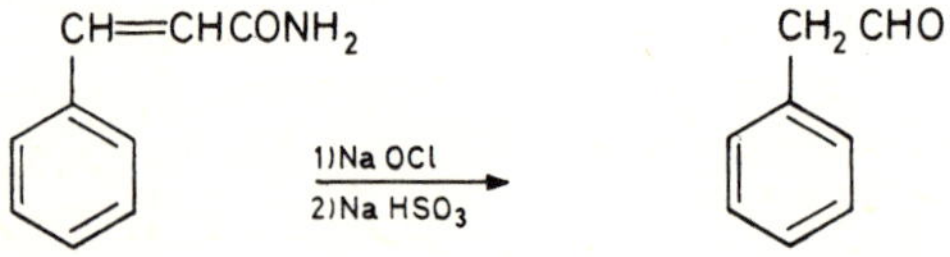

$$C_6H_5-CH=CHCONH_2 \xrightarrow{\substack{1) NaOCl \\ 2) NaHSO_3}} C_6H_5-CH_2CHO$$

WEERMAN DEGRADATION

Further examination of the reaction[2] established the method for degrading the amides of α-hydroxy acids obtained from carbohydrates, the amides being easily accessible from aldoses by oxidizing to the lactones and treating with alcoholic ammonia.

Weerman's procedure enabled aldose sugars to be degraded by one carbon atom at a time. Thus, arabinose was obtained in a 50 per cent yield from glucose

$$\begin{array}{c}\text{CHO}\\\text{H}-\text{C}-\text{OH}\\\text{HO}-\text{C}-\text{H}\\\text{H}-\text{C}-\text{OH}\\\text{H}-\text{C}-\text{OH}\\\text{CH}_2\text{OH}\end{array} \rightarrow \begin{array}{c}\text{O}\\\|\\\text{C}-\\\text{H}-\text{C}-\text{OH}\\\text{HO}-\text{C}-\text{H}\quad\text{O}\\\text{H}-\text{C}-\\\text{H}-\text{C}-\text{OH}\\\text{CH}_2\text{OH}\end{array} \rightarrow \begin{array}{c}\text{O}\\\|\\\text{C}-\text{NH}_2\\\text{H}-\text{C}-\text{OH}\\\text{HO}-\text{C}-\text{H}\\\text{H}-\text{C}-\text{OH}\\\text{H}-\text{C}-\text{OH}\\\text{CH}_2\text{OH}\end{array} \rightarrow \left[\begin{array}{c}\text{NCO}\\\text{H}-\text{C}-\text{OH}\\\text{HO}-\text{C}-\text{H}\\\text{H}-\text{C}-\text{OH}\\\text{H}-\text{C}-\text{OH}\\\text{CH}_2\text{OH}\end{array}\right] \rightarrow \begin{array}{c}\text{CHO}\\\text{HO}-\text{C}-\text{H}\\\text{H}-\text{C}-\text{OH}\\\text{H}-\text{C}-\text{OH}\\\text{CH}_2\text{OH}\end{array}$$

Structural relationships in sugars were investigated by the sequential degradation of mannose to erythrose by the route mannose→mannonamide→arabinose→arabonamide→erythrose.

The yields from this procedure are usually above 50 per cent, and Weerman felt that despite the steps involved it was superior to Ruff's method (No. 30). It is not now frequently employed, however, although the formation of isocyanate in the reaction is a test for the presence of unsubstituted α-hydroxyamides.

Mechanism

Haworth and his collaborators[3] confirmed Weerman's observation that with α-hydroxyamides the reaction led to the formation of isocyanates. With α-methoxyamides, however, ammonia was formed in addition to the aldehyde, and they concluded that the overall reaction in this case takes the form

$$\begin{array}{c}\text{CONH}_2\\\text{CHO CH}_3\\\text{R}\end{array} \longrightarrow \begin{array}{c}\text{NCO}\\\text{CHO CH}_3\\\text{R}\end{array} \longrightarrow \begin{array}{c}\text{CHO}+\text{CH}_3\text{OH}+\text{NH}_3+\text{Na}_2\text{CO}_3\\\text{R}\end{array}$$

Arcus and Greenwood[4] suggested that the difference between the two routes is due to the presence of the α-hydroxy or α-methoxy group. In the former case the anion (*A*) may be formed, as shown in

NAMED ORGANIC REACTIONS

their proposed mechanism, but this cannot be the case with the α-methoxyamides

$$\begin{array}{c}CONH_2\\|\\CHOH\\|\\R\end{array} \xrightarrow{ClO^-} \begin{array}{c}NCO\\|\\CHOH\\|\\R\end{array} \xrightarrow{HO^-} \begin{array}{c}\overset{+}{N}=C-\bar{O}\\|\\CH\underset{}{-}\bar{O}^-\\|\\R\\(A)\end{array} \longrightarrow \begin{array}{c}\bar{N}CO\\|\\CH=O\\|\\R\end{array}$$

It appears that the α-methoxyamides undergo reaction involving the intermediate formation of cyclic urethans, since Haworth *et al.*[3] isolated these in their work with tetramethylgluconamides

$$\begin{array}{c}CONH_2\\|\\H-C-OCH_3\\|\\CH_3O-C-H\\|\\H-C-OH\\|\\H-C-OCH_3\\|\\CH_2OCH_3\end{array} \longrightarrow \begin{array}{c}N=CO\\|\\H-C-OCH_3\\|\\CH_3O-C-H\\|\\H-C-OH\\|\\H-C-OCH_3\\|\\CH_2OCH_3\end{array} \longrightarrow \begin{array}{c}NH\text{------}\\|\quad\quad\quad|\\H-C-OCH_3\;|\\|\quad\quad\quad|\\CH_3O-C-H\;\;CO\\|\quad\quad\quad|\\H-C-O\text{-----}\\|\\H-C-OCH_3\\|\\CH_2OCH_3\end{array} \longrightarrow \begin{array}{c}CHO\\|\\CH_3O-C-H\\|\\H-C-OH\\|\\H-C-OCH_3\\|\\CH_2OCH_3\end{array}$$

General reaction conditions

The preparation is carried out by adding an aqueous alkaline hypochlorite solution to an ice-cold aqueous solution of the amide[5]. The mixture is sometimes warmed to dissolve the amide and then acidified with hydrochloric acid. After heating to 35° for a short time, the acidic solution is neutralized with calcium carbonate. Aldehydic products are usually isolated by forming semicarbazones or, alternatively, by evaporating the aqueous solution and extracting with acetone.

Applications

The reaction also takes place with α-hydroxyamides other than carbohydrates. Weerman[2] illustrated this by preparing benzaldehyde from mandelamide

$$\text{Ph-CHOH-CONH}_2 \longrightarrow \text{Ph-CHO}$$

WEERMAN DEGRADATION

Mason and Nord[6] used the procedure to prepare 2-thiophene-acetaldehyde but achieved a yield of only 19 per cent

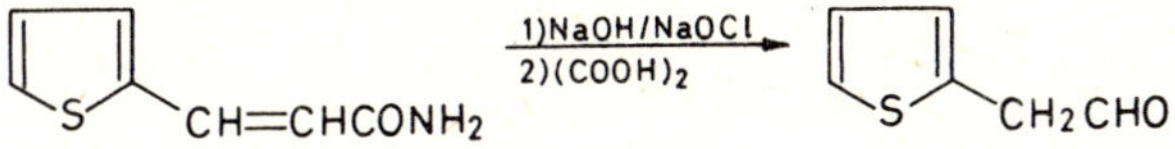

[thiophene-CH=CHCONH$_2$] →(1)NaOH/NaOCl, 2)(COOH)$_2$)→ [thiophene-CH$_2$CHO]

2-Thiopheneacrylamide

O'Colla and co-workers[7] have used Weerman's conditions to cleave polyimides prepared from polysaccharides such as starch. They found this procedure to be comparable with other accepted methods for cleaving these macromolecules.

REFERENCES

[1] Weerman, R. A. *Justus Liebigs Annln Chem.* 401 (1913) 1 [*Chem. Abstr.* 8 (1914) 64]
[2] Weerman, R. A. *Proc. Sect. Sci. ned. Akad. Wet.* 17 (1915) 1163 [*J. chem. Soc.* (A) 108 (1915) [1] 387]; *Recl Trav. chim. Pays-Bas Belg.* 37 (1918) 16 [*Chem. Abstr.* 12 (1918) 1463]
[3] Ault, R. G., Haworth, W. N. and Hirst, E. L. *J. chem. Soc.* (1934) 1722; Haworth, W. N., Peat, S. and Whetstone, J. (1938) 1975
[4] Arcus, C. L., with Greenwood, D. B. *J. chem. Soc.* (1953) 1937; with Prydal, B. S. (1954) 4018
[5] Irvine, J. C. and Pryde, J. *J. chem. Soc.* 125 (1924) 1045
[6] Mason, C. D. and Nord, F. F. *J. org. Chem.* 16 (1951) 1869
[7] O'Colla, P. S., O'Donnell, J. J. and Mulloy, J. A. *Proc. chem. Soc.* (1961) 300

32

WOHL DEGRADATION

Nature of the reaction

The terminal carbon atom is eliminated from an aldose sugar by the intermediate formation of a nitrile, followed by treatment with ammoniacal silver oxide

$$\begin{array}{c}\text{CHO}\\|\\\text{CHOH}\\|\\(\text{CHOH})_n\\|\\R\end{array} \xrightarrow{NH_2OH} \begin{array}{c}\text{CH=NOH}\\|\\\text{CHOH}\\|\\(\text{CHOH})_n\\|\\R\end{array} \xrightarrow[\text{Pyridine}]{(CH_3CO)_2O} \begin{array}{c}\text{CN}\\|\\\text{CHOCOCH}_3\\|\\(\text{CHOCOCH}_3)_n\\|\\R\end{array} \xrightarrow{NH_3/Ag_2O} \begin{array}{c}\text{CH(NHCOCH}_3)_2\\|\\\text{CHOH}\\|\\(\text{CHOH})_n\\|\\R\end{array} \xrightarrow[\text{2)Ba CO}_3]{1)H_2SO_3} \begin{array}{c}\text{CHO}\\|\\(\text{CHOH})_n\\|\\R\end{array}$$

Historical development

Wohl's ideas[1] for degrading aldoses originated in 1891 and were applied[2] in 1893 to the elimination of the terminal carbon atom from glucose to give arabinose. Later he converted arabinose to erythrose[3]

$$\begin{array}{c}\text{CHO}\\|\\\text{H}-\text{C}-\text{OH}\\|\\\text{HO}-\text{C}-\text{H}\\|\\\text{H}-\text{C}-\text{OH}\\|\\\text{H}-\text{C}-\text{OH}\\|\\\text{CH}_2\text{OH}\end{array} \longrightarrow \begin{array}{c}\text{CHO}\\|\\\text{HO}-\text{C}-\text{H}\\|\\\text{H}-\text{C}-\text{OH}\\|\\\text{H}-\text{C}-\text{OH}\\|\\\text{CH}_2\text{OH}\end{array} \longrightarrow \begin{array}{c}\text{CHO}\\|\\\text{H}-\text{C}-\text{OH}\\|\\\text{H}-\text{C}-\text{OH}\\|\\\text{CH}_2\text{OH}\end{array}$$

D(+)-Glucose   D(−)-Arabinose   D(−)-Erythrose

This degradation is the reverse of the Kiliani–Fischer method for increasing the length of aldose sugar chains (No. 28).

Wohl employed sodium acetate and acetic anhydride to form the acetyl derivatives, used to protect the hydroxy groups during the process, but Behrend[4] found pyridine and acetic anhydride to be superior for this purpose.

The overall yield for the conversion is between 30 and 40 per cent. A comparison between this and other methods of degradation has been made by Deulofeu[5].

Mechanism

A great deal of interest has revolved around the diacetamide formed by treating the acetylated nitrile with silver oxide[6]. It is

WOHL DEGRADATION

believed to arise from an intramolecular rearrangement of the acetyl groups from the second and third carbon atoms[7]. The rearrangement is initiated by attack of ammonia on the carbonyl formed after loss of the nitrile group

$$
\begin{array}{c}
\text{CN} \\
| \\
\text{H--C--O--COCH}_3 \\
| \\
\text{H--C--O--COCH}_3 \\
| \\
\text{H--C--O--COCH}_3 \\
| \\
R
\end{array}
\quad
\begin{array}{c}
\text{H--C=O} \\
| \\
\text{H--C--O--COCH}_3 \\
| \\
\text{H--C--O--COCH}_3 \\
| \\
R
\end{array}
\xrightarrow{\text{NH}_3}
\begin{array}{c}
\text{OH} \\
| \\
\text{H--C--NH}_2 \\
| \\
\text{H--C--O--COCH}_3 \\
| \\
\text{H--C--O--COCH}_3 \\
| \\
R
\end{array}
$$

$$
\longrightarrow
\begin{array}{c}
\text{OH} \\
| \\
\text{H--C--NH}\diagdown \\
| \quad\quad\quad \text{C}\diagup\text{CH}_3 \\
\text{H--C--O}\diagup\quad\diagdown\text{OH} \\
| \\
\text{H--C--O--COCH}_3 \\
| \\
R
\end{array}
\longrightarrow
\begin{array}{c}
\text{OH} \\
| \\
\text{H--C--NH--COCH}_3 \\
| \\
\text{H--C--OH} \\
| \\
\text{H--C--OCOCH}_3 \\
| \\
R
\end{array}
\xrightarrow{\text{NH}_3}
$$

$$
\begin{array}{c}
\quad\quad\text{OH} \\
\quad\quad | \\
\quad\text{H--C--NH--COCH}_3 \\
\text{NH}_2\; | \text{HCOH} \\
| \quad | \\
\text{CH}_3\text{--C--O--CH} \\
| \quad\quad | \\
\text{OH} \quad R
\end{array}
\xrightarrow{-\text{H}_2\text{O}}
\begin{array}{c}
\text{HC--NHCOCH}_3 \\
\text{HN}\diagup \quad\text{HCOH} \\
| \quad\quad | \\
\text{CH}_3\text{--C--O--CH} \\
| \quad\quad | \\
\text{OH} \quad R
\end{array}
\longrightarrow
\begin{array}{c}
\text{CH}_3\text{CONH--CH--NHCOCH}_3 \\
| \\
\text{HCOH} \\
| \\
\text{HO--C--H} \\
| \\
R
\end{array}
$$

General reaction conditions

The several stages are executed by straightforward procedures.

(a) The oxime—this may be formed by adding the aldose to a solution of hydroxylamine hydrochloride in sodium ethoxide and warming the mixture. The oxime is deposited after standing.

(b) The acetyl-nitrile derivative[8]—the oxime is dissolved in pyridine and acetic anhydride is added. After warming between 50° and 70° for a short time, and standing at room temperature for 24 h, the solution is poured onto iced water. The product is often isolated as a syrup.

(c) Rearrangement to the diacetamide[9]—this can be accomplished by heating the acetyl nitrile with concentrated aqueous ammonia (30 per cent). The silver oxide originally used has been found to be unnecessary. After cooling and standing, the solution is evaporated and the product recrystallized from ethanol.

(d) Hydrolysis—A solution of the diacetamide in 0·5N sulphuric acid is prepared and left to stand for 1 h. It is then neutralized with barium hydroxide, filtered and the aqueous solution evaporated to give the degraded carbohydrate.

Modification

Zemplén[10] found that the acetyl nitrile could be converted directly to the lower aldose by dissolving in chloroform and adding sodium methoxide to the cold stirred solution. After neutralization by acetic acid, the aldose was obtained by evaporation of the aqueous phase. The method was originally developed for use on the degradation of disaccharides for which Wohl's procedure is unsuitable.

Application

Other acyl derivatives of carbohydrates have been studied under Wohl and Zemplen conditions. It has been found[11] that the pentabenzoyl nitriles derived from aldohexoses could be degraded to the aldopentoses as readily as could the penta-acetyl nitriles

```
      CHO                    HC=NOH                      CN
       |                       |                          |
  HO—C—H                   HO—C—H               C₆H₅COO—C—H                   CHO
       |                       |                          |                     |
  HO—C—H                   HO—C—H               C₆H₅COO—C—H               HO—C—H
       |      NH₂OH·HCl         |     C₆H₅COCl              |       CHCl₃         |
   H—C—OH   ————————→      H—C—OH   ————————→      H—C—OCOC₆H₅  ————————→   H—C—OH
       |                       |      Pyridine              |       NaOCH₃         |
   H—C—OH                   H—C—OH                 H—C—OCOC₆H₅             H—C—OH
       |                       |                          |                     |
     CH₂OH                   CH₂OH                  CH₂OCOC₆H₅               CH₂OH

   D(+)-Mannose                                                        D(-)-Arabinose
```

REFERENCES

[1] Wohl, A. *Ber. dt. chem. Ges.* 24 (1891) 994 [*J. chem. Soc.* (A) 60 (1891) 813]
[2] Wohl, A. *Ber. dt. chem. Ges.* 26 (1893) 730; with List, E. 30 (1897) 3101 [*J. chem. Soc.* (A) 64 (1893) [1] 292; 74 (1898) [1] 168]
[3] Wohl, A. *Ber. dt. chem. Ges.* 32 (1899) 3666 [*J. chem. Soc.* (A) 78 (1900) [1] 140]
[4] Behrend, R. *Justus Liebigs Annln Chem.* 353 (1907) 106 [*Chem. Abstr.* 1 (1907) 2234]
[5] Deulofeu, V. *Adv. Carbohyd. Chem.* 4 (1949) 119
[6] Sowden, J. C. *The Carbohydrates* (ed. Pigman, W. W.), p. 119, New York (Academic Press) 1957
[7] Isbell, H. S. and Frush, H. L. *J. Am. chem. Soc.* 71 (1949) 1579
[8] Deulofeu, V. *J. chem. Soc.* (1932) 2973
[9] Hockett, R. C. *J. Am. chem. Soc.* 57 (1935) 2266
[10] Zemplén, G. *Ber. dt. chem. Ges.* 59 (1926) 1254; with Kiss, D. 60 (1927) 165 [*Chem. Abstr.* 20 (1926) 2988; 21 (1927) 1633]
[11] Labriola, E. R. de and Deulofeu, V. *J. org. Chem.* 12 (1947) 726

33

MEERWEIN–PONNDORF–VERLEY REDUCTION

Nature of the reaction
Carbonyl compounds are reduced to alcohols by the action of aluminium alkoxides in alcohols

$$RCHO + R'CH_2OH \xrightleftharpoons{(R'CH_2O)_3Al} RCH_2OH + R'CHO$$

Historical development
From independent researches, Meerwein[1], Ponndorf[2] and Verley[3] published almost simultaneously this reaction which now bears their names. Their work had developed from Claisen's[4] observation that aromatic aldehydes were converted into esters by the action of sodium alkoxides

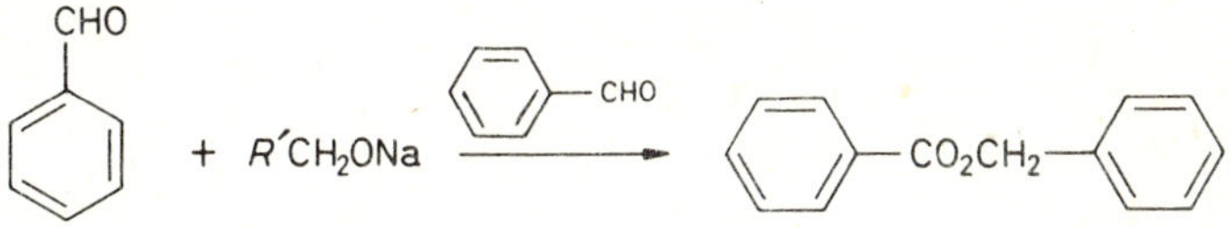

$$\text{PhCHO} + R'CH_2ONa \longrightarrow \text{Ph-CO}_2CH_2\text{-Ph}$$

Meerwein and Verley employed aluminium ethoxide in their reductions, while Ponndorf found aluminium benzyloxide also suitable. The main advantages of the method lie in the mild conditions, high yields (60–90 per cent) and the lack of reactivity with unsaturated systems. One of its earliest successes was the reduction of cinnamaldehyde to cinnamyl alcohol

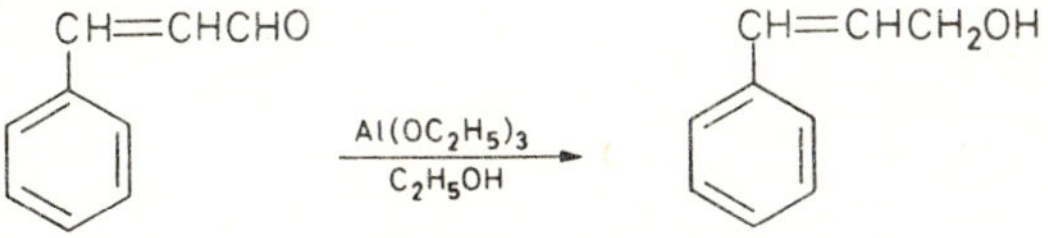

$$\text{PhCH=CHCHO} \xrightarrow[C_2H_5OH]{Al(OC_2H_5)_3} \text{PhCH=CHCH}_2OH$$

The early workers were aware that the reaction was reversible but did not investigate this facet which was later developed by Oppenauer (No. 34).

NAMED ORGANIC REACTIONS

The full value of the reduction was not appreciated until 1937 when Lund[5] described his work on the reduction of ketones with aluminium isopropoxide[6]: in most cases, yields of 90–100 per cent were obtained with this reagent

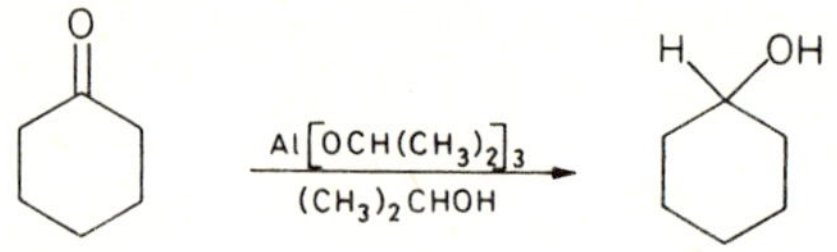

Cyclohexanone $\xrightarrow[(CH_3)_2CHOH]{Al[OCH(CH_3)_2]_3}$ Cyclohexanol

The alcohol employed in the reaction is simultaneously oxidized to the corresponding ketone. The value and scope of the reaction has been reviewed by Wilds[7].

Mechanism

The work of Meerwein[8] and Jackman[9] has suggested that the mechanism involves a cyclic transition state formed between the carbonyl compound and the alkoxide. Recent work by Yager and Hancock[10] on the equilibrium constants of methyl ketones, $R\text{COCH}_3$, under M.P.V. conditions lends further support to the generally accepted mechanism.

The initial step is considered to be an equilibrium between the alcohol and the aluminium alkoxide

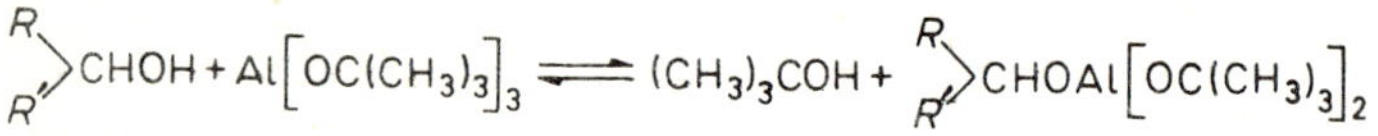

$$\underset{R'}{\overset{R}{>}}\text{CHOH} + \text{Al}[\text{OC}(CH_3)_3]_3 \rightleftharpoons (CH_3)_3\text{COH} + \underset{R'}{\overset{R}{>}}\text{CHOAl}[\text{OC}(CH_3)_3]_2$$

The second step is reaction with the ketone to give the cyclic transition state. (Structures are simplified by employing Al/3 to represent that portion of alkoxide involved)

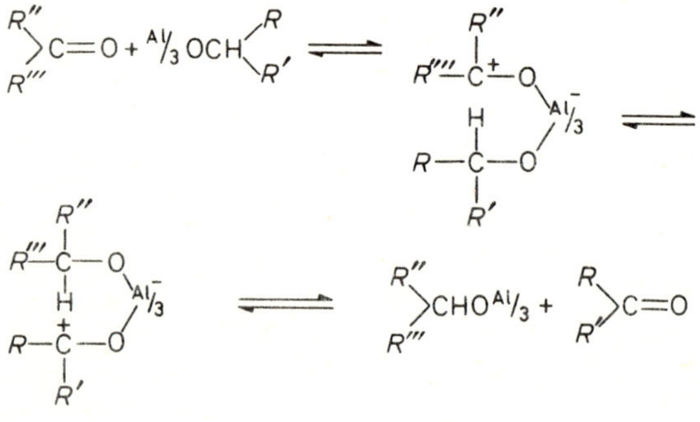

By distilling off the ketone formed, the equilibrium is thrown to the right to enable the high yields of reduction product to be obtained.

General reaction conditions

Aluminium isopropoxide is the alkoxide most commonly employed for this purpose, as it effects rapid reactions with very few side reactions; in many cases, particularly the reduction of aldehydes, it also gives superior yields compared with other alkoxides.

Irrespective of the alkoxide employed, reactions are generally carried out in isopropyl alcohol, as the acetone formed is readily removed.

Young and co-workers[11] established conditions for the reduction of crotonaldehyde which are of wide application. The aldehyde in isopropyl alcohol was added to an excess of aluminium isopropoxide in the same alcohol and the mixture slowly distilled for 9 h, using a column-condenser unit. Excess solvent was removed under reduced pressure, and hydrolysis of the residue with sulphuric acid led to the separation of the crude product as an oil

$$CH_3CH=CHCHO \longrightarrow CH_3CH=CHCH_2OH$$

For ketones which may be resistant to reduction under these conditions, it is common to make a solution from solid aluminium isopropoxide and the ketone in a high-boiling solvent such as toluene or xylene to which only a small quantity of isopropyl alcohol is added. By very slow distillation a suitable reduction may be effected.

Modifications

Amongst the various by-products in M.P.V. reductions, hydrocarbons have been found to arise from further reduction of the main product. By working at 250° with an excess of aluminium isopropoxide, Hoffsommer and co-workers[12] have managed to reduce aromatic ketones to hydrocarbons with very high yields (95 per cent)

Applications

The wide acceptance of this method of reduction as a synthetic procedure, particularly in the realms of steroid chemistry, has been due to the mild conditions which are employed. Marker and Rohrmann[13] prepared both α- and β-oestradiols by reducing oestrone with aluminium isopropoxide in isopropyl alcohol

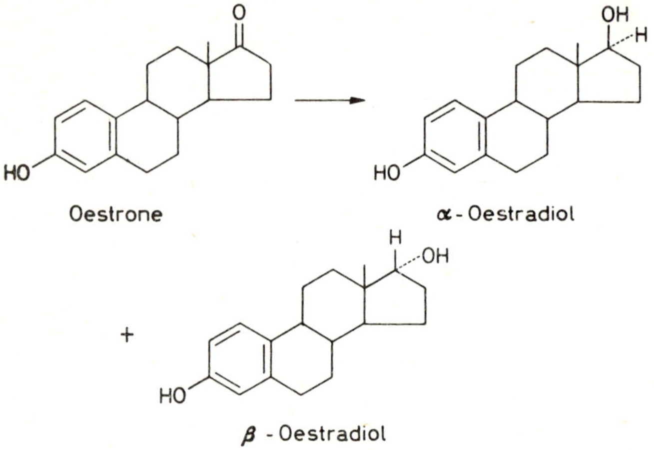

Oestrone → α-Oestradiol

+ β-Oestradiol

REFERENCES

[1] Meerwein, H. and Schmidt, R. *Justus Liebigs Annln Chem.* 444 (1925) 221 [*Chem. Abstr.* 19 (1925) 3251]
[2] Ponndorf, W. *Angew. Chem.* 39 (1926) 138 [*Chem. Abstr.* 20 (1926) 1611]
[3] Verley, A. *Bull. Soc. chim. Fr.* 37 (1925) 537; 41 (1927) 788 [*Chem. Abstr.* 19 (1925) 2635; 21 (1927) 3346]
[4] Claisen, L. *Ber. dt. chem. Ges.* 20 (1887) 646 [*J. chem. Soc.* (A) 52 (1887) 574]
[5] Lund, H. *Ber. dt. chem. Ges.* 70 (1937) 1520 [*Chem. Abstr.* 31 (1937) 6611]
[6] Linstead, R. P. *Rep. Prog. Chem.* 34 (1938) 228
[7] Wilds, A. L. *Org. React.* 2 (1944) 178
[8] Meerwein, H. *et al.*, *J. prakt. Chem.* 147 (1936) 211 [*Chem. Abstr.* 31 (1937) 656]
[9] Jackman, L. M., with Mills, J. A. *Nature, Lond.* 164 (1949) 789; with Macbeth, A. K. *J. chem. Soc.* (1952) 3252
[10] Yager, B. J. and Hancock, C. K. *J. org. Chem.* 30 (1965) 1174
[11] Young, W. G., Harting, W. H. and Crossley, F. S. *J. Am. chem. Soc.* 58 (1936) 100
[12] Hoffsommer, R. D., Taub, D. and Wendler, N. L. *Chemy Ind.* (1964) 482
[13] Marker, R. E. and Rohrmann, E. *J. Am. chem. Soc.* 60 (1938) 2927

34

OPPENAUER OXIDATION

Nature of the reaction

Alcohols may be dehydrogenated to the corresponding aldehyde or ketone by the action of an aluminium alkoxide in the presence of a ketone

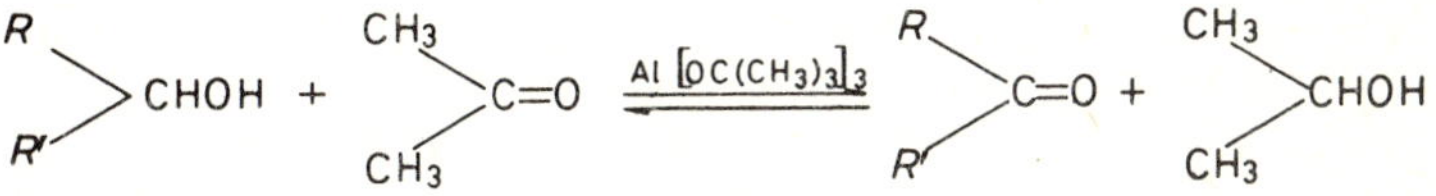

$$\underset{R'}{\overset{R}{>}}CHOH + \underset{CH_3}{\overset{CH_3}{>}}C=O \xrightarrow{Al[OC(CH_3)_3]_3} \underset{R'}{\overset{R}{>}}C=O + \underset{CH_3}{\overset{CH_3}{>}}CHOH$$

Historical development

The reversible nature of the Meerwein–Ponndorf–Verley reaction (No. 33) was demonstrated by both Ponndorf and Verley. Application of the reversibility to the preparation of ketones was not shown until Oppenauer[1] reported his investigations on unsaturated sterols. Oxidation of cholesterol necessitated the use of these mild conditions to give cholestenone[2], with a 70–81 per cent yield

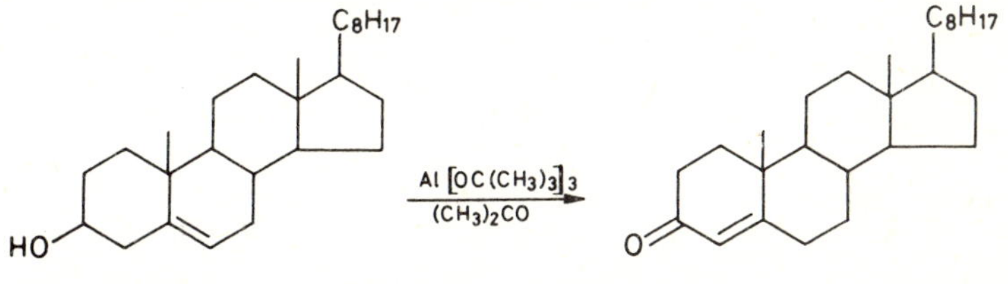

Robinson and co-workers[3] developed the procedure further in the preparation of ketonic derivatives of naphthalene and phenanthrene.

Although the reaction is generally applied to secondary alcohols, some primary groups have been oxidized; however, yields for these are generally lower, between 50 and 90 per cent.

The Oppenauer oxidation has been reviewed in detail by Djerassi[4] and considered in conjunction with the Meerwein–Ponndorf–Verley reduction by Bersin[5].

Mechanism

The Oppenauer oxidation proceeds by the same cyclic transition state mechanism as the M.P.V. reduction and involves the formation of a mixed alkoxide between the alcohol and the aluminium t-butoxide

$$\underset{R'}{\overset{R}{>}}CHOH + Al_{/3}OC(CH_3)_3 \rightleftharpoons (CH_3)_3COH + \underset{R'}{\overset{R}{>}}CHOAl_{/3}$$

$$\underset{R'}{\overset{R}{>}}CHOAl_{/3} + \underset{CH_3}{\overset{CH_3}{>}}C=O \rightleftharpoons \begin{array}{c} CH_3-\overset{CH_3}{\underset{+}{C}}-O \\ \diagdown \\ Al^-_{/3} \\ \diagup \\ R-\underset{R'}{\overset{H}{C}}-O \end{array} \rightleftharpoons$$

$$\begin{array}{c} CH_3-\overset{CH_3}{\underset{H}{C}}-O \\ \diagdown \\ Al^-_{/3} \\ \diagup \\ R-\overset{+}{\underset{R'}{C}}-O \end{array} \rightleftharpoons \underset{CH_3}{\overset{CH_3}{>}}CHO\,Al_{/3} + \underset{R'}{\overset{R}{>}}C=O$$

General reaction conditions

Since the Oppenauer oxidation represents one side of an equilibrium reaction, conditions must be chosen such that the process will be forced in the required direction. Because of this, procedures necessitate either the use of a large excess of carbonyl starting material or of a carbonyl compound that produces an alcohol boiling at a higher temperature than the ketone required from the reaction.

Adkins and Franklin[6] found that acetone, methyl ethyl ketone, cyclohexanone, benzil and *p*-benzoquinone were of particular value as hydrogen acceptors for the reaction.

This is carried out by mixing the alcohol with aluminium t-butoxide and an excess of the hydrogen acceptor, usually acetone; benzene and toluene are often used as additional solvents. The mixture is gently heated for several hours and the product, if a low-boiling ketone, distilled off. Alternatively, acidification of the mixture, followed by water-washing the organic layer and removal of the solvent, enables the product to be obtained.

Modifications

The Oppenauer oxidation fails with molecules containing basic groups in close proximity to the hydroxyl group.

Woodward[7] found that aromatic alcohols possessing basic groups in the benzene ring could be oxidized by using potassium t-butoxide in benzene solution with benzophenone as the hydrogen acceptor;

a 40 per cent yield of 2-dimethylamino-5-methylbenzaldehyde was obtained

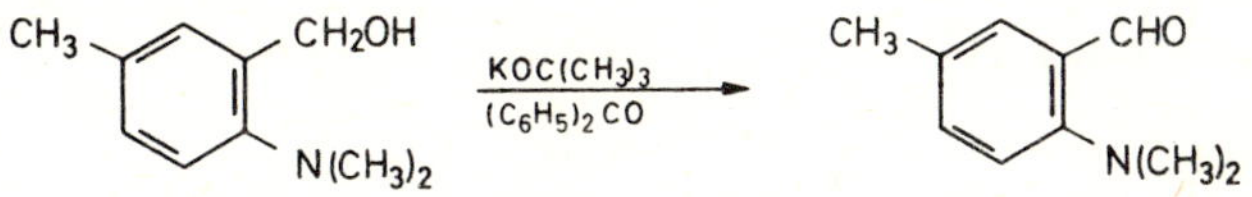

$$\text{CH}_3\text{-C}_6\text{H}_3(\text{N}(\text{CH}_3)_2)\text{-CH}_2\text{OH} \xrightarrow[\text{(C}_6\text{H}_5)_2\text{CO}]{\text{KOC(CH}_3)_3} \text{CH}_3\text{-C}_6\text{H}_3(\text{N}(\text{CH}_3)_2)\text{-CHO}$$

Warnhoff[8] found this method unsuitable for sensitive alcohols but was able to carry out oxidations under mild conditions by using the rapid hydrogen acceptor fluorenone,

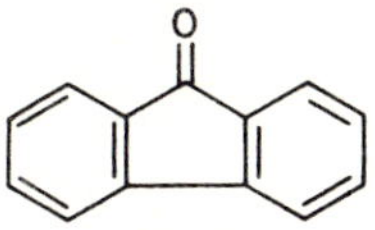

Fluorenone

Applications

Most applications of the Oppenauer oxidation have been in the field of natural product chemistry, particularly with steroids. As in the case of cholesterol, oxidation of $\triangle^5$-3-hydroxy steroids leads not only to formation of the ketone but also to migration of the double bond to the $\triangle^4$ position

In their work on the structure of irone, Ruzicka et al.[9] used aluminium isopropoxide in acetone and benzene to convert α- and β-ionols to the corresponding ionones

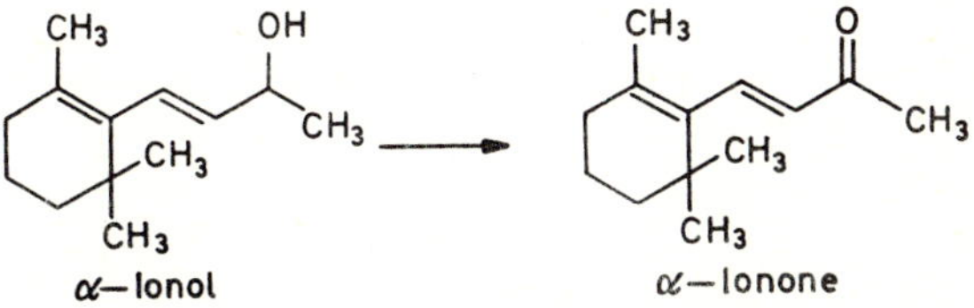

α—Ionol   α—Ionone

Lee and Bhardway[10] have employed cinnamaldehyde as the hydrogen acceptor to prepare cyclopropane carboxaldehyde from cyclopropyl carbinol (20–25 per cent yield)

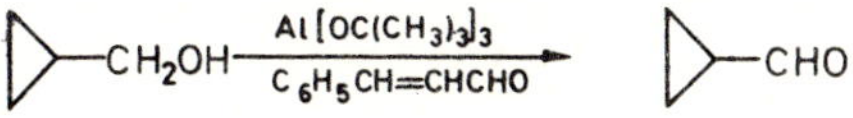

$$\triangleright\text{-CH}_2\text{OH} \xrightarrow[\text{C}_6\text{H}_5\text{CH=CHCHO}]{\text{Al[OC(CH}_3)_3]_3} \triangleright\text{-CHO}$$

REFERENCES

[1] Oppenauer, R. V. *Recl Trav. chim. Pays-Bas Belg.* 56 (1937) 137 [*Chem. Abstr.* 31 (1937) 3061]
[2] Oppenauer, R. V. *Org. Synth., Coll. Vol.* 3 (1955) 207
[3] Robinson, R. and Walker, J. *J. chem. Soc.* (1938) 185; Cornforth, J. W. and Robinson, R. (1949) 1855
[4] Djerassi, C. *Org. React.* 6 (1951) 207
[5] Bersin, T. *Newer Methods of Preparative Chemistry*, p. 125, New York (Interscience) 1948
[6] Adkins, H. and Franklin, R. C. *J. Am. chem. Soc.* 63 (1941) 2381
[7] Woodward, R. B. and Kornfield, E. C. *J. Am. chem. Soc.* 70 (1948) 2511
[8] Warnhoff, E. W. and Reynolds-Warnhoff, P. *J. org. Chem.* 28 (1963) 1431
[9] Ruzicka, L. *et al.*, *Helv. chim. Acta* 31 (1948) 257 [*Chem. Abstr.* 42 (1948) 5427]
[10] Lee, C. C. and Bhardway, I. S. *Can. J. Chem.* 41 (1963) 1031

35
BARBIER–WIELAND DEGRADATION

Nature of the reaction

Acids are converted into the next lower homologue by the formation of an olefin from the ester, followed by oxidation of the double bond

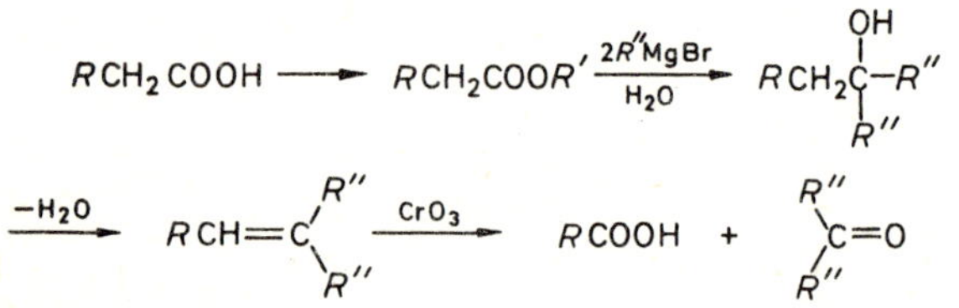

$$RCH_2COOH \longrightarrow RCH_2COOR' \xrightarrow[H_2O]{2R''MgBr} RCH_2\underset{R''}{\overset{OH}{\underset{|}{\overset{|}{C}}}}-R''$$

$$\xrightarrow{-H_2O} RCH=C\underset{R''}{\overset{R''}{\diagup}} \xrightarrow{CrO_3} RCOOH + \underset{R''}{\overset{R''}{\diagup}}C=O$$

Historical development

The method of degrading monobasic and dibasic acids, introduced by Barbier and Locquin[1] in 1913, involved treating the ester with a double molar proportion of methyl magnesium bromide per carbonyl group to give a tertiary alcohol, oxidized in turn to the shorter-chain acid

BARBIER–WIELAND DEGRADATION

$$\text{(CH}_3\text{)}_2\text{CH–CH}_2\text{–CO}_2\text{C}_2\text{H}_5 \xrightarrow{2\,\text{CH}_3\text{MgI}} \text{(CH}_3\text{)}_2\text{CH–CH}_2\text{–C(OH)(CH}_3\text{)–CH}_3$$

$$\xrightarrow{\text{CrO}_3} \text{(CH}_3\text{)}_2\text{CH COOH} + \text{(CH}_3\text{)}_2\text{C=O}$$

The value of the procedure was not fully recognized until Wieland and his collaborators[2] applied it to the stepwise degradation of the side chain in cholanic acid.

The method was later improved[3] by dehydrating the tertiary alcohol to an olefin prior to the oxidation step. The reaction is generally employed in structural determinations, particularly in the field of steroids. Overall yields vary from less than 10 to over 40 per cent. A number of methods for improving the yield have been discussed in outline by Shoppee[4].

Mechanism

Chromium trioxide can give rise to a wide variety of products when used to oxidize carbon–carbon double bonds. Under different conditions ketones, acids and epoxides have all been isolated, but the mechanism for oxidations under Barbier–Wieland conditions is still obscure[5].

The extensive work of Hickinbottom and his collaborators[6] suggests that the initial step is electrophilic attack by chromium trioxide or chromic acid on the olefinic double bond to give either a carbanion or a cyclic intermediate[7]. The latter is cleaved to give an aldehyde and a ketone, the aldehyde being oxidized further to the required acid[8]

$$R_2C=CHR' + \text{CrO}_2(\text{OH})_2 \longrightarrow R_2C(\text{O–})\text{–CH}(R')(\text{–O–})\text{Cr(OH)}_2$$

$$\longrightarrow R_2C=O + R'CH=O$$

$$R'CH=O + H_2\text{CrO}_4 \rightleftharpoons R\text{–CH(OH)(OCrO}_3\text{H)} \longrightarrow R\text{–C(OH)=O} + H_2\text{CrO}_3$$

NAMED ORGANIC REACTIONS

General reaction conditions

The methyl ester of the acid starting material is produced by standard esterification procedures.

Formation of the tertiary alcohol is accomplished by refluxing the ester with a slight excess of a double molar proportion of phenyl magnesium bromide in ether. After the complex has been formed, the bulk of the solution is reduced and the tertiary alcohol obtained by treating the residue with iced dilute hydrochloric acid.

The crude alcohol is dehydrated by refluxing for about 1 h with acetic anhydride. Excess anhydride is removed and the crude olefin washed with water and dried.

Oxidation of the olefin is carried out below 50° in glacial acetic acid to which a solution of chromium trioxide in 70 per cent acetic acid is slowly added. After a total time of 30–45 min, excess chromium trioxide is destroyed by careful addition of methanol. The solution is concentrated by evaporation and treated with dilute hydrochloric acid, the solid product being filtered off, washed and recrystallized.

The preparation of 3-methyl-3-phenylhexanoic acid by Lane and Wallis[9] following this procedure gave an overall yield of about 35 per cent

$$CH_3(CH_2)_3 - \underset{\underset{C_6H_5}{|}}{\overset{\overset{CH_3}{|}}{C}} - CH_2CO_2CH_3 \quad \xrightarrow[\substack{2)\ (CH_3CO)_2O \\ 3)\ CrO_3}]{1)\ C_6H_5MgBr} \quad CH_3(CH_2)_3 - \underset{\underset{C_6H_5}{|}}{\overset{\overset{CH_3}{|}}{C}} - CO_2H$$

Riegel and co-workers[10] have given a detailed procedure for the degradation of steroids which is of wide application.

Modification

The major modification has been due to Miescher and his collaborators[11] who developed a procedure for the elimination of three carbon atoms at a time from the carbon chain. After formation of the double bond, a Wohl–Ziegler reaction (No. 25) with *N*-bromosuccinimide introduces a bromine atom into the allylic position of the chain. Dehydrobromination leads to the formation of another double bond which can be oxidized to eliminate the three carbon atoms

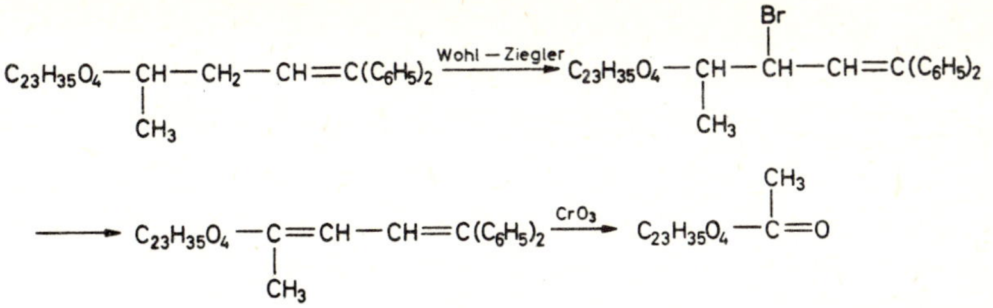

BARBIER–WIELAND DEGRADATION

$$C_{23}H_{35}O_4-\underset{\underset{CH_3}{|}}{CH}-CH_2-CH=C(C_6H_5)_2 \xrightarrow{Wohl-Ziegler} C_{23}H_{35}O_4-\underset{\underset{CH_3}{|}}{CH}-\underset{\underset{}{|}}{\overset{\overset{Br}{|}}{CH}}-CH=C(C_6H_5)_2$$

$$\longrightarrow C_{23}H_{35}O_4-\underset{\underset{CH_3}{|}}{C}=CH-CH=C(C_6H_5)_2 \xrightarrow{CrO_3} C_{23}H_{35}O_4-\underset{\underset{CH_3}{|}}{C}=O$$

The oxidation step in Barbier–Wieland reactions has been improved in a number of cases[12] by using sodium metaperiodate with ruthenium tetroxide catalyst in place of chromic oxide.

Application

Typical of the use of the degradation method was the determination of the position of the hydroxyl group and a methyl group in cholesterol[13]. This was oxidized to an unknown keto acid, (A), that could be reduced to an acid (B) which, after two Barbier–Wieland steps, gave a third acid which would not undergo further degradation and was, therefore, attached to a tertiary carbon atom

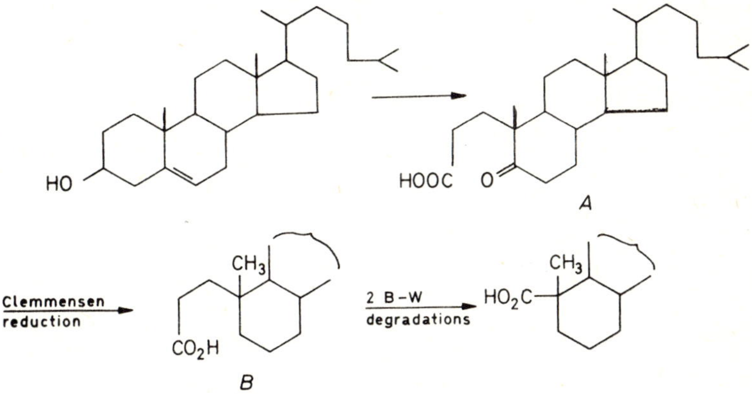

Pattison and Buchanan[14] have successfully applied the Barbier–Wieland degradation to fluoroalkanoic acids.

REFERENCES

[1] Barbier, P. and Locquin, R. *C. r. hebd. Séanc. Acad. Sci., Paris* **156** (1913) 1443 [*Chem. Abstr.* **7** (1913) 3110]

[2] Wieland, H., Schlichting, O. and Jacobi, R. *Hoppe-Seyler's Z. physiol. Chem.* **161** (1926) 80 [*Chem. Abstr.* **21** (1927) 590]

3 Hoehn, W. M. and Mason, H. L. *J. Am. chem. Soc.* 60 (1938) 1493
4 Shoppee, C. W. *Rep. Prog. Chem.* 44 (1947) 184
5 House, H. O. *Modern Synthetic Reactions*, p. 92, New York (Benjamin) 1965
6 Hickinbottom, W. J. *et al., J. chem. Soc.* (1954) 4400; Davis, M. A. and Hickinbottom, W. J. (1958) 2205
7 Waters, W. A. *Q. Rev. chem. Soc.* 12 (1958) 287
8 Stewart, R. *Oxidation Mechanisms*, pp. 48, 54, New York (Benjamin) 1964
9 Lane, J. F. and Wallis, E. S. *J. Am. chem. Soc.* 63 (1941) 1674
10 Riegel, B., Moffett, R. B. and McIntosh, A. V. *Org. Synth., Coll. Vol.* 3 (1955) 234, 237
11 Miescher, K. *et al., Helv. chim. Acta* 27 (1944) 1815 [*Chem. Abstr.* 40 (1946) 884]
12 Berkowitz, L. M. and Rylader, P. N. *J. Am. chem. Soc.* 80 (1958) 6682; Sarel, S. and Yanuka, Y. Y. *J. org. Chem.* 24 (1959) 2018
13 Tschesche, R. *Justus Liebigs Annln Chem.* 498 (1932) 185 [*Chem. Abstr.* 27 (1933) 100]; Heilbron, I. M., Simpson, J. C. E. and Spring, F. S. *J. chem. Soc.* (1933) 626
14 Pattison, F. L. M. and Buchanan, R. L. *Biochem. J.* 92 (1964) 100

36

ELBS PERSULPHATE OXIDATION

Nature of the reaction

Potassium persulphate in alkaline solution is used to oxidize monohydric to dihydric phenols

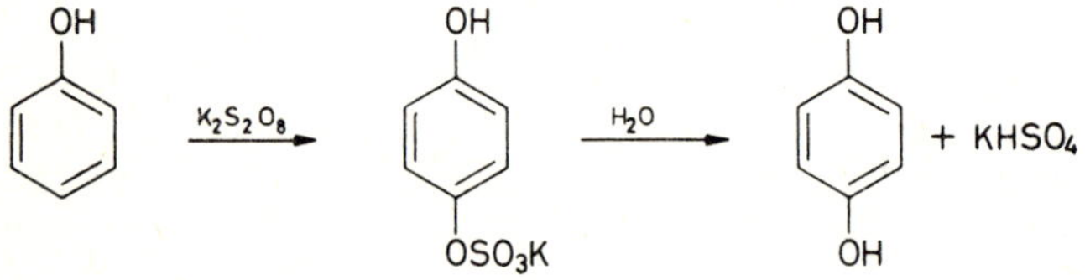

Historical development

The first reaction of this type, described by Elbs[1] in 1893, was the conversion of *o*-nitrophenol to nitroquinol

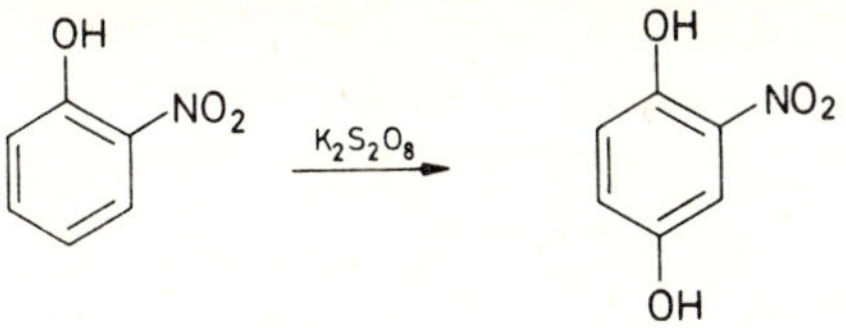

ELBS PERSULPHATE OXIDATION

Despite the fact that yields are rarely above 50 per cent, the method has been found useful for introducing a second hydroxyl group *para* to the one already present. If the *para* position is already occupied, an hydroxyl group is introduced in the *ortho* position to give a catechol derivative, generally in low yield.

A review[2] of the reaction, covering the literature to 1951, illustrates the value of the procedure, particularly in the field of natural product chemistry. A short section on the Elbs persulphate oxidation has also been included in a review of organic reactions involving electrophilic oxidation[3].

Mechanism

An ionic mechanism for the reaction has been proposed by Baker and Brown[4], suggesting that the persulphate ion is converted to an ion radical by interaction with a trace metal

$$S_2O_8^{2-} + Fe^{2+} \longrightarrow Fe^{3+} + SO_4^{2-} + \cdot OSO_3^-$$

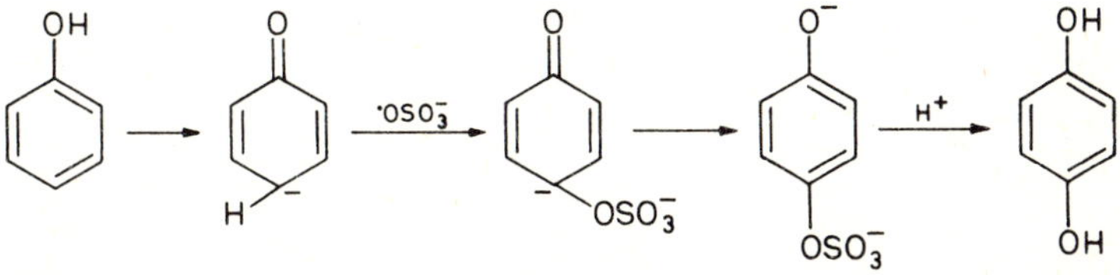

Dermer and Edmunsons' discussion of this mechanism[5] drew attention to the fact that the $\cdot OSO_3^-$ ion radical causes *ortho* substitution in arylamines but does not affect a number of other aromatic systems.

Further mechanistic studies by Behrmann and Walker[6] indicated the rate-determining step to be electrophilic attack of the persulphate ion on the phenolate. Behrmann[7] has suggested that two kinetically indistinguishable routes are possible involving the persulphate ion rather than a sulphate ion radical

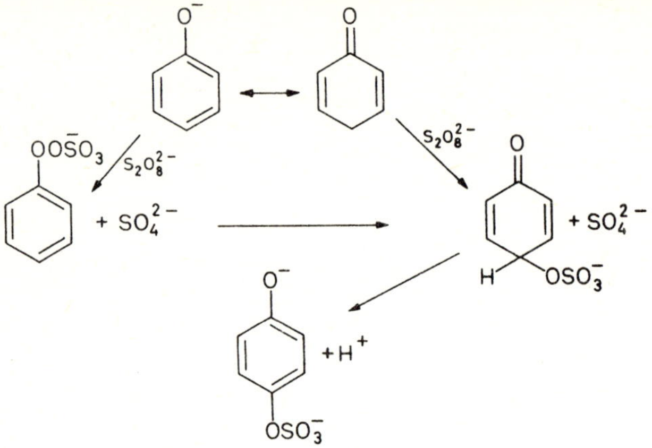

General reaction conditions

Optimum conditions for persulphate oxidations of substituted phenols were determined by Baker and Brown[4]. The phenolic compound (1 mol) is dissolved in 10 per cent aqueous sodium hydroxide (5 mol), and a saturated aqueous solution of potassium persulphate (1 mol) is added gradually with stirring during 3–4 h. The temperature is controlled below 20° for the period of the addition, and then the mixture is left to stand overnight. After acidification and removal of any solid by filtration, the aqueous solution is extracted with ether. The aqueous layer is treated further with hydrochloric acid and re-extracted. Evaporation of the combined ether extracts affords the dihydric phenol.

For dihydric phenols which are readily soluble in water, it is necessary to acidify by passing carbon dioxide into the aqueous solution. The bulk of the solution is then evaporated to a smaller volume prior to ether extraction.

By employing excess potassium persulphate, it is possible to prepare polyhydric phenols. Schock and Tabern[8] used this approach to obtain 2,3,5-trihydroxybenzoic acid from salicylic acid

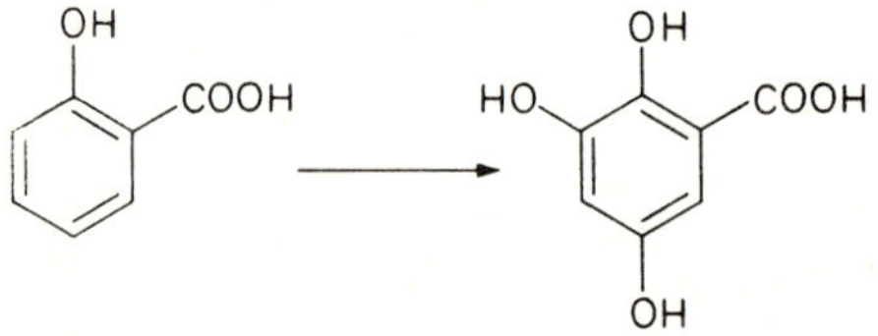

ELBS PERSULPHATE OXIDATION

Applications

Elbs persulphate oxidation has been used extensively, with various minor modifications, by Seshadri and co-workers[9] in flavone syntheses. The conversion of 5,7-dihydroxy-4'-methoxyflavone to 5,7,8-trihydroxy-4'-methoxyflavone is a typical example:

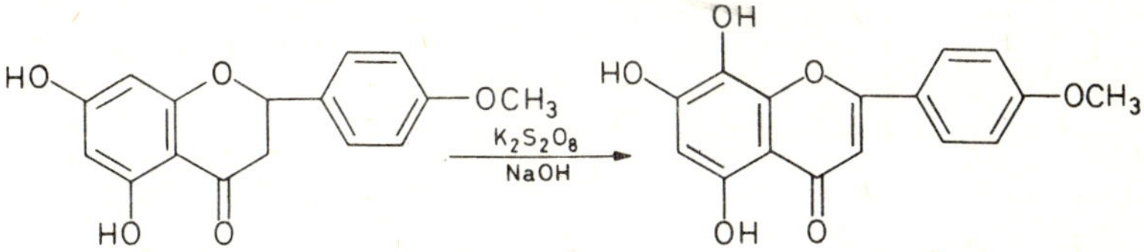

The scope of the reaction has been extended to the oxidation of aromatic amines[10], giving rise to *ortho* substitution in preference to the *para* substitution which occurs with phenols; in these preparations, the intermediate potassium salt is isolated prior to the hydrolysis step

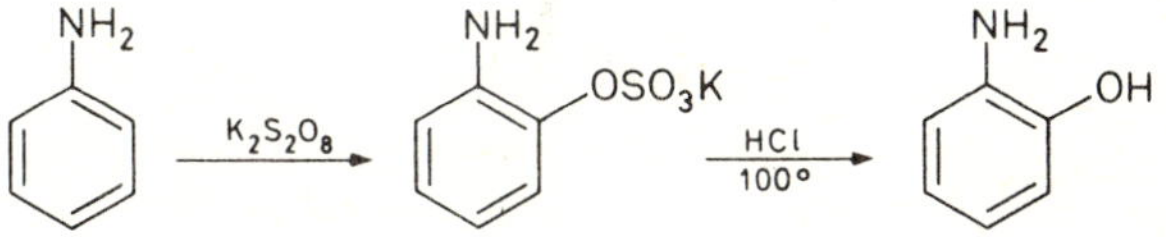

REFERENCES

[1] Elbs, K. *J. prakt. Chem.* 48 (1893) 179 [*J. chem. Soc. (A)* 64 (1893) [1] 640]
[2] Sethna, S. M. *Chem. Rev.* 49 (1951) 91
[3] Lee, J. B. and Uff, B. C. *Q. Rev. chem. Soc.* 21 (1967) 453
[4] Baker, W. and Brown, N. C. *J. chem. Soc.* (1948) 2303
[5] Dermer, O. C. and Edmison, M. T. *Chem. Rev.* 57 (1957) 103
[6] Behrmann, E. J. and Walker, P. P. *J. Am. chem. Soc.* 84 (1962) 3454
[7] Behrmann, E. J. *J. Am. chem. Soc.* 85 (1963) 3478
[8] Schock, R. U. and Tabern, D. L. *J. org. Chem.* 16 (1951) 1772
[9] Rao, K. V., Seshadri, T. R. and Viswanadham, M. *Proc. Indian Acad. Sci.* 29A (1949) 72 [*Chem. Abstr.* 44 (1950) 3985]
[10] Boyland, E., with Manson, D. and Sims, P. *J. chem. Soc.* (1953) 3623; with Sims, P. (1958) 4198

37

ÉTARD REACTION

Nature of the reaction
Aromatic alkyl groups, particularly methyl, are oxidized to carbonyl groups by the action of chromyl chloride

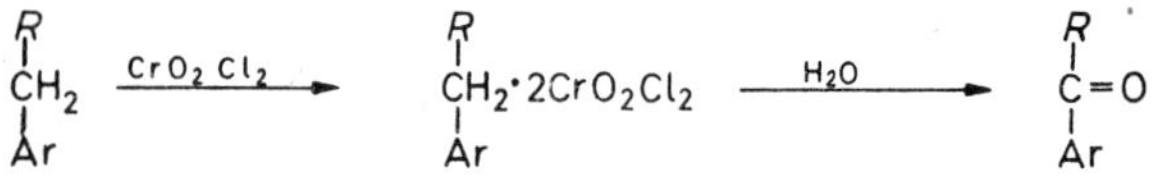

$$\underset{Ar}{\overset{R}{\underset{|}{CH_2}}} \xrightarrow{CrO_2Cl_2} \underset{Ar}{\overset{R}{\underset{|}{CH_2 \cdot 2CrO_2Cl_2}}} \xrightarrow{H_2O} \underset{Ar}{\overset{R}{\underset{|}{C=O}}}$$

$R = H$ or alkyl

Historical development

Étard[1] first reported the use of chromyl chloride for this reaction in 1877 and subsequently developed it further. Toluene was oxidized to benzaldehyde, and longer carbon chains gave ketones

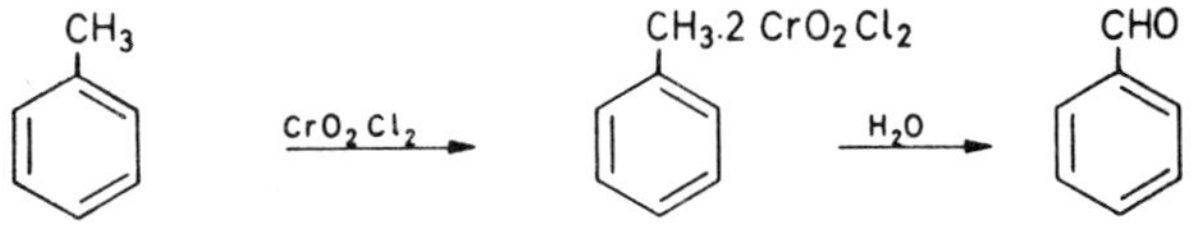

$$\text{PhCH}_3 \xrightarrow{CrO_2Cl_2} \text{PhCH}_3 \cdot 2\,CrO_2Cl_2 \xrightarrow{H_2O} \text{PhCHO}$$

Ferguson[2] included the Étard reaction in his review of synthetic methods for preparing aldehydes, and Hartford and Darrin[3] have covered the preparative and mechanistic studies in their review of the literature up to 1957.

Yields from the reaction vary greatly, as side products often result. Thus *o*- and *m*-tolualdehydes are prepared in 70–80 per cent yields by oxidizing the corresponding xylenes[4], whilst oxidation of ethylbenzene gives a mixture of phenylacetaldehyde, acetophenone and side-chain cleavage products[5].

Mechanism

Since its discovery, the process has been frequently investigated and is considered[3] to proceed through the formation of an hydrogen-bonded complex between the chromyl chloride and the hydrocarbon

ÉTARD REACTION

$$\text{Ar}-\underset{\underset{H}{|}}{\overset{\overset{H}{|}}{C}}-R + 2\,CrO_2Cl_2 \longrightarrow \text{Ar}-\underset{\underset{H\cdots\cdots O=\underset{Cl}{\overset{Cl}{|}}=O}{|}}{\overset{\overset{H\cdots\cdots O=\underset{Cl}{\overset{Cl}{|}}=O}{|}}{C}}R \xrightarrow{H_2O}$$

$$2\,HO-\underset{\underset{Cl}{|}}{\overset{\overset{Cl}{|}}{C}r}-OH + \left[\text{Ar}\underset{\underset{OH}{|}}{\overset{\overset{OH}{|}}{C}}R\right] \longrightarrow \text{Ar}\overset{\overset{O}{\|}}{C}R + H_2O$$

Whilst this satisfies much of the experimental evidence, there are cases where analyses of the complexes do not show simple ratios of 2 molecules of chromyl chloride to 1 molecule of hydrocarbon. This has led to a great deal of recent activity in this field and, on the basis of magnetic susceptibility measurements, Wheeler[6] proposed that the chromium atoms are in a tetravalent state and the complex may be represented as

$$\text{Ar}-CH \underset{O-\underset{\underset{Cl}{|}}{\overset{\overset{Cl}{|}}{C}r}-OH}{\overset{O-\underset{\underset{Cl}{|}}{\overset{\overset{Cl}{|}}{C}r}-OH}{\diagup\hspace{-2pt}\diagdown}}$$

This conflicts with the electron-spin resonance studies of Nenitzescu, Necsoiu et al.[7] who claim that one of the chromium atoms is in an hexavalent state.

Wiberg and Eisenthal[8] have summarized the conflicting evidence and on the basis of this and their own investigations support Wheeler's structure for the complex.

General reaction conditions

The aromatic compound is dissolved in carbon tetrachloride or carbon disulphide, and a solution of chromyl chloride in the same solvent is added gradually with stirring at a rate such that the temperature does not exceed 40°; the addition is preferably carried out under an inert atmosphere. The explosive brown complex is insoluble and precipitates during the course of the addition. When addition is complete, the solution is left to stand for 2–3 days before the complex is filtered off. It is washed with carbon tetrachloride and decomposed by the use of a large quantity of water or dilute sulphurous acid.

In some cases alcohols have been employed for the hydrolysis step, giving hydrolysis products of lower acidity and lessening the possibility of further oxidation occurring.

Applications

Standard Étard reaction conditions were used by Pettit[9] to obtain 35 per cent yields of 2-nitro-6-methoxybenzaldehyde from 2-nitro-6-methoxytoluene

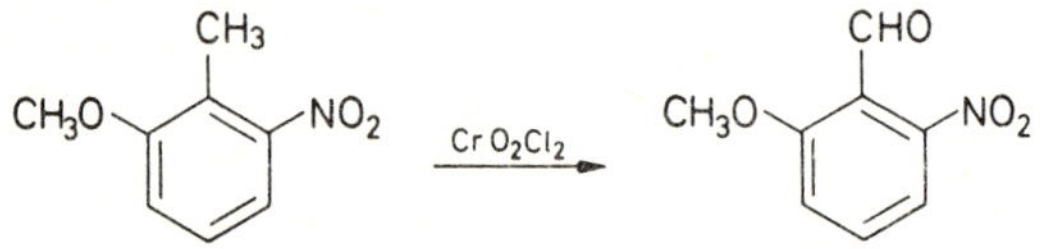

Where complex formation was slow, Tillotson and Houston[10] found the addition of 1 per cent of an alkene catalysed the reaction. A 25 per cent yield of hexahydrobenzaldehyde was obtained from methylcyclohexane when a small quantity of 2-methylbutene was added to the mixture

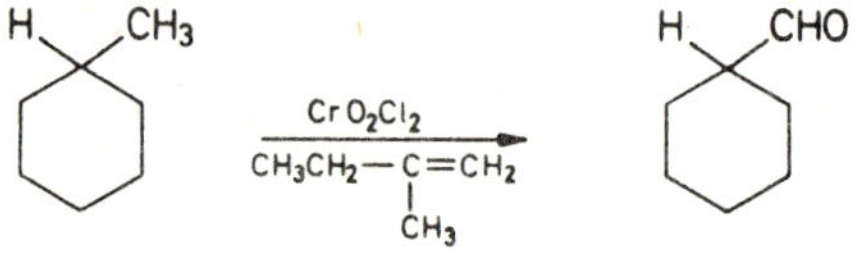

REFERENCES

[1] Étard, A. L. *C. r. hebd. Séanc. Acad. Sci., Paris* 84 (1877) 127; *Annls Chim. Phys.* 22 (1881) [5] 218 [*J. chem. Soc.* (A) 31 (1877) [1] 584; 40 (1881) 581]
[2] Ferguson, L. N. *Chem. Rev.* 38 (1946) 237
[3] Hartford, W. F. and Darrin, M. *Chem. Rev.* 58 (1958) 25
[4] Bornemann, E. *Ber. dt. chem. Ges.* 17 (1884) 1462 [*J. chem. Soc.* (A) 46 (1884) 1161]
[5] von Miller, W. and Rohde, G. *Ber. dt. chem. Ges.* 23 (1890) 1070 [*J. chem. Soc.* (A) 58 (1890) 978]
[6] Wheeler, O. H. *Can. J. Chem.* 38 (1960) 2137; 42 (1964) 706
[7] Necsoiu, I., Nenitzescu, C. D. *et al.*, *Tetrahedron* 19 (1963) 1133
[8] Wiberg, K. B. and Eisenthal, R. *Tetrahedron* 20 (1964) 1151
[9] Pettit, G. R. *J. org. Chem.* 24 (1959) 866
[10] Tillotson, A. and Houston, B. *J. Am. chem. Soc.* 73 (1951) 211

38

BLANC REACTION (BLANC'S RULE)

Nature of the reaction
Cyclic ketones are formed by the action of acetic anhydride upon 1 : 6 and 1 : 7 dicarboxylic acids.

Blanc's rule states that dicarboxylic acids with 4 or 5 carbon atoms in the chain give cyclic anhydrides on treatment with acetic anhydride, whilst those with 6 or 7 carbon atoms give cyclic ketones

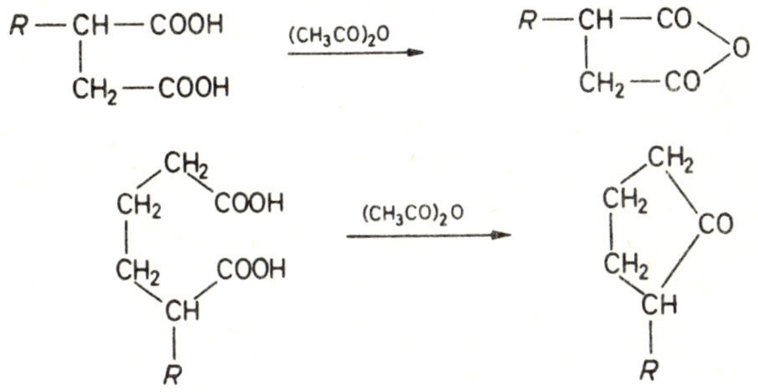

Historical development

The reaction as originally reported by Blanc[1] was applied to a number of substituted and unsubstituted dicarboxylic acids, particularly methyl substituted adipic and pimelic acids.

Ketone formation from the unsubstituted acids is achieved with less than 50 per cent yields. With alkyl groups α and β to the acid group, much greater yields of the corresponding ketones are obtained[2].

The validity of Blanc's rule was confirmed by Windaus and his colleagues[3]; their work on hydroaromatic dicarboxylic acids showed that 2-carboxycyclohexaneacetic acid gave only an anhydride but 2-carboxycyclohexanepropionic acid formed a ketone

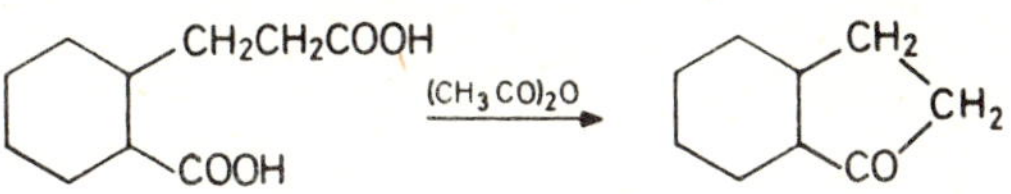

125

It has been demonstrated[4], however, that exceptions to Blanc's rule do occur, and this has led to the suggestion[5] that while the rule is valid for most systems it is not applicable where the carbon chain is incorporated into more than one ring system

General reaction conditions

The dicarboxylic acid is refluxed with acetic anhydride for several hours and the excess reagent then distilled off. The temperature is raised to between 200° and 300° to cause decarboxylation to occur after which the ketone distils off.

Applications

Blanc's rule has been of value in terpene chemistry for ascertaining the proximity of carboxylic acid groups[6]. Haller and Blanc[7] found camphoric acid to give an anhydride while homocamphoric acid gave a ketone. Their established structures show that this would be expected

Camphoric acid

Homocamphoric acid   Camphor

Bachman and Deno[8] employed the reaction to prepare 2-(1'-naphthylmethyl)cyclopentanone with a 71 per cent yield

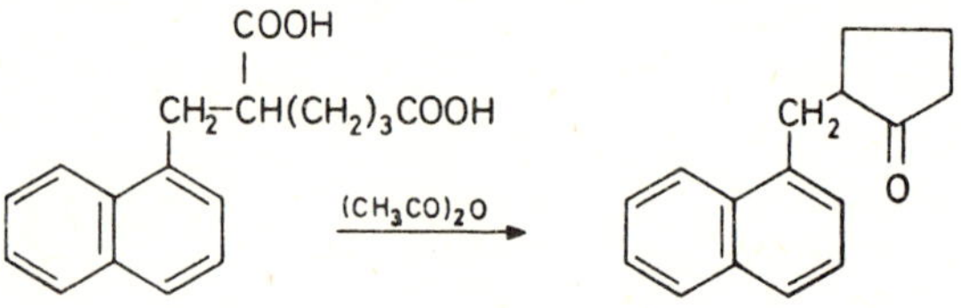

REFERENCES

[1] Blanc, G. *C. r. hebd. Séanc. Acad. Sci., Paris* 144 (1907) 1356; *Bull. Soc. chim. Fr.* 3 (1908) 778 [*Chem. Abstr.* 1 (1907) 2561; 2 (1908) 2949]
[2] Gilman, H. *Organic Chemistry* 1, 80, New York (Wiley) 1943
[3] Windaus, A., Hückel, W. and Reverey, G. *Ber. dt. chem. Ges.* 56B (1923) 91 [*Chem. Abstr.* 17 (1923) 1221]
[4] Wieland, H. and Dane, E. *Hoppe-Seyler's Z. physiol. Chem.* 210 (1932) 268 [*Chem. Abstr.* 26 (1932) 5961]
[5] Fieser, L. F. and M. *Natural Products Related to Phenanthrene*, 3rd edn., p. 140, New York (Reinhold) 1949
[6] Gilman, H. *Organic Chemistry* 4, 648, New York (Wiley) 1953
[7] Haller, A. and Blanc, G. *C. r. hebd. Séanc. Acad. Sci., Paris* 130 (1900) 376 [*J. chem. Soc. (A)* 78 (1900) [1] 202]
[8] Bachman, W. E. and Deno, N. C. *J. Am. chem. Soc.* 71 (1949) 3540

39

DIECKMANN REACTION

Nature of the reaction

Cyclic β-keto esters are formed by the base-catalysed intramolecular condensation of esters of dicarboxylic acids

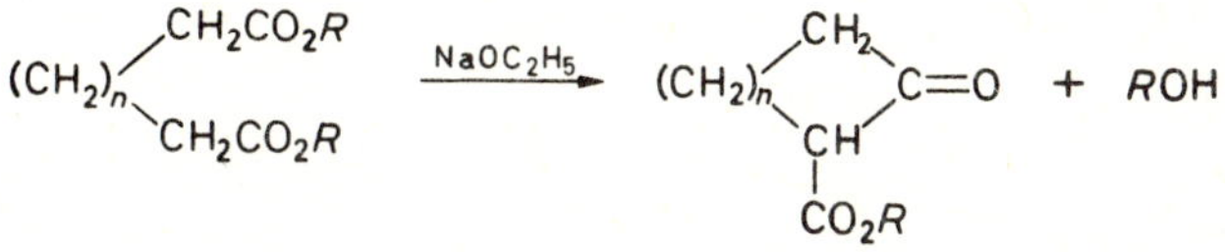

$$(CH_2)_n \begin{matrix} CH_2CO_2R \\ CH_2CO_2R \end{matrix} \xrightarrow{NaOC_2H_5} (CH_2)_n \begin{matrix} CH_2 \\ CH \\ CO_2R \end{matrix} C{=}O \ + \ ROH$$

Historical development

The reaction constitutes an intramolecular Claisen condensation and was originally reported concerning the cyclization of the esters of adipic and pimelic acids[1]

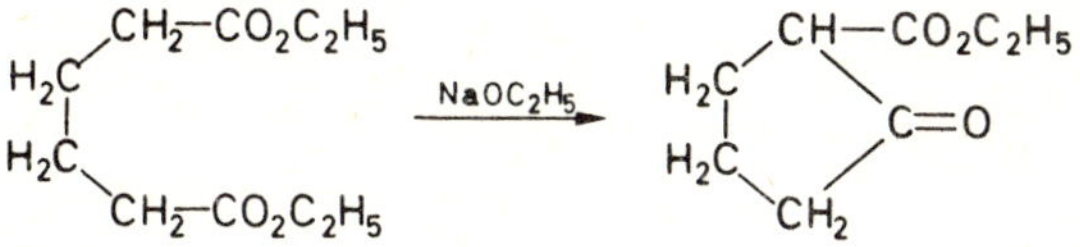

$$\begin{matrix} H_2C \\ | \\ H_2C \end{matrix} \begin{matrix} CH_2-CO_2C_2H_5 \\ CH_2-CO_2C_2H_5 \end{matrix} \xrightarrow{NaOC_2H_5} \begin{matrix} H_2C \\ | \\ H_2C \end{matrix} \begin{matrix} CH-CO_2C_2H_5 \\ | \\ CH_2 \end{matrix} C{=}O$$

Yields for the cyclization vary greatly, depending upon the length of the carbon chain, but are generally below 60 per cent. The reaction was discussed by Hauser and Hudson[2] in their review of acetoacetic ester-type condensations. It has recently been reviewed in more detail by Schaefer and Bloomfield[3].

Mechanism

By direct relationship to the sodium ethoxide-catalysed Claisen condensation, the reaction may be represented as

$$(CH_2)_n \begin{matrix} CH_2C(=O)-OC_2H_5 \\ CH_2C(=O)-OC_2H_5 \end{matrix} \xrightarrow{^-OC_2H_5} (CH_2)_n \begin{matrix} CH_2-C(=O)-OC_2H_5 \\ CH-C(=O)-OC_2H_5 \end{matrix} \longrightarrow$$

$$(CH_2)_n \begin{matrix} CH_2 \\ CH \end{matrix} \begin{matrix} O^- \\ C \\ OC_2H_5 \end{matrix} \quad O=C-OC_2H_5 \longrightarrow (CH_2)_n \begin{matrix} CH_2 \\ CH \end{matrix} \begin{matrix} C=O \\ CO_2C_2H_5 \end{matrix} + {}^-OC_2H_5$$

A

Carrick and Fry[4] established that the rate-determining step is the formation of the carbon ring, *A*.

General reaction conditions

The cyclization procedure has been well documented[5, 6] for the formation of simple cyclic keto esters. To ensure success, reactions are carried out with specially dried reagents under inert atmospheres. Sodium powder is stirred in toluene or benzene, and the diester, with a small quantity of ethanol, added gradually during 1–2 h. Refluxing is continued for several hours after addition has been completed, then the mixture is allowed to cool. The organic solution is poured into a cold aqueous acid solution to effect the hydrolysis and the product isolated from the organic layer. As an alternative, sodium hydride may be used as the base in the condensation[3].

Titley[7] employed the sodium ethoxide procedure to prepare a number of compounds incorporating benzene in the alicyclic ring

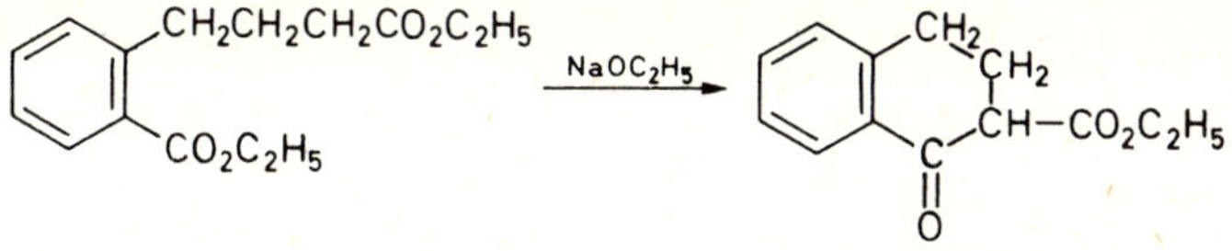

Ethyl-γ-o-carbethoxyphenyl
n-butyrate

Ethyl-1-keto-1,2,3,4-
tetrahydronaphthalene-2-
carboxylate

Applications

The generally held view that the Dieckmann reaction is only applicable to 5-, 6- and 7-membered rings has been discredited by Leonard and Schimelpfenig[8]. By working at high dilution and using potassium-t-butoxide in place of the conventional sodium ethoxide, they made cyclic ketones possessing more than 12 carbon atoms in yields as high as 50 per cent. The procedure was later[9] used to prepare *para*cyclophanes

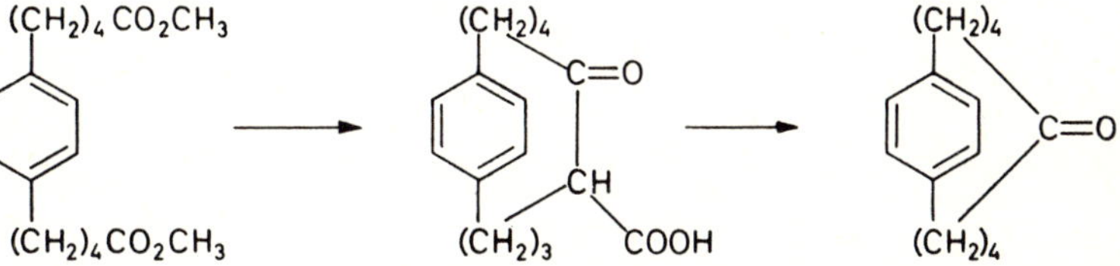

The dilution approach enabled Overberger and his collaborators[10] to prepare the heterocyclic 1-thiacyclo-octane-5-one

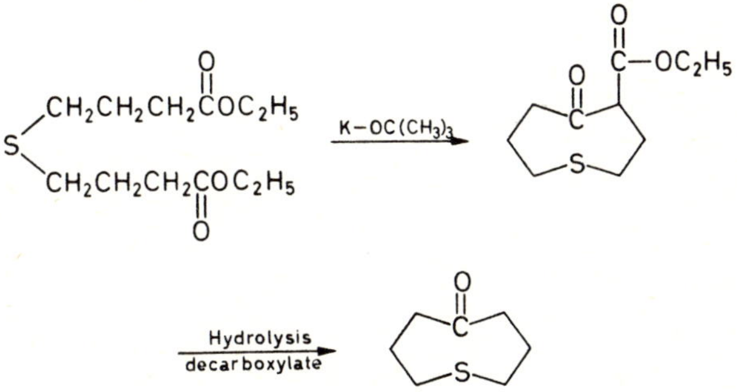

REFERENCES

[1] Dieckmann, W. *Ber. dt. chem. Ges.* 27 (1894) 102, 965; with Groeneveld, A. 33 (1900) 595 [*J. chem. Soc.* (A) 66 (1894) [1] 173, 324; 78 (1900) [1] 297]
[2] Hauser, C. R. and Hudson, B. E. *Org. React.* 1 (1942) 274

[3] Schaefer, J. P. and Bloomfield, J. J. *Org. React.* 15 (1967) 1
[4] Carrick, W. J. and Fry, A. *J. Am. chem. Soc.* 72 (1955) 2395
[5] Pinkney, P. S. *Org. Synth., Coll. Vol.* 2 (1943) 116
[6] Vogel, A. I. *A Textbook of Practical Organic Chemistry*, 3rd edn., p. 856, London (Longmans) 1956
[7] Titley, A. F. *J. chem. Soc.* (1928) 2571
[8] Leonard, N. J. and Schimelpfenig, C. W. *J. org. Chem.* 23 (1958) 1708
[9] Schimelpfenig, C. W., Lin, Y. T. and Waller, J. F. *J. org. Chem.* 28 (1963) 805
[10] Overberger, C. G. et al., *J. Am. chem. Soc.* 84 (1962) 2814

40

RUZICKA RING SYNTHESIS

Nature of the reaction

Large-ring alicyclic ketones are formed by the thermal decomposition of salts (calcium, thorium) of dicarboxylic acids

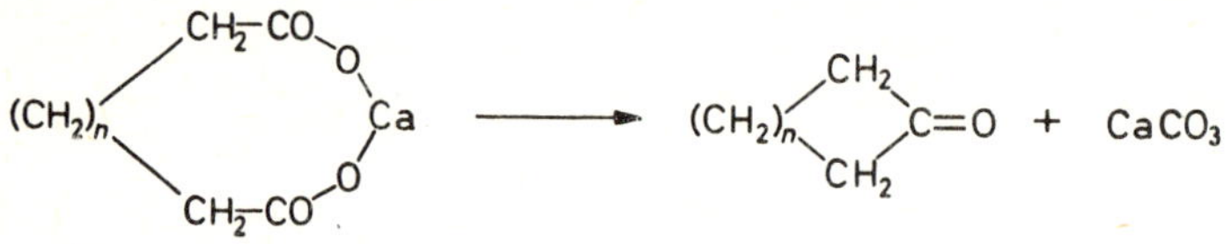

$$(CH_2)_n\begin{matrix}CH_2-CO-O\\ \\CH_2-CO-O\end{matrix}Ca \longrightarrow (CH_2)_n\begin{matrix}CH_2\\ \\CH_2\end{matrix}C=O + CaCO_3$$

Historical development

In 1836, Boussingault[1] reported the formation of a liquid product when suberic acid was heated with calcium carbonate. Mager[2] subsequently examined this reaction further and established optimum conditions for the formation of suberone from calcium suberate

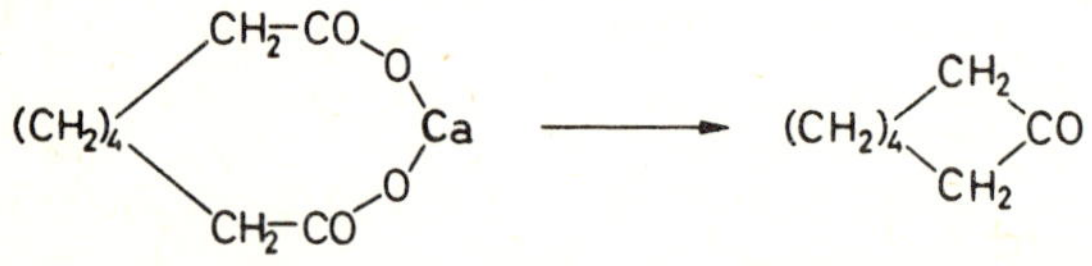

$$(CH_2)_4\begin{matrix}CH_2-CO-O\\ \\CH_2-CO-O\end{matrix}Ca \longrightarrow (CH_2)_4\begin{matrix}CH_2\\ \\CH_2\end{matrix}CO$$

Suberone (cycloheptanone)

This work was almost completely ignored for thirty years until investigations on the structure of civetone led to the necessity of examining methods for preparing large-ring compounds

$$\begin{array}{c} CH-(CH_2)_7 \\ \| \qquad\qquad\quad C=O \\ CH-(CH_2)_7 \end{array}$$

Civetone

Ruzicka and co-workers[3] carried out a very extensive study into the conditions for cyclization and prepared alicyclic ketones possessing up to 34 carbon atoms. The background and growth of this subject was discussed in an article by Ruzicka[4]. His researches had shown that the use of thorium, cerium or yttrium salts in place of calcium salts led to increased yields. These vary with the size of the ketone ring, a maximum of 80 per cent being obtained with cyclohexanone decreasing to 0·2 per cent for the 11-carbon rings, rising to 8 per cent for the 17-carbon rings and decreasing again to a constant 2 per cent yield for larger rings.

Mechanism

The mechanism of the elimination and cyclization is obscure, as it is influenced by many factors including the nature of substituents, ring strain and metal salt used[5].

Neunhöffer and Paschke[6] have suggested that the reaction involves the formation of the salt of a keto acid

$$(CH_2)_n\!\!\begin{array}{c}CH_2-CO_2M \\ CH_2-CO_2M\end{array} \longrightarrow (CH_2)_n\!\!\begin{array}{c}CH_2 \\ CH-CO_2M\end{array}\!\!CO \longrightarrow (CH_2)_n\!\!\begin{array}{c}CH_2 \\ CH_2\end{array}\!\!CO$$

(M = metal)

General reaction conditions

Thorium salts are made by dissolving the dicarboxylic acid in dilute sodium hydroxide solution and adding thorium tetrachloride. The precipitated salt is filtered off and dried above 100°. Cyclization is performed by dry distilling the salt under reduced pressure (<12 mm) up to about 500°. To obtain the product (or products), the collected liquid fraction is washed with sodium bicarbonate to remove any free acid and then carefully fractionated.

Application

Thorpe and Kon[7] have described a procedure for the preparation of cyclopentanone in which adipic acid is heated with barium hydroxide (80 per cent yield)

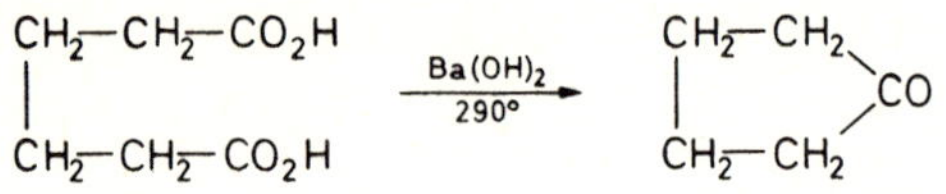

$$\begin{array}{c} CH_2\text{--}CH_2\text{--}CO_2H \\ | \\ CH_2\text{--}CH_2\text{--}CO_2H \end{array} \xrightarrow[290°]{Ba(OH)_2} \begin{array}{c} CH_2\text{--}CH_2 \\ | \quad\quad\;\; \diagdown \\ \quad\quad\quad\;\; CO \\ | \quad\quad\;\; \diagup \\ CH_2\text{--}CH_2 \end{array}$$

REFERENCES

[1] Boussingault, M. *Justus Liebigs Annln Chem.* 19 (1836) 308
[2] Mager, H. *Justus Liebigs Annln Chem.* 275 (1893) 357 [*J. chem. Soc.* (*A*) 64 (1893) [1] 557]
[3] Ruzicka, L., with Stoll, M. and Schinz, H. *Helv. chim. Acta* 9 (1926) 249; 11 (1928) 670; with Stoll, M. *ibid.* 1159; with Brugger, W. 9 (1926) 399 [*Chem. Abstr.* 20 (1926) 1792; 22 (1928) 4482; 23 (1929) 1110; 20 (1926) 2150]
[4] Ruzicka, L. *Chemy Ind.* 54 (1935) 2
[5] Fuson, R. C. *Organic Chemistry, an Advanced Treatise* (ed. Gilman, H.), 1, 78, New York (Wiley) 1943
[6] Neunhöffer, O. and Paschke, P. *Ber. dt. chem. Ges.* 72B (1939) 919 [*Chem. Abstr.* 33 (1939) 5367]
[7] Thorpe, J. F. and Kon, G. A. R. *Org. Synth., Coll. Vol.* 1 (1932) 187

41
ZIEGLER–THORPE LARGE-RING SYNTHESIS

Nature of the reaction
Large alicyclic rings are formed by treating alkyl dinitriles in highly dilute solution with metal amides

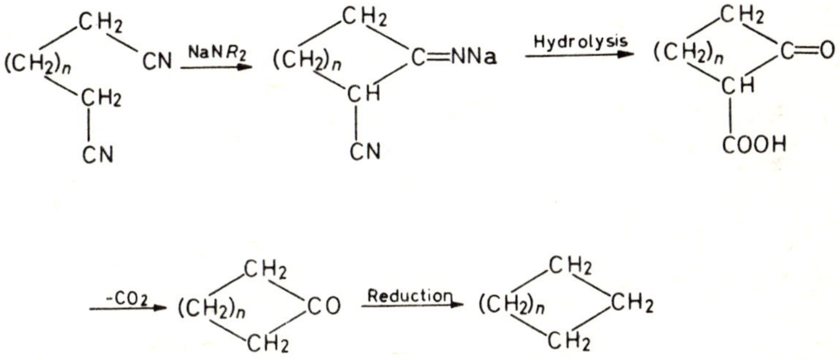

$$(CH_2)_n \begin{pmatrix} CH_2-CN \\ CH_2-CN \end{pmatrix} \xrightarrow{NaNR_2} (CH_2)_n \begin{pmatrix} CH_2 \\ CH-CN \end{pmatrix} C=NNa \xrightarrow{Hydrolysis} (CH_2)_n \begin{pmatrix} CH_2 \\ CH-COOH \end{pmatrix} C=O$$

$$\xrightarrow{-CO_2} (CH_2)_n \begin{pmatrix} CH_2 \\ CH_2 \end{pmatrix} CO \xrightarrow{Reduction} (CH_2)_n \begin{pmatrix} CH_2 \\ CH_2 \end{pmatrix} CH_2$$

Historical development
    Until 1926, the largest alicyclic ring that had been prepared was cyclo-octane. Within a short period after this date, several methods for preparing large rings were developed (Ruzicka, No. 40).
    Ziegler's method, introduced in 1933, was based upon Thorpe's[1] observation that 1,4-dicyanobutane gave 1-imino-2-cyanocyclopentane on boiling in ethanol with sodium ethoxide

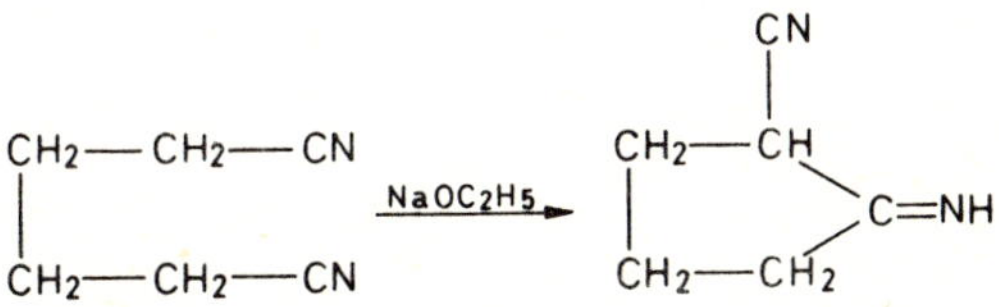

$$\begin{array}{c} CH_2-CH_2-CN \\ | \\ CH_2-CH_2-CN \end{array} \xrightarrow{NaOC_2H_5} \begin{array}{c} CN \\ | \\ CH_2-CH \\ | \diagdown \\ CH_2-CH_2 \end{array} C=NH$$

    Ziegler[2] applied this intramolecular condensation to longer aliphatic dinitrile chains and effected cyclization in very dilute solutions to prevent intermolecular polymerization. Over a period of years[3] alicyclic rings possessing more than 30 carbon atoms were prepared

$$(CH_2)_{12}\begin{smallmatrix}CH_2CN\\ \\CH_2CN\end{smallmatrix} \longrightarrow (CH_2)_{12}\begin{smallmatrix}CH_2\\ \\CH\\|\\CN\end{smallmatrix}C=NH \longrightarrow (CH_2)_{12}\begin{smallmatrix}CH_2\\ \\CH\\|\\COOH\end{smallmatrix}C=O$$

$$\longrightarrow (CH_2)_{12}\begin{smallmatrix}CH_2\\ \\CH_2\end{smallmatrix}C=O \longrightarrow (CH_2)_{12}\begin{smallmatrix}CH_2\\ \\CH_2\end{smallmatrix}CH_2$$

The method gives almost quantitative yields for rings with 5–8 carbon atoms but low ones of the 9- to 13-membered rings. For larger rings, the yields improve to a constant figure of 70–80 per cent. Ziegler[4] compared his method to that of Ruzicka and showed that the use of dinitriles gave superior yields to the distillation of carboxylic acids.

Prelog[5] has discussed the different methods available for preparing large-ring compounds in the light of improved approaches. A short section on the Ziegler–Thorpe reaction has been included in a recent review of the Dieckmann condensation[6].

Mechanism

Little work has been carried out on the mechanism of this reaction. It appears that the metal amide (usually sodium *N*-methylanilide) serves to replace a methylene hydrogen atom by the metal[7], the resulting intermediate rearranging to give the cyclic structure

$$(CH_2)_n\begin{smallmatrix}CH_2CN\\ \\CH_2CN\end{smallmatrix} \xrightarrow{NaN\begin{smallmatrix}CH_3\\C_6H_5\end{smallmatrix}} (CH_2)_n\begin{smallmatrix}\overset{Na}{\overset{|}{C}H}-CN\\ \\CH_2-C\equiv N\end{smallmatrix}$$

$$\longrightarrow (CH_2)_n\begin{smallmatrix}CHCN\\ \\CH_2\end{smallmatrix}C=NNa \xrightarrow{Hydrolysis} (CH_2)_n\begin{smallmatrix}CHCOOH\\ \\CH_2\end{smallmatrix}C=O$$

General reaction conditions

Reactions are carried out in ether solution and necessitate the use of an ether-soluble metal amine. For this purpose the lithium or sodium salts of N-methyl aniline are employed. A very dilute solution of the dinitrile in ether solution is slowly added to the metal compound dissolved in the same solvent, and the mixture is stirred at room temperature for at least 10 h. Then the ethereal solution is washed with water and dilute hydrochloric acid. Evaporation of the ethereal solution affords the cyano-imine which is hydrolysed to the ketone by refluxing with 50 per cent sulphuric acid until evolution of carbon dioxide ceases. Ketone reduction may be brought about by the standard procedures (Clemmensen or Wolff–Kishner).

This method was used by Rapoport and Williams[8] to prepare dibenzo[a.c.][1·3]cycloheptadiene

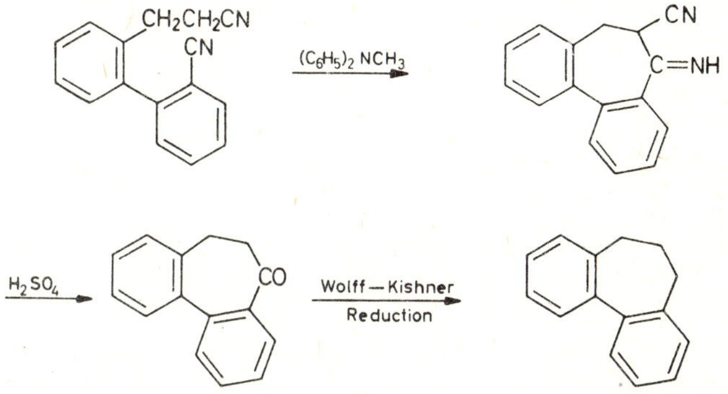

Other bases have been employed to obtain ring closure of dinitriles. Thompson[9] achieved an 85 per cent yield of 2-amino-1-cyano-1-cyclopentene by refluxing adiponitrile with sodium-t-butoxide

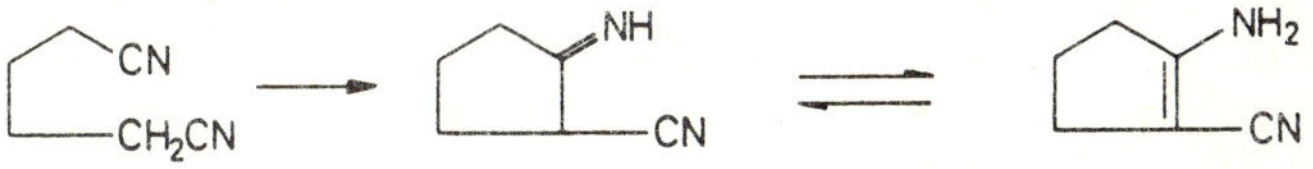

Applications

Very early in the development of the method, Ziegler and Lüttringhaus[10] showed that it could be applied to give more complicated systems than simple carbon–carbon chains

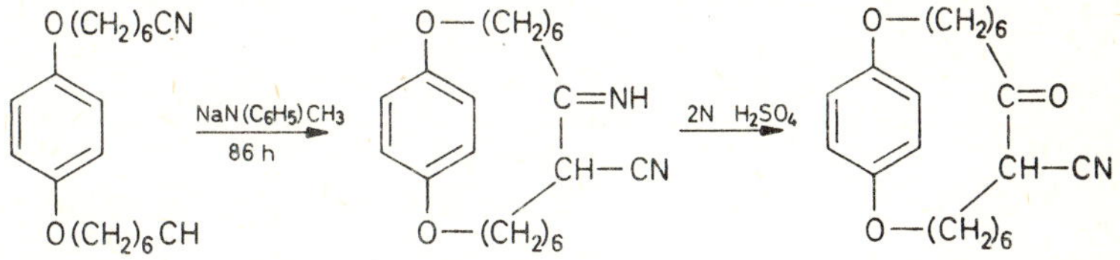

Irie and his collaborators[11] cyclized γ-cyano-γ-(3 : 4 methylenedioxyphenyl)pimelonitrile, using sodium amide in toluene, as part of their work on the structure of the alkaloid Tazettine

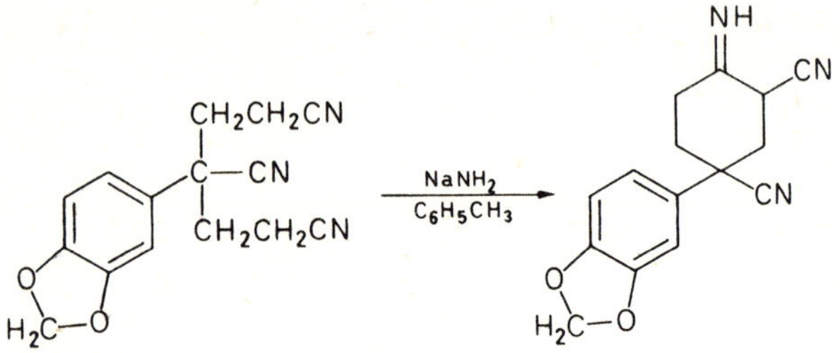

REFERENCES

[1] Thorpe, J. F. *J. chem. Soc.* 95 (1909) 1901
[2] Ziegler, K., Eberle, H. and Ohlinger, H. *Justus Liebigs Annln Chem.* 504 (1933) 94 [*Chem. Abstr.* 28 (1934) 117]
[3] Ziegler, K., with Holl, H. *Justus Liebigs Annln Chem.* 528 (1937) 143; with Hechelhammer, W. *ibid.* 114 [*Chem. Abstr.* 31 (1937) 6624, 6623]
[4] Ziegler, K. and Aurnhammer, R. *Justus Liebigs Annln Chem.* 513 (1934) 43 [*Chem. Abstr.* 29 (1935) 746]
[5] Prelog, V. *J. chem. Soc.* (1950) 421
[6] Schaefer, J. P. and Bloomfield, J. J. *Org. React.* 15 (1967) 28
[7] Finar, I. L. *Organic Chemistry*, 4th edn., 1, 493, London (Longmans) 1963
[8] Rapoport, H. and Williams, A. R. *J. Am. chem. Soc.* 71 (1949) 1774
[9] Thompson, Q. E. *J. Am. chem. Soc.* 80 (1958) 5483
[10] Ziegler, K. and Lüttringhaus, A. *Justus Liebigs Annln Chem.* 511 (1934) 1 [*Chem. Abstr.* 28 (1934) 5422]
[11] Irie, H., Tsuda, Y. and Uyeo, S. *J. chem. Soc.* (1959) 1446

42

FISCHER INDOLE SYNTHESIS

Nature of the reaction
Indole derivatives are formed by intramolecular condensations of arylhydrazones of aldehydes or ketones in the presence of Lewis acids

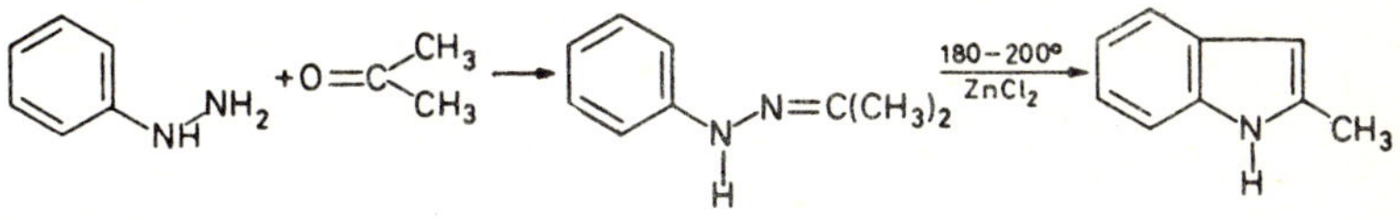

Historical development
The first reaction of this type, reported by Fischer and Jourdan[1], was the cyclization of the methyl phenyl hydrazone of pyruvic acid in 5 per cent yield. The actual nature of the product, however, was not established until later[2]

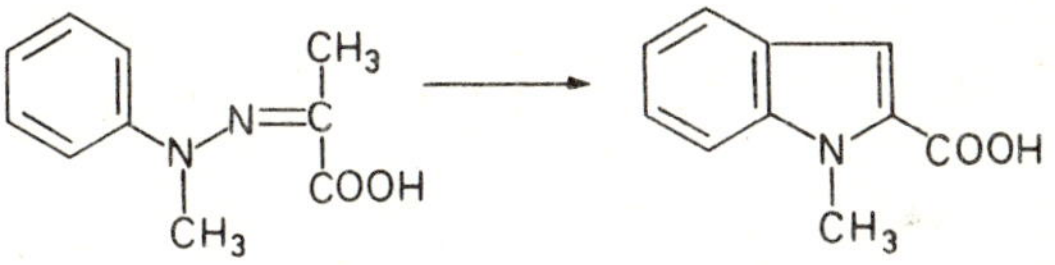

The growth and development of the procedure has been reviewed by van Order and Lindwall[3], many substituted indoles being listed.

The use of catalysts such as zinc chloride in the reaction led to substantial increases in yields. Varying catalysts have been employed: with calcium hydride, Endler and Becker[4] obtained a 50 per cent yield of 3-methyloxindole from β-propionylphenyl-hydrazone

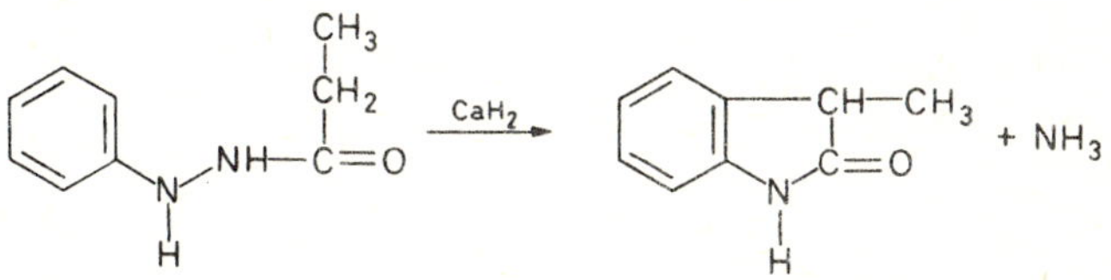

Yields as high as 95 per cent have often been achieved with a suitable catalyst.

137

Mechanism

Several mechanisms have been proposed for the cyclization; the currently accepted one was set forth by Robinson and Robinson[5]. This is consistent with observations such as those by Allen and Wilson[6] and Clusius and Weisser[7] who showed, by ^{15}N labelling, that the retained nitrogen is attached to the aromatic ring. The mechanism has been discussed[8] in the light of current concepts; it involves a rearrangement before cyclization takes place

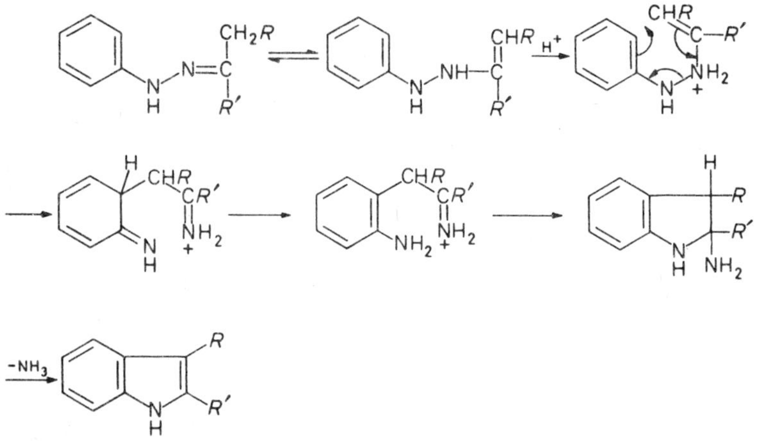

Successful attempts to isolate intermediates formed in this process lend support to the accepted mechanism[9, 10].

General reaction conditions

The Fischer indole reaction is normally a straightforward process, easy to carry out. The phenylhydrazone is mixed with a large excess of a Lewis acid catalyst, stirred and heated to 170° for a short period. On cooling, the mixture solidifies and is broken up and treated with dilute hydrochloric acid. The organic product is filtered off and recrystallized[11].

In many cases the indole is formed in a single continuous procedure by refluxing the ketone and the aryl hydrazine in the presence of acetic acid or sulphuric acid for 1–2 h. On cooling, the indole crystallizes out and can be filtered off. This type of procedure was employed by Rogers and Corson[12] to give 1,2,3,4-tetrahydrocarbazole in 85 per cent yields from cyclohexanone and phenylhydrazine

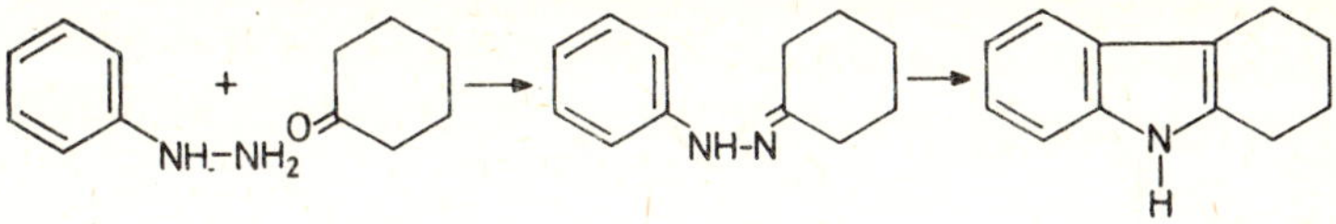

FISCHER INDOLE SYNTHESIS

Application

It has been found that phenols will often react in the keto form when treated with hydrazines, and this can be employed as the initial step to a Fischer indolization[13]. 3,4-Benzocarbazole has been prepared from 2-naphthol[14] by this approach

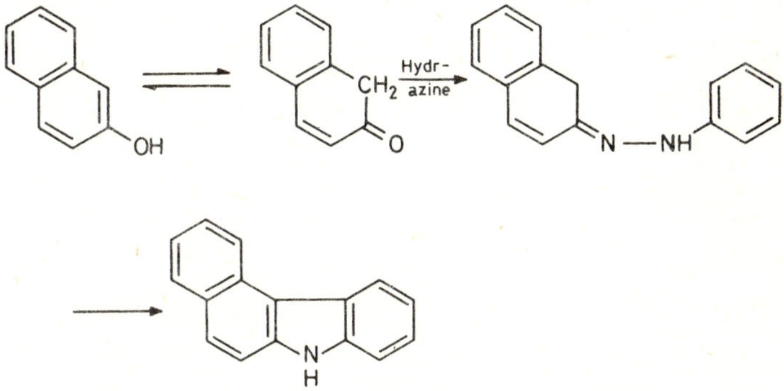

REFERENCES

[1] Fischer, E. and Jourdan, F. *Ber. dt. chem. Ges.* 16 (1883) 2241 [*J. chem. Soc. (A)* 46 (1884) 52]
[2] Fischer, E. and Hess, O. *Ber. dt. chem. Ges.* 17 (1884) 559 [*J. chem. Soc. (A)* 46 (1884) 1180]
[3] van Order, R. B. and Lindwall, H. G. *Chem. Rev.* 30 (1942) 78
[4] Endler, A. S. and Becker, E. I. *Org. Synth., Coll. Vol.* 4 (1963) 657
[5] Robinson, R. and G. M. *J. chem. Soc.* 113 (1918) 639; 125 (1924) 827
[6] Allen, C. F. H. and Wilson, C. V. *J. Am. chem. Soc.* 65 (1943) 611
[7] Clusius, K. and Weisser, H. R. *Helv. chim. Acta* 35 (1952) 400 [*Chem. Abstr.* 46 (1952) 11179]
[8] Robinson, B. *Chem. Rev.* 63 (1963) 373
[9] Conroy, H. and Firestone, R. A. *J. Am. chem. Soc.* 78 (1956) 2290
[10] Robinson, F. P. and Brown, R. K. *Can. J. Chem.* 42 (1964) 1940
[11] Shriner, R. L., Ashley, W. C. and Welch, E. *Org. Synth., Coll. Vol.* 3 (1955) 725
[12] Rogers, C. U. and Corson, B. B. *Org. Synth., Coll. Vol.* 4 (1963) 884
[13] Sumpter, W. C. and Miller, F. M. *Heterocyclic Compounds with Indole and Carbazole Systems* (ed. Weissberger, A.), 9, 77, New York (Interscience) 1954
[14] Japp, F. R. and Maitland, W. *J. chem. Soc.* 83 (1903) 269

43

SKRAUP SYNTHESIS

Nature of the reaction

Quinolines are formed by the reaction between primary aromatic amines and glycerol in the presence of sulphuric acid and an oxidizing agent

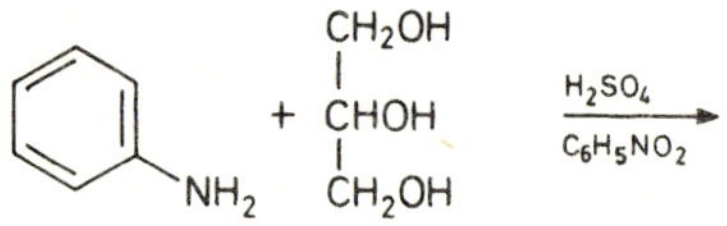

C₆H₅NH₂ + CH₂OH-CHOH-CH₂OH $\xrightarrow{H_2SO_4, C_6H_5NO_2}$ quinoline

Historical development

Koenigs's[1] synthesis of quinoline, first from allyl aniline and later[2] from aniline and glycerol, preceded the work of Skraup[3]. However, the latter procedure gave better yields and has lent itself to extensive development. By employing substituted anilines, the corresponding substituted quinolines are produced

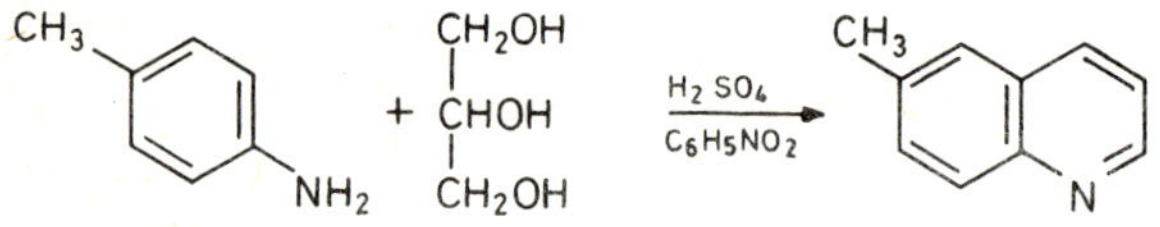

p-Toluidine + CH₂OH-CHOH-CH₂OH $\xrightarrow{H_2SO_4, C_6H_5NO_2}$ 6-Methyl quinoline

When more than one primary amino group is present in the molecule, additional cyclization reactions take place

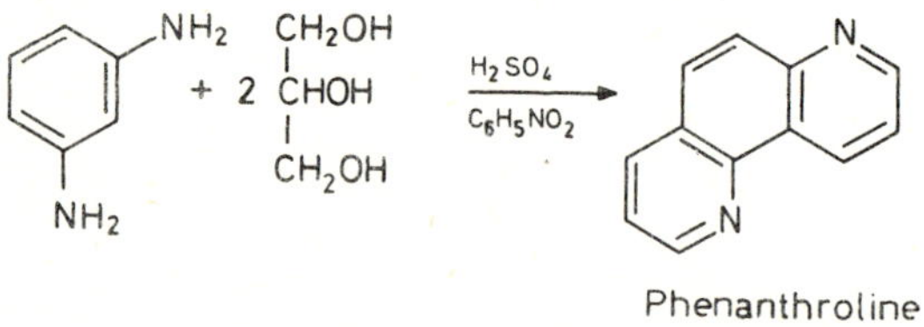

m-phenylenediamine + 2 CH₂OH-CHOH-CH₂OH $\xrightarrow{H_2SO_4, C_6H_5NO_2}$ Phenanthroline

By using α,β-unsaturated carbonyl compounds in place of glycerol, the reaction can be made to form quinolines with substituents in the hetero ring

140

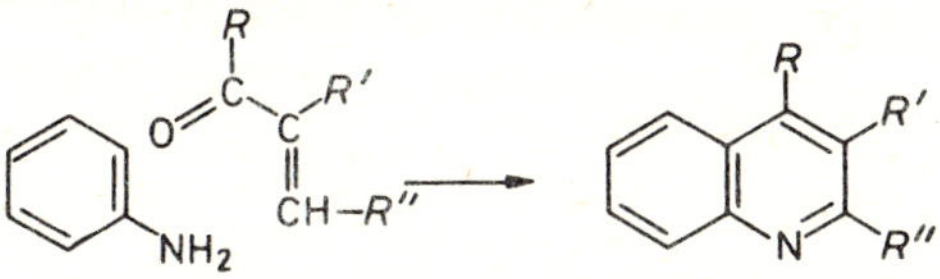

SKRAUP SYNTHESIS

Yields are for the greater part below 60 per cent but depend a great deal upon how the reaction is carried out[4]. It has been reviewed by Manske and Kulka[5].

Mechanism

The early investigators quickly realized that the glycerol used is initially dehydrated to acrolein prior to condensation with the primary amine. The β-aminoaldehyde thus formed undergoes subsequent acid-catalysed cyclization and oxidation to the final product

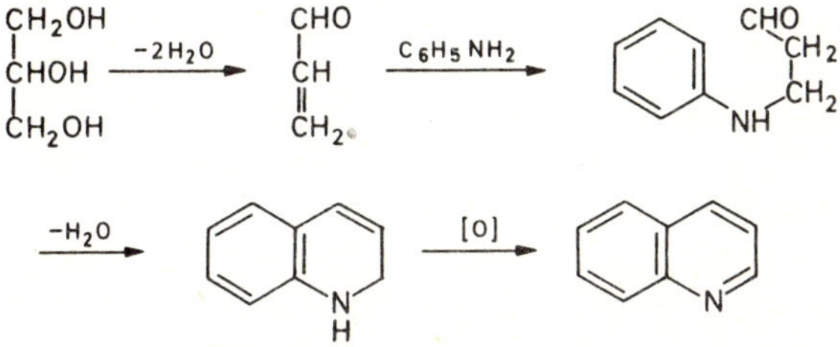

This mechanism is supported by the work of Badger *et al.*[6] in which β-aryl amino ketones have been isolated as intermediates in reactions between substituted anilines and α,β-unsaturated carbonyl compounds

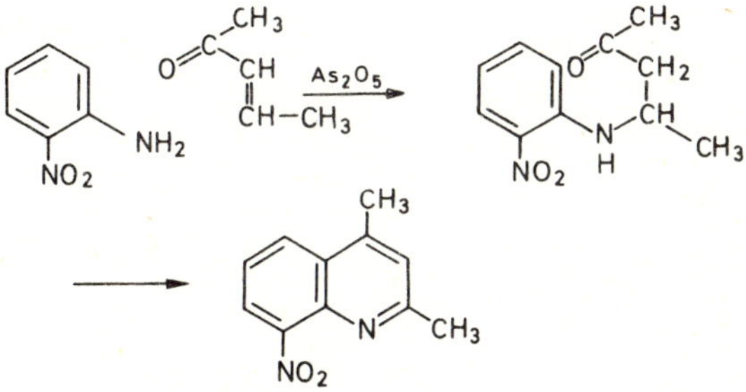

2,4-Dimethyl-8-nitroquinoline

General reaction conditions

The Skraup reaction is often very violent, and although substances such as ferrous sulphate, acetic and boric acid have been added to moderate it, reports on their efficacy are conflicting.

Aniline, or a substituted aniline, is mixed with glycerol (or the α,β-unsaturated ketone) and the oxidizing agent, usually nitrobenzene. Concentrated sulphuric acid is then added dropwise to the stirred solution. Gentle heating may be needed to initiate the reaction which should be exothermic enough to maintain steady boiling. When it is complete, the solution is made alkaline and steam-distilled. Volatile quinolines pass over in the distillate and may be extracted.

Clarke and Davis[7] have reported the preparation of quinoline with an 84–91 per cent yield by this type of reaction employing a ferrous sulphate moderator.

Modifications

It has been stated[8] that the best way to control the reaction is to add gradually the hot sulphuric acid, glycerol, aniline solution to a mixture of the oxidizing agent and ferrous sulphate. By this means the reaction proceeds steadily at 160°–170°.

An alternative oxidizing agent that has been extensively employed is arsenic acid or arsenic oxide; the preparation of 6-methoxy-8-nitroquinoline by this means, with 65–76 per cent yield, has been described by Mosher et al.[9]

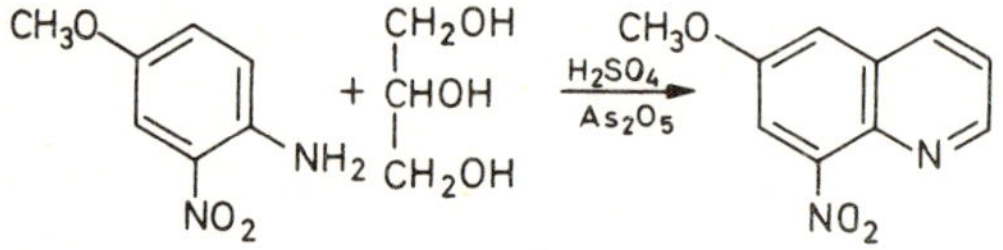

This compound is an intermediate in the synthesis of the antimalarial drug Plasmoquin[10].

Denton and Suschitzky[11] have found better yields to be obtained if polyphosphoric instead of sulphuric acid is used.

Applications

The Skraup synthesis may be applied to aromatic systems other than benzene. Finar and Hurlock[12] successfully reacted 4-amino-1-phenylpyrazole with glycerol to obtain 1′-phenylpyrazole-(4′ : 5′-2 : 3)pyridine (26 per cent yield)

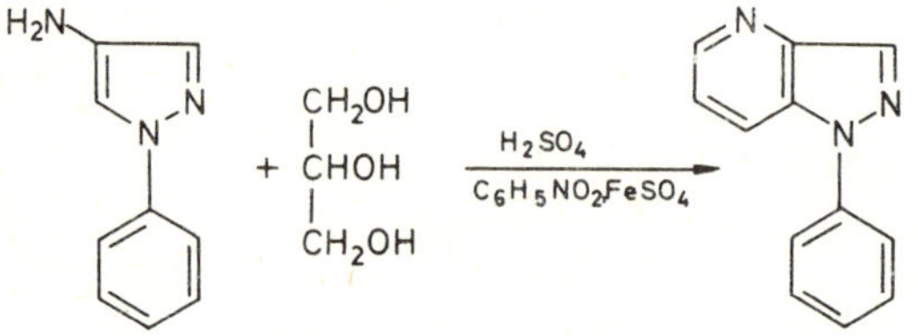

REFERENCES

[1] Koenigs, W. *Ber. dt. chem. Ges.* 12 (1879) 453 [*J. chem. Soc.* (*A*) 36 (1879) 540]

[2] Koenigs, W. *Ber. dt. chem. Ges.* 13 (1880) 911 [*J. chem. Soc.* (*A*) 38 (1880) 672]

[3] Skraup, Z. H. *Ber. dt. chem. Ges.* 13 (1880) 2086; 15 (1882) 893; *Sber. Akad. Wiss. Wien* 83 (1880) [2] 434 [*J. chem. Soc.* (*A*) 38 (1880) 287; 42 (1882) 1111; 38 (1880) 919]

[4] Yale, H. L. and Bernstein, J. *J. Am. chem. Soc.* 70 (1948) 254

[5] Manske, R. H. F. and Kulka, M. *Org. React.* 7 (1953) 59

[6] Badger, G. M. *et al.*, *Aust. J. Chem.* 16 (1963) 814; with Crocker, H. P. and Ennis, B. C. *ibid.* 840

[7] Clarke, H. T. and Davis, A. W. *Org. Synth., Coll. Vol.* 1 (1932) 478

[8] Manske, R. H. F., Ledingham, A. E. and Ashford, W. R. *Can. J. Res.* 27F (1949) 359

[9] Mosher, H. S., Yanko, W. H. and Whitmore, F. C. *Org. Synth., Coll. Vol.* 3 (1955) 568

[10] Finar, I. L. *Organic Chemistry*, 3rd edn., 2, 630, London (Longmans) 1964

[11] Denton, D. A. and Suschitzky, H. *J. chem. Soc.* (1963) 4741

[12] Finar, I. L. and Hurlock, R. J. *J. chem. Soc.* (1958) 3259

44

FRIEDEL–CRAFTS REACTION

Nature of the reaction

The term Friedel–Crafts reaction is now used to cover a wide range of alkylation and acylation reactions of both aromatic and aliphatic compounds that take place under the catalytic influence of Lewis acids:

Alkylation $\quad RH + R'Br \xrightarrow{MX_n} R-R' + HBr$

Acylation $\quad RH + R'COCl \xrightarrow{MX_n} R-CO-R' + HCl$

R, R' = alkyl or aryl

NAMED ORGANIC REACTIONS

Historical development

The reaction was first reported by Friedel and Crafts[1] in connection with the formation of 1-phenylpentane from benzene and n-amyl chloride

$$C_6H_6 + CH_3CH_2CH_2CH_2CH_2Cl \xrightarrow{AlCl_3} C_6H_5CH_2CH_2CH_2CH_2CH_3 + HCl$$

The procedure was soon extended[2] to the preparation of ketones from acid chlorides and hydrocarbons; benzophenones were easily formed by this method

$$C_6H_5-COCl + C_6H_6 \xrightarrow{AlCl_3} C_6H_5-CO-C_6H_5 + HCl$$

Where more than one halogen atom exists in the alkylating agent, all may be displaced if sufficient hydrocarbon is employed

$$CCl_4 + 4\ C_6H_6 \xrightarrow{AlCl_3} C(C_6H_5)_4 + 4\ HCl$$

The Friedel–Crafts reaction is often misrepresented as being applicable only to replacement of aromatic hydrogen atoms. In fact, its extensive scope also embraces the preparation of aliphatic compounds, as has been shown by the work of Taylor and his collaborators[3] on alkenes

$$CH_3O_2C(CH_2)_2COCl + CH_2=CH_2 \xrightarrow{AlCl_3} CH_2=CHCO(CH_2)_2CO_2CH_3$$

Numerous reviews on different aspects of the reaction have been published[4], but the most extensive coverage has been achieved in a series of volumes edited by Olah[5].

Although aluminium chloride is the catalyst most frequently used, many other substances have also been found to be suitable. Catalysts which initiate Friedel–Crafts reactions are all electron acceptors, within the classification of acids defined by Lewis[6], and include metal halides such as $FeCl_3$, $TiCl_4$, $SnCl_4$ and $ZnCl_2$.

Mechanism

The traditional mechanism for the reaction involves the formation of a complex between the alkyl halide and the catalyst

FRIEDEL–CRAFTS REACTION

$$RCl + AlCl_3 \rightleftharpoons [R\text{—}ClAlCl_3] \rightleftharpoons R^+ + \bar{A}lCl_4$$

[benzene + R⁺ → arenium ion intermediate → substituted benzene + H⁺]

$$H^+ + \bar{A}lCl_4 \rightleftharpoons AlCl_3 + HCl$$

The work of Brown and Grayson[7] is not in agreement with this simple ionization mechanism: they suggest that attack involves an addition compound between the organic and the metal halide

$$RCl + AlCl_3 \rightleftharpoons RCl:AlCl_3$$

$$ArH + RCl:AlCl_3 \longrightarrow \left[Ar{\overset{H}{\underset{R}{\diagup}}} \right]^+ \bar{A}lCl_4$$

$$\rightleftharpoons ArR + HCl + AlCl_3$$

An acyl carbonium ion mechanism has been proposed[8] for Friedel–Crafts acylation reactions; there is little doubt that some reactions do proceed by the formation of RCO^+ ions[9]

$$RCOCl + AlCl_3 \rightleftharpoons \bar{A}lCl_4 + RCO^+$$

$$\xrightarrow{ArH} RCO\overset{+}{A}rH \xrightarrow{\bar{A}lCl_4} R\text{—}\overset{\overset{O:AlCl_3}{\|}}{C}\text{—}Ar + HCl$$

Interest has been aroused by the observation that free radicals exist in some Friedel–Crafts alkylation reactions[10]. Thus that of benzene with either benzyl chloride or dichloromethane in the presence of aluminium chloride gave electron-spin resonance spectra showing the existence of the anthracene positive radical ion.

General reaction conditions

Lewis acids require anhydrous conditions, since small quantities of water impair their activities as catalysts. Reactions are carried out in specially dried solvents such as carbon disulphide or nitrobenzene, or in an excess of the hydrocarbon being acylated.

In general, alkylation reactions demand only small quantities of catalyst, while for acylation molar proportions of it are needed. This is due to the stable complexes formed between the catalyst and the carbonyl group[11].

The catalyst is dissolved, with stirring, in the dried solvent and the two reagents added slowly. Solids are dissolved in some of the solvent prior to addition. The reaction mixture is maintained at about 0° during the addition and kept at this temperature while any gases (usually HCl) are evolved. Finally, after gas evolution is virtually complete, the mixture may be heated for a short period before being cooled and treated with dilute acid. The product is extracted, washed and isolated by evaporation.

Numerous procedures are listed in the standard literature, most of them using similar methods.

Applications

Reactions occur just as readily with aromatic systems other than benzene. Villani and King[12] made 3-benzoylpyridine in 96 per cent yields by employing 2·5 mol of aluminium chloride

[Reaction scheme: pyridine-COCl + benzene →(AlCl₃) pyridyl-CO-phenyl + HCl]

By contrast, the alkylation of benzene by propylene in the presence of ferric chloride, to give a 91 per cent yield of cumene, was accomplished by Potts and Carpenter[13] with only 0·3 mol of catalyst.

One of the many substances employed as a catalyst is polyphosphoric acid (PPA). This is used extensively for both inter- and intramolecular acylation reactions involving carboxylic acids and eliminates the necessity of forming acid chlorides. Ayres and Denney[14] used this approach to prepare a number of substituted benzophenones as precursors for lignans

[Reaction scheme: HO-/CH₃O-substituted benzene-COOH + OCH₃/OCH₃-substituted benzene →(PPA) HO-/CH₃O-substituted benzene-CO-benzene-OCH₃/OCH₃]

A review of the use of PPA for alkylation and acylation reactions has been given by Popp and McEwen[15].

A particular aspect of the Friedel–Crafts reaction is the large number of intramolecular condensations that have been accomplished. Acylation by this route leads to cyclic ketones, which are dealt with in the review by Johnson[4]. The cyclization of 1-phenylbutyryl chloride to α-tetralone is an example of the standard procedure[16]

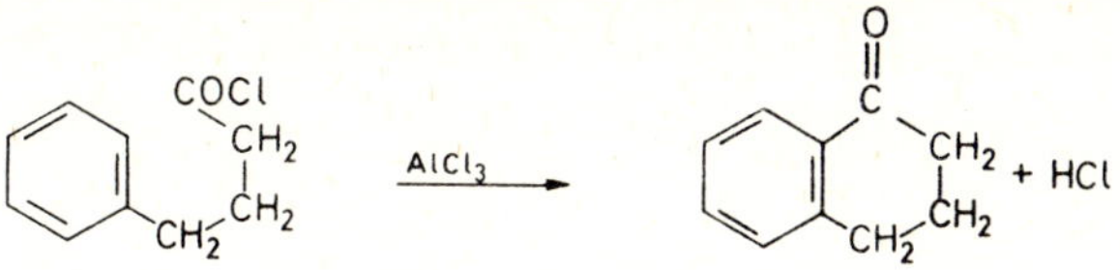

FRIEDEL–CRAFTS REACTION

REFERENCES

[1] Friedel, C. and Crafts, J. M. *C. r. hebd. Séanc. Acad. Sci., Paris* 84 (1877) 1392 [*J. chem. Soc.* 32 (1877) 725]
[2] Friedel, C. and Crafts, J. M. *C. r. hebd. Séanc. Acad. Sci., Paris* 84 (1887) 1450 [*J. chem. Soc.* 32 (1877) 864]
[3] Taylor, H. T. *J. chem. Soc.* (1958) 3922; with Jones, N. and Rudd, E. J. (1963) 2354
[4] Johnson, W. S. *Org. React.* 2 (1944) 130; Price C. C. 3 (1946) 1; Berliner, E. 5 (1949) 229; Gore, P. H. *Chem. Rev.* 55 (1955) 229; Roberts, R. M. *Chem. Engng News* 43 (1965) [4] 96
[5] Olah, G. A. (ed.), *Friedel Crafts and Related Reactions*, 4 vols., New York (Interscience) 1965
[6] Lewis, G. N. *J. Franklin Inst.* 226 (1938) 293
[7] Brown, H. C. and Grayson, M. *J. Am. chem. Soc.* 75 (1953) 6285
[8] Hine, J. *Physical Organic Chemistry*, p. 357, New York (McGraw-Hill) 1962
[9] Burton, H. and Praill, P. F. G. *J. chem. Soc.* (1950) 1203; (1953) 837
[10] Haszeldine, R. N. *et al.*, *Proc. chem. Soc.* (1964) 396
[11] Vogel, A. I. *A Textbook of Practical Organic Chemistry*, 3rd edn., p. 725, London (Longmans) 1956
[12] Villani, F. J. and King, M. S. *Org. Synth., Coll. Vol.* 4 (1963) 88
[13] Potts, W. M. and Carpenter, L. L. *J. Am. chem. Soc.* 61 (1939) 663
[14] Ayres, D. C. and Denney, R. C. *J. chem. Soc.* (1961) 4506
[15] Popp, F. D. and McEwen, W. E. *Chem. Rev.* 58 (1958) 377
[16] Martin, E. L. and Fieser, L. F. *Org. Synth., Coll. Vol.* 2 (1943) 569

45

GATTERMANN ALDEHYDE SYNTHESIS

Nature of the reaction

An aldehyde group is introduced into the aromatic system of phenols and their ethers by the action of hydrogen cyanide and hydrogen chloride in the presence of a metal halide catalyst

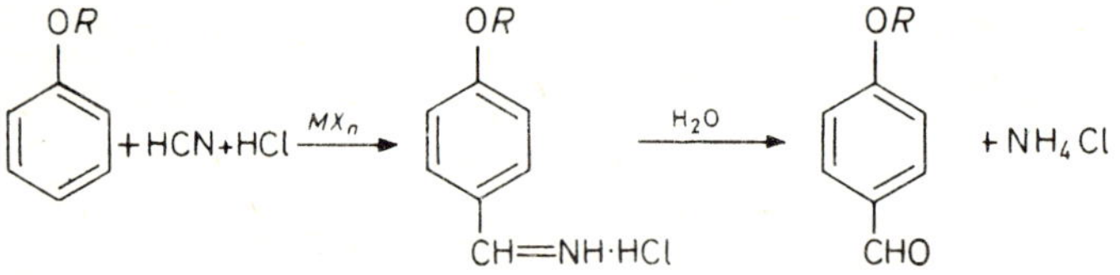

$$\text{ArOR} + \text{HCN} + \text{HCl} \xrightarrow{MX_n} \text{RO-C}_6\text{H}_4\text{-CH=NH·HCl} \xrightarrow{H_2O} \text{RO-C}_6\text{H}_4\text{-CHO} + \text{NH}_4\text{Cl}$$

Historical development

Gattermann[1] reported this method in 1898. It was his second approach to the formation of aldehydes, his first having been in collaboration with Koch (No. 46). The wide applicability of this procedure was demonstrated by the formation of a number of aromatic aldehydes from phenyl ethers, with yields of at least 80 per cent

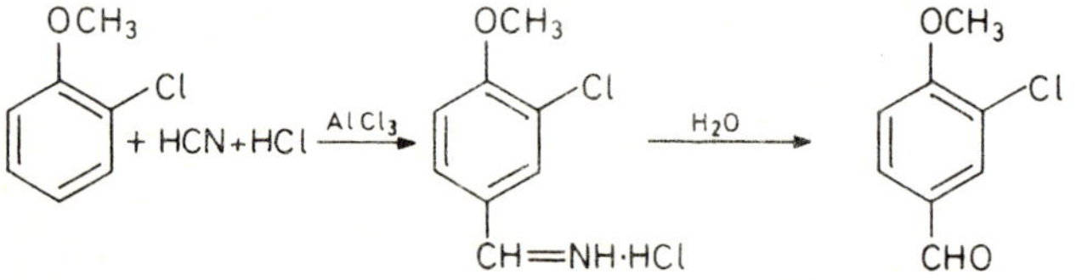

$$\text{2-Cl-C}_6\text{H}_4\text{-OCH}_3 + \text{HCN} + \text{HCl} \xrightarrow{AlCl_3} \text{CH=NH·HCl derivative} \xrightarrow{H_2O} \text{CHO derivative}$$

A detailed review of the reaction, covering the literature to 1956, has been given by Truce[2]. Its scope also includes related procedures in which the aldehyde group is introduced into aromatic hydrocarbons[3] at elevated temperatures. Aluminium chloride catalyst is normally used in Gattermann reactions, but other Lewis acids such as zinc chloride are also effective.

Mechanism

The nature of the mechanism has not been entirely clarified, although it is often represented as involving the formation of form-imino chloride as the first step

GATTERMANN ALDEHYDE SYNTHESIS

$$HCN + HCl \longrightarrow ClCH=NH$$

$$ArH + ClCH=NH \xrightarrow{AlCl_3} ArCH=NH \cdot HCl \xrightarrow{H_2O} ArCHO + NH_4Cl$$

The extensive work of Hinkel and his co-workers[4], however, indicated that the mechanism varies under different reaction conditions. It appears that two molecules of hydrogen cyanide are involved with every one of hydrogen chloride to form chloromethyleneformamidine

$$2\,HCN + HCl \longrightarrow ClCH=NCH=NH$$

$$ArH + ClCH=NCH=NH \xrightarrow{AlCl_3} HCl + ArCH=NCH=NH \xrightarrow{H_2O}$$

$$ArCH=NH + NH_3 + HCO_2H \xrightarrow{H_2O} ArCHO + NH_3$$

Further investigation in this field is still required.

General reaction conditions

All reactions are carried out under fume hoods with good ventilation! The phenol or ether is stirred into anhydrous hydrogen cyanide cooled by an ice bath. An inert organic solvent is also often used. Gaseous hydrogen chloride is passed through the mixture and the metal halide catalyst added gradually while the temperature is maintained below 40°. After passage of hydrogen chloride for several hours the mixture is poured on to iced dilute hydrochloric acid, heated for a short time, and any organic solvent removed. The aldehyde formed is isolated by forming the sodium bisulphite addition product.

Temperatures below 40° are adequate for most reactions, but to bring about reactions with aromatic hydrocarbons such as toluene, 70° and an excess of catalyst are required.

Applications

Gattermann's aldehyde synthesis has been applied to many different systems including pyrroles and furans. Fischer[5] found the method useful for introducing the aldehyde group into substituted indoles, although this could not be achieved with indole itself

2-Methyl indole → (HCl/HCN) → [3-CHNH·HCl intermediate] → 2-Methyl indole-3-aldehyde

NAMED ORGANIC REACTIONS

To avoid large quantities of hydrogen cyanide, Adams and co-workers[6] employed zinc cyanide as an *in situ* source of both hydrogen cyanide and zinc chloride catalyst by its reaction with hydrogen chloride. Fuson *et al.*[7] used this procedure, with tetrachloroethane as solvent, for the preparation of mesitaldehyde, in 80 per cent yields

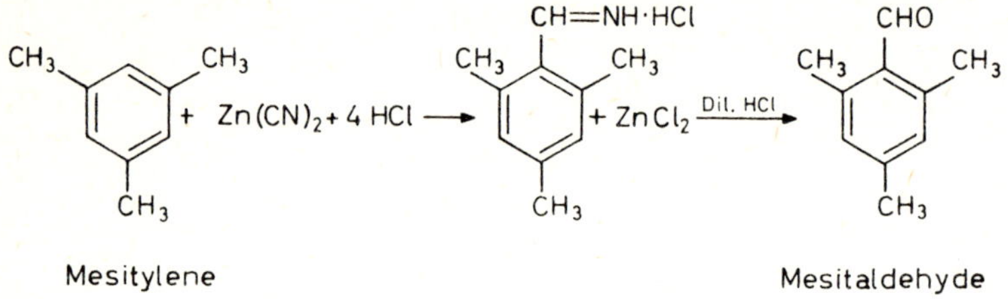

$$\text{Mesitylene} + Zn(CN)_2 + 4\ HCl \longrightarrow [\text{intermediate with } CH=NH\cdot HCl] + ZnCl_2 \xrightarrow{\text{Dil. HCl}} \text{Mesitaldehyde}$$

REFERENCES

[1] Gattermann, L. *Ber. dt. chem. Ges.* 31 (1898) 1149; with Berchelmann, W. *ibid.* 1765; *Justus Liebigs Annln Chem.* 357 (1907) 313 [*J. chem. Soc.* (A) 74 (1898)[1] 476, 581; *Chem. Abstr.* 2 (1908) 820]

[2] Truce, W. E. *Org. React.* 9 (1957) 37

[3] Hinkel, L. E., Ayling, E. E. and Morgan, W. H. *J. chem. Soc.* (1932) 2793

[4] Hinkel, L. E., with Ayling, E. E. and Beynon, J. H. *J. chem. Soc.* (1935) 674; with Watkins, T. I. and Jones, K. M. (1944) 647

[5] Fischer, H. and Pistor, K. *Ber. dt. chem. Ges.* 56B (1923) 2313 [*Chem. Abstr.* 18 (1924) 1126]

[6] Adams, R., with Levine, I. *J. Am. chem. Soc.* 45 (1923) 2373; with Montgomery, E. 46 (1924) 1518

[7] Fuson, R. C. *et al.*, *Org. Synth., Coll. Vol.* 3 (1955) 549

46

GATTERMANN–KOCH REACTION

Nature of the reaction
Aromatic aldehydes are formed by the action of carbon monoxide and hydrogen chloride on aromatic hydrocarbons in the presence of Lewis acid catalysts

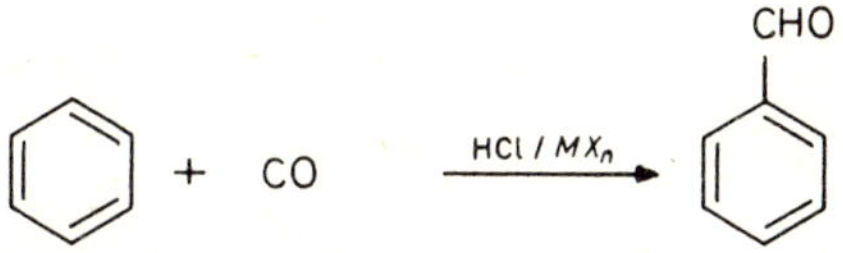

$$\text{C}_6\text{H}_6 + \text{CO} \xrightarrow{\text{HCl}/MX_n} \text{C}_6\text{H}_5\text{CHO}$$

Historical development
The original reactions described by Gattermann and Koch[1], carried out under atmospheric conditions, were applied to a wide variety of substituted benzenes; the aldehyde group usually enters in the *para* position in monosubstituted compounds

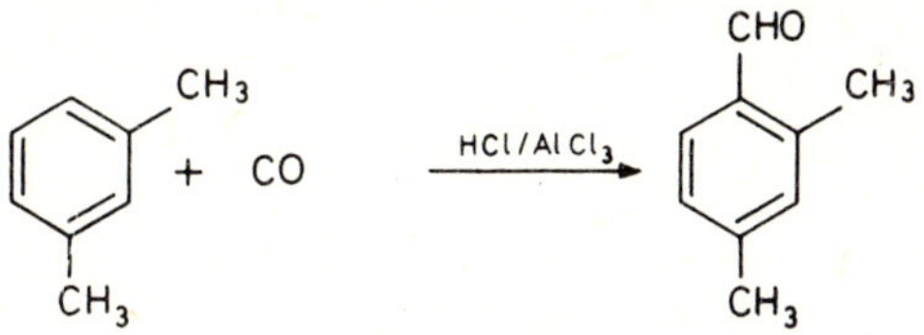

$$m\text{-xylene} + \text{CO} \xrightarrow{\text{HCl}/\text{AlCl}_3} \text{2,4-dimethylbenzaldehyde}$$

The reaction, now used in modified forms mainly for industrial purposes, has been reviewed by Crounse[2]. The review deals with its application to alkyl benzenes and covers the literature to 1949. Phenols and aromatic ethers cannot be substituted by this method.

Aluminium chloride is normally the catalyst employed, often with cuprous or nickel chloride as a support[3]. Yields of aldehydes from alkyl benzenes are usually about 40–60 per cent under atmospheric conditions; however, by working at increased pressures, yields between 80 and 90 per cent are more common.

Mechanism
Gattermann and Koch[1] considered the reaction to be an extension of the Friedel–Crafts reaction (No. 44) involving the intermediate

formation of formyl chloride from carbon monoxide and hydrogen chloride

$$CO + HCl \longrightarrow HC{\overset{\displaystyle O}{\underset{\displaystyle Cl}{\diagup\!\!\!\diagdown}}}$$

Later investigations showed that formyl chloride exists only as an intermediate complex with the catalyst[4]. On the basis of its relationship to the Friedel–Crafts reaction, a simple ionic mechanism may be postulated

$$HCl + CO + AlCl_3 \rightleftharpoons [HCO \cdot AlCl_4] \rightleftharpoons HCO^+ + \bar{A}lCl_4$$

$$C_6H_6 + HCO^+ \rightleftharpoons [C_6H_6 \cdot CHO]^+ \longrightarrow C_6H_5CHO + H^+$$

$$H^+ + \bar{A}lCl_4 \rightleftharpoons HCl + AlCl_3$$

The work of Dilkes and Eley[5], involving different metallic halide catalysts, showed that HCO^+ was in fact the active ion, and the rate-determining step its reaction with the aromatic system. Existence of formyl chloride or its complex is unnecessary, as the formyl ion may be produced by direct protonation of carbon monoxide[6]. This was found to occur[5] when carbon monoxide was complexed with the cuprous ion under acidic conditions

$$Cu^+ + CO \longrightarrow Cu^+CO \longrightarrow H\overset{+}{C}O + Cu^+$$

General reaction conditions

The main difference between reactions under atmospheric conditions and under pressure is that, in the former, a cuprous chloride carrier is used in addition to the catalyst.

Dry carbon monoxide and hydrogen chloride gases are passed continuously for several hours into a vigorously stirred suspension of catalyst and carrier in an excess of the hydrocarbon; alternatively, this may be dissolved in a less reactive solvent. The reaction is kept at 30°–50° and left to stand overnight before being added to an equal volume of iced water, causing separation into two phases. The aldehyde is isolated and purified by distillation or by the formation of a bisulphite addition compound.

In high-pressure reactions, the organic solution is saturated with hydrogen chloride before being placed with the catalyst in the autoclave. Carbon monoxide is added at a pressure between 500 and 1,000 lb/in^2 until no further absorption occurs. Most pressure reactions do not require heating, so the reaction is just left to stand for several hours in the closed vessel before isolating the product.

Applications

A procedure for preparing *p*-tolualdehyde in 50 per cent yields, devised by Coleman and Craig[7], included an effective mixing of the two gases before passing them into the mixture; their rates of flow were followed by passage through wash bottles

C$_6$H$_4$(CH$_3$) $\xrightarrow{AlCl_3/Cu_2Cl_2}$ C$_6$H$_3$(CHO)(CH$_3$)

Dry carbon monoxide and hydrogen chloride suitable for laboratory procedures can be prepared in equimolar proportions by reacting formic with chlorosulphonic acid[8]

$$HCO_2H + ClSO_3H \longrightarrow H_2SO_4 + CO + HCl$$

The industrial manufacture of benzaldehyde by the Gattermann–Koch reaction has been extensively studied[9]. Under optimum high-pressure conditions, yields of 66 per cent are possible.

Gresham and Tabet[10] found that gaseous hydrogen chloride can be avoided in industrial processes by the use of a boron trifluoride–hydrogen fluoride catalyst; the method has the additional advantage that the catalyst can be re-used

naphthalene + CO $\xrightarrow{BF_3/HF}$ 1-naphthaldehyde (CHO)

REFERENCES

[1] Gattermann, L., with Koch, J. A. *Ber. dt. chem. Ges.* 30 (1897) 1622; *Justus Liebigs Annln Chem.* 347 (1906) 347 [*J. chem. Soc.* (*A*) 72 (1897) [1] 519; 90 (1906) [1] 589]

[2] Crounse, N. N. *Org. React.* 5 (1949) 290

[3] Korczynski, A. and Mrozinski, W. *Bull. Soc. chim. Fr.* 29 (1921) [4] 459 [*Chem. Abstr.* 15 (1921) 3640]
[4] Hopff, H. et al., *Ber. dt. chem. Ges.* 69 (1936) 2244 [*Chem. Abstr.* 31 (1937) 366]
[5] Dilke, M. H. and Eley, D. D. *J. chem. Soc.* (1949) 2601, 2613
[6] Olah, G. A. *Friedel Crafts and Related Reactions*, 1, 115, New York (Interscience) 1963
[7] Coleman, G. H. and Craig, D. *Org. Synth., Coll. Vol.* 2 (1943) 583
[8] Bert, L. *C. r. hebd. Séanc. Acad. Sci., Paris* 221 (1945) 77 [*Chem. Abstr.* 40 (1946) 6061]
[9] Holloway, J. H. and Krase, N. W. *Ind. Engng Chem.* 25 (1933) 497
[10] Gresham, W. F. and Tabet, G. E. *U.S. Pat.* 2 485 237 (1949) [*Chem. Abstr.* 44 (1950) 2027]

47

GRIGNARD REAGENTS (GRIGNARD REACTIONS)

Nature of the reaction

Organo-magnesium halides are formed by reactions between magnesium and organic halides in ethereal solutions; the reagents are employed in reactions with double and triple bond systems

$$RX + Mg \longrightarrow RMgX$$

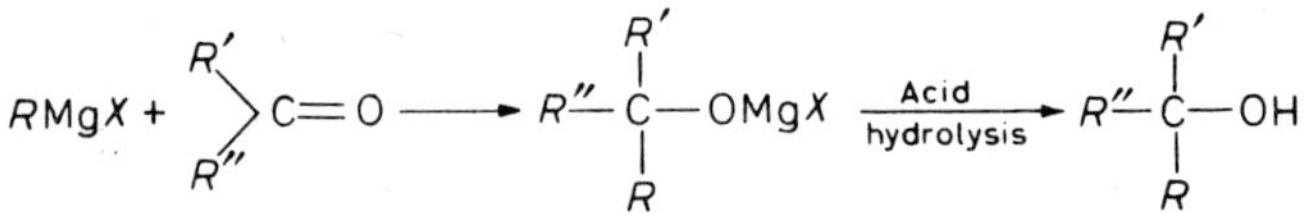

$$RMgX + \underset{R''}{\overset{R'}{>}}C=O \longrightarrow R''-\underset{R}{\overset{R'}{\underset{|}{C}}}-OMgX \xrightarrow{\text{Acid hydrolysis}} R''-\underset{R}{\overset{R'}{\underset{|}{C}}}-OH$$

Historical development

The complete reaction of magnesium with an organic halide and a ketone in diethyl ether was first reported by Barbier[1], the procedure being carried out with all the reagents mixed together

GRIGNARD REAGENTS (GRIGNARD REACTIONS)

$$CH_3-\underset{\underset{CH_3}{|}}{C}=CH-CH_2CH_2CH_2-\underset{\underset{CH_3}{|}}{CO}+CH_3I+Mg \longrightarrow CH_3-\underset{\underset{CH_3}{|}}{C}=CHCH_2CH_2CH_2-\underset{\underset{CH_3}{|}}{\overset{\overset{OMgI}{|}}{C}}-CH_3$$

$$\xrightarrow{\text{Hydrolysis}} CH_3-\underset{\underset{CH_3}{|}}{C}=CHCH_2CH_2CH_2-\underset{\underset{CH_3}{|}}{\overset{\overset{OH}{|}}{C}}-CH_3$$

Grignard[2] investigated the role played by magnesium and found that the metal initially reacted with the alkyl halide to give an intermediate organo-magnesium halide which could be reacted with aldehydes, ketones and similar unsaturated systems. This was illustrated by the preparation of 1-phenyl-3-methyl butanol

$$\underset{CH_3}{\overset{CH_3}{>}}CH-CH_2Br \xrightarrow{Mg/(C_2H_5)_2O} \underset{CH_3}{\overset{CH_3}{>}}CH-CH_2 \, MgBr \longrightarrow$$ (with PhCHO)

$$H-\underset{\underset{Ph}{|}}{\overset{\overset{OMgBr}{|}}{C}}-CH_2-CH\underset{CH_3}{\overset{CH_3}{<}} \xrightarrow{\text{Hydrolysis}} \underset{\underset{Ph}{|}}{\overset{\overset{OH}{|}}{CH}}-CH_2-CH\underset{CH_3}{\overset{CH_3}{<}}$$

Further development of the procedure was rapid, and the use of Grignard reagents has now become common laboratory practice. The early progress in this field was reviewed by Grignard[3]; since then, it has become the subject for books rather than for reviews. Kharasch and Reinmuth[4] have carried out an extensive coverage of the literature to 1950, dealing with reactions of Grignard reagents with most organic systems.

Yields of the organo-magnesium halides are usually well over 85 per cent for simple molecules, but lower where steric hindrance and electronegativity factors apply. For different halogen atoms in the alkyl halide it has been found that the order of reactivity with magnesium is $RI > RBr > RCl$, when R is identical throughout.

Mechanism and Structure

The mechanism involved in the formation of Grignard reagents has been studied for many years. Gomberg and Bachman[5] suggested that the formation of the organo-magnesium halide is a radical reaction. Evidence in favour of such a mechanism, summarized by Kharasch and Reinmuth[4], has been the basis of a reaction scheme proposed by Walborsky and Young[6], involving surface absorption of the alkyl halide on the magnesium

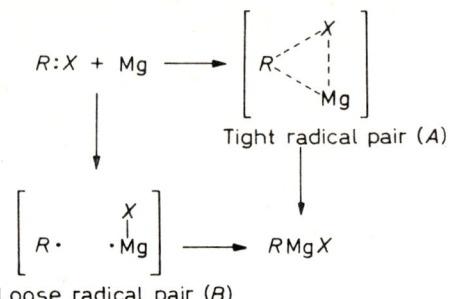

They suggest that the extent of formation of either (A) or (B) will depend upon the strength of the carbon–halogen bond.

Tremendous controversy has raged for many years over the constitution of Grignard reagents in ethereal solutions. Jolibois[7] proposed a dimeric structure represented as $R_2Mg \cdot MgX_2$ and later supported by evidence from the tracer study by Dessy et al.[8].

Ashby and Smith[9], however, claim that monomeric species exist at low concentrations while dimeric species occur at high ones and suggested an equilibrium between different forms of the reagent

$$(RMgX)_2 \rightleftharpoons 2RMgX \rightleftharpoons R_2Mg + MgX_2 \rightleftharpoons R_2Mg \cdot MgX_2$$

Dessy and co-workers[10] have outlined the background to the controversy and discussed the theoretical aspects of contemporary ideas. A review by Ashby[11] deals with both the mechanism of formation and the structure of Grignard reagents, covering the literature to 1966.

General reaction conditions

Most Grignard reagents are prepared in either diethyl ether or tetrahydrofuran and used without isolation; other ethers, such as monoglyme and diglyme, have also been used. Sodium-dried ether is placed in a flask fitted with a stirrer and reflux condenser, and clean, dry, granulated magnesium is added. The organic halide, dissolved in ether, is added dropwise to the stirred mixture. The reaction often commences immediately and has to be controlled by cooling or slowing the rate of addition. Sometimes the formation of

GRIGNARD REAGENTS (GRIGNARD REACTIONS)

the organo-magnesium halide needs to be initiated by the addition of a crystal of iodine or, alternatively, by gently warming the mixture on a steam bath.

Many different Grignard reagents have been prepared and used, and are easily accessible in the standard chemical literature. The procedure given by Gilman and Catlin[12] for the preparation of cyclohexyl magnesium chloride and its conversion to cyclohexyl carbinol is a fairly standard one, suitable for adaptation to other compounds

$$\text{C}_6\text{H}_{11}\text{-Cl} + \text{Mg} \longrightarrow \text{C}_6\text{H}_{11}\text{-MgCl} \xrightarrow{H_2C=O} \text{C}_6\text{H}_{11}\text{-CH}_2\text{OMgCl} \xrightarrow{H^+} \text{C}_6\text{H}_{11}\text{-CH}_2\text{OH}$$

Grignard reagents in solution are used by stirring with a corresponding solution of the appropriate organic compound. After a short time, the magnesium complex is decomposed by either water or dilute hydrochloric acid.

Summary of Standard Grignard Reactions

Formaldehyde $\longrightarrow$ Primary alcohols
$$R\text{MgX} + H_2\text{CO} \longrightarrow R\text{CH}_2\text{OH}$$

Aldehydes $\longrightarrow$ Secondary alcohols
$$R\text{MgX} + R'\text{CHO} \longrightarrow {R \atop R'}\!\!>\!\!\text{CH}_2\text{OH}$$

Ketones $\longrightarrow$ Tertiary alcohols
$$R\text{MgX} + {R' \atop R''}\!\!>\!\!\text{C=O} \longrightarrow {R' \atop R''}\!\!>\!\!\text{C}\!<\!\!{R \atop \text{OH}}$$

Acid chlorides (and anhydrides) $\longrightarrow$ Tertiary alcohols
$$R\text{MgX} + R'\text{COX} \longrightarrow {R \atop R'}\!\!>\!\!\text{C=O} \xrightarrow{R\text{MgX}} {R \atop R'}\!\!>\!\!\text{C}\!<\!\!{\text{OH} \atop R}$$

Carbon dioxide $\longrightarrow$ Carboxylic acids
$$R\text{MgX} + \text{C}{<\!\!\!{O \atop O}} \longrightarrow R-\text{C}{<\!\!\!{O \atop \text{OH}}}$$

Nitriles $\longrightarrow$ Ketones
$$R\text{MgX} + R'\text{CN} \longrightarrow {R \atop R'}\!\!>\!\!\text{C=O}$$

*Ortho*formates $\longrightarrow$ Aldehydes
$$R\text{MgX} + \text{H-C}(\text{OC}_2\text{H}_5)_3 \longrightarrow R\text{CHO}$$

*Ortho*esters $\longrightarrow$ Ketones
$$R\text{MgX} + R'\text{C}(\text{OC}_2\text{H}_5)_3 \longrightarrow {R \atop R'}\!\!>\!\!\text{C=O}$$

Yields from reactions involving Grignard reagents can often be improved by the use of metal halide catalysts[13]; Rabjohn and co-workers[14] found that cuprous iodide added to the reagents increased them in reactions with esters by as much as 30 per cent.

Fieser and Fieser[15] have given a useful summary of the Grignard reactions that have been described in the volumes of *Organic Syntheses*. Examples of most forms of elaboration occur in the list.

The Zerewitinoff Determination

An important application of methyl magnesium halides is for determining the number of active hydrogen atoms in organic compounds. The active hydrogen atoms react to give a quantitative yield of methane which is collected and measured

$$CH_3MgI + ROH \longrightarrow CH_4 + ROMgI$$

This method, pioneered by Zerewitinoff[16] soon after Grignard reagents were discovered, has become a very refined and precise procedure[17]. With amines and amides, the active hydrogen atoms react progressively, the second hydrogen giving methane only on heating with the reagent.

REFERENCES

[1] Barbier, P. A. *C. r. hebd. Séanc. Acad. Sci., Paris* 128 (1899) 110 [*J. chem. Soc. (A)* 76 (1899) [1] 323]
[2] Grignard, V., *C. r. hebd. Séanc. Acad. Sci., Paris* 130 (1900) 1322 [*J. chem. Soc. (A)* 78 (1900) [1] 382]
[3] Grignard, V. *Bull. Soc. chim. Fr.* 13 (1913) [4] Conf. IX; 39 (1926) [4] 1285 [*Chem. Abstr.* 21 (1927) 731]
[4] Kharasch, M. S. and Reinmuth, O. *Grignard Reactions of Non-metallic Substances*, New York (Prentice-Hall) 1954
[5] Gomberg, M. and Bachman, W. E. *J. Am. chem. Soc.* 49 (1927) 236
[6] Walborsky, H. M. and Young, A. E. *J. Am. chem. Soc.* 86 (1964) 3288
[7] Jolibois, P. *C. r. hebd. Séanc. Acad. Sci., Paris* 155 (1912) 353 [*Chem. Abstr.* 6 (1912) 2741]
[8] Dessy, R. E., with Handler, G. S. *J. Am. chem. Soc.* 80 (1958) 5824; *et al.*, 79 (1957) 3476
[9] Ashby, E. C., with Smith, M. B. *J. Am. chem. Soc.* 86 (1964) 4363; *Trans. N.Y. Acad. Sci.* 27 (1964) 29
[10] Dessy, R. E., Green, S. E. I. and Salinger, R. M. *Tetrahedron Lett.* 21 (1964) 1369
[11] Ashby, E. C., *Q. Rev. chem. Soc.* 21 (1967) 259
[12] Gilman, H. and Catlin, W. E. *Org. Synth., Coll. Vol.*, 2nd edn., 1 (1941) 188
[13] Hook, W. H. and Robinson, R. *J. chem. Soc.* (1944) 1952

[14] Rabjohn, N., Phillips, L. V. and De Feo, R. J. *J. org. Chem.* 24 (1959) 1964
[15] Fieser, L. F. and M. *Reagents for Organic Synthesis*, p. 415, New York (Wiley) 1967
[16] Zerewitinoff, T. *Ber. dt. chem. Ges.* 40 (1907) 2023 [*Chem. Abstr.* 2 (1908) 2810]
[17] Niederl, J. B. and V., *Micromethods of Quantitative Organic Analysis*, p. 263, New York (Wiley) 1946

48

REFORMATSKY REACTION

Nature of the reaction

Organo-zinc derivatives of α-halogeno esters are reacted with carbonyl compounds to give β-hydroxyesters

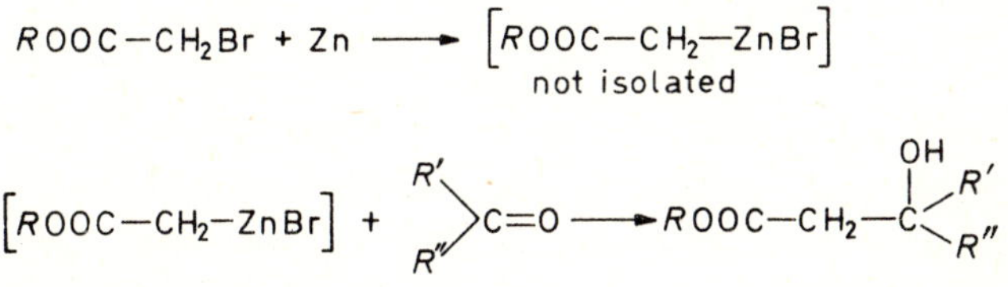

$$ROOC-CH_2Br + Zn \longrightarrow [ROOC-CH_2-ZnBr]$$
$$\text{not isolated}$$

$$[ROOC-CH_2-ZnBr] + \underset{R''}{\overset{R'}{>}}C=O \longrightarrow ROOC-CH_2-\underset{R''}{\overset{OH}{\underset{|}{C}}}\overset{R'}{\underset{R''}{<}}$$

Historical development

Reformatsky[1] found the reaction of ethyl monochloroacetate with acetone in the presence of zinc to lead to the formation of ethyl-β-dimethylethylenelactic acid

$$ClCH_2CO_2C_2H_5 + Zn + \underset{CH_3}{\overset{CH_3}{>}}C=O \longrightarrow \underset{CH_3}{\overset{CH_3}{>}}\underset{}{\overset{OH}{C}}-CH_2CO_2C_2H_5$$

Further researches[2] into this field with a variety of aldehydes and ketones laid the basis for what is now a widely used synthetic method; it is of particular value in the formation of substituted acrylic acids

$$ClCH_2CO_2C_2H_5 + Zn + \underset{CH_3CH_2CH_2}{\overset{CH_3}{>}}CO \longrightarrow \underset{CH_3CH_2CH_2}{\overset{CH_3}{>}}\underset{}{\overset{OH}{C}}=CH_2CO_2H$$

$$\xrightarrow{H_2SO_4} \underset{CH_3CH_2CH_2}{\overset{CH_3}{>}}C=CHCO_2H$$

Advantages of the method are: (a) the organo-zinc compound is prepared and used *in situ*; (b) it has a wide application for lengthening carbon chains; (c) it enables preparation of numerous branched-chain compounds; and (d) the organo-zinc intermediates do not react readily with themselves.

Yields from the reaction, normally between 50 and 75 per cent, depend a great deal upon the skill of the operator, but those above 90 per cent are quite common.

A detailed review of the Reformatsky reaction has been given by Shriner[3], and some of its applications have been covered by Gensler[4] and Draper and Kuksis[5].

Mechanism

The formation of an intermediate between zinc and an α-halogeno-ester was postulated by Reformatsky. Dain[6] confirmed this by isolating the organometallic compound from the interaction of zinc and ethyl α-bromoisobutyrate.

Dippy and Parkins[7] investigated a number of systems and concluded that the intermediate formed can react further by two possible routes to give either the β-hydroxy ester (route A) or the β-ketonic ester (route B) often found as a by-product

REFORMATSKY REACTION

$$RCHBrCO_2C_2H_5 + Zn \longrightarrow BrZn \cdot CHR \cdot CO_2C_2H_5$$

$$BrZn \cdot CHR \cdot CO_2C_2H_5 \;\xrightarrow{R'_2CO}\; BrZnO \cdot CR'_2 \cdot CHR \cdot CO_2C_2H_5 \quad \text{\textcircled{A}}$$

$$\xrightarrow{BrZn \cdot CHR \cdot CO_2C_2H_5}\; \underset{\underset{ZnBr}{|}\;\underset{OZnBr}{|}}{CHR-\overset{OC_2H_5}{\underset{|}{C}}-CHR-CO_2C_2H_5} \quad \text{\textcircled{B}}$$

$$A \xrightarrow{H^+} HO-CR'_2-CHR-CO_2C_2H_5$$

$$B \xrightarrow{H^+} CH_2R-CO-CHR-CO_2C_2H_5$$

The predominance of one over the other depends upon the nature of the carbonyl compound: with aldehydes and simple ketones *A* is favoured; with bulky ketones, *B* predominates.

General reaction conditions

Reformatsky's original conditions required mixing the ester and carbonyl compound, allowing the mixture to stand over granulated zinc for several days, then heating for a short period. The general method now employs a low-boiling solvent (benzene, toluene) which helps to control the initial exothermic reaction. Anhydrous conditions need to be maintained as moisture leads to lower yields. As an additional precaution, the zinc dust may be added portionwise. After the initial reaction has subsided, total reaction time is shortened by refluxing the mixture for a few hours, and the product is isolated by removal of the solvent and acidification.

Hauser and Breslow[8] have given a detailed procedure for the preparation of ethyl β-phenyl-β-hydroxypropionate, with 64 per cent yield

$$Ph-CHO + BrCH_2CO_2C_2H_5 + Zn \longrightarrow Ph-\underset{|}{\overset{OZnBr}{CH}}CH_2CO_2C_2H_5$$

$$\xrightarrow{H_2SO_4} Ph-\underset{|}{\overset{OH}{CH}}CH_2CO_2C_2H_5$$

Rinehart and Perkins[9] found an excess of ester and zinc to improve yields (87 per cent) in the preparation of ethyl 4-ethyl-2-methyl-3-hydroxyoctanoate

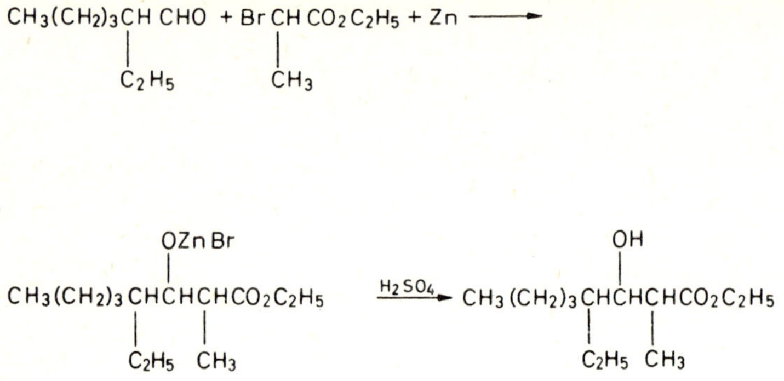

$$CH_3(CH_2)_3\underset{C_2H_5}{CH}CHO + Br\underset{CH_3}{CH}CO_2C_2H_5 + Zn \longrightarrow$$

$$CH_3(CH_2)_3\underset{C_2H_5}{\underset{|}{CH}}\overset{OZnBr}{\underset{CH_3}{\underset{|}{CH}}}CHCO_2C_2H_5 \xrightarrow{H_2SO_4} CH_3(CH_2)_3\underset{C_2H_5}{\underset{|}{CH}}\overset{OH}{\underset{CH_3}{\underset{|}{CH}}}CHCO_2C_2H_5$$

Applications

Barbier and Bouveault[10] employed a Reformatsky reaction in their conversion of methylheptenone to geranic acid, later extended by Tiemann[11] to give citral

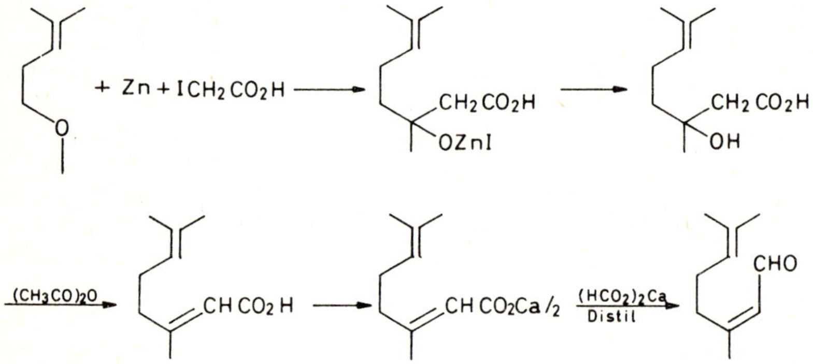

Geranic acid   Citral

REFERENCES

[1] Reformatsky, S. N. *Ber. dt. chem. Ges.* **20** (1887) 1210 [*J. chem. Soc.* (A) 52 (1887) 717]
[2] Reformatsky, S. N. *Zh. russk. fiz.-khim. Obshch.* **22** (1890) 44; *Ber. dt. chem. Ges.* **28** (1895) 2842; *J. prakt. Chem.* **54** (1896) [2] 469 [*J. chem. Soc.* (A) 60 (1891) 169; 70 (1896) [1] 128; 72 (1897) [1] 212]
[3] Shriner, R. L. *Org. React.* **1** (1942) 1
[4] Gensler, W. J. *Chem. Rev.* **57** (1957) 265
[5] Diaper, D. G. M. and Kuksis, A. *Chem. Rev.* **59** (1959) 89
[6] Dain, G. *Zh. russk. fiz.-khim. Obshch.* **28** (1896) 543
[7] Dippy, J. F. J. and Parkins, J. C. *J. chem. Soc.* (1951) 1570

[8] Hauser, C. R. and Breslow, D. S. *Org. Synth., Coll. Vol.* 3 (1955) 408
[9] Rinehart, K. L. and Perkins, E. G. *Org. Synth., Coll. Vol.* 4 (1963) 444
[10] Barbier, P. and Bouveault, L. *C. r. hebd. Séanc. Acad. Sci., Paris* 122 (1896) 393 [*J. chem. Soc. (A)* 70 (1896) [1] 445]
[11] Tiemann, F. *Ber. dt. chem. Ges.* 31 (1898) 2899 [*J. chem. Soc. (A)* 76 (1899) [1] 190]

49

GATTERMANN REACTION

Nature of the reaction

Diazonium salts are converted to aryl halides by the action of heat on a solution of this salt in the presence of freshly prepared copper powder and mineral acid

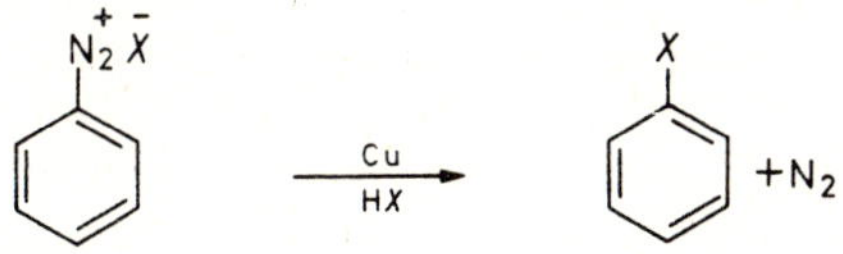

$$\text{Ph-N}_2^+ X^- \xrightarrow[HX]{Cu} \text{Ph-}X + N_2$$

Historical development

The procedure, introduced by Gattermann[1] in 1890, is closely related to the Sandmeyer reaction (No. 50), and the two have been jointly reviewed[2,3]. Gattermann's discovery was accidental, as he had hoped to prepare diphenyl from benzene diazonium chloride; instead, he found the action of the copper powder to lead to the formation of chlorobenzene

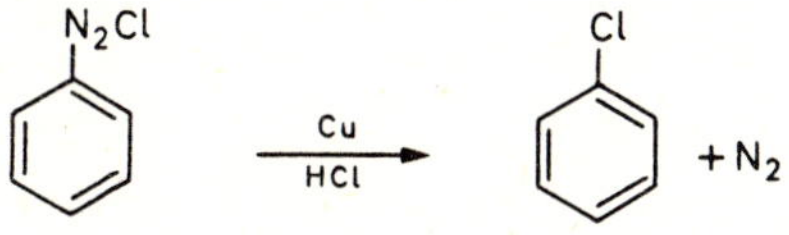

$$\text{Ph-N}_2\text{Cl} \xrightarrow[HCl]{Cu} \text{Ph-Cl} + N_2$$

The main advantage of the method is that it is carried out at a lower temperature than the Sandmeyer reaction. Better yields are also often obtained, usually between 40 and 50 per cent, and frequently greater.

Mechanism

The mechanism has not been clearly established. Gattermann considered it as fundamentally different from the Sandmeyer reaction, but it is difficult to decide whether this is so or whether the reaction actually involves the formation of a superficial layer of halide upon the copper catalyst[2].

A great deal of the complexity involving replacement of the diazo group by a halogen has been discussed by Saunders[4]. Waters[5] has suggested that the copper catalyst becomes a cuprous ion, as the ionization potential for the change $Cu \rightarrow Cu^+ + e^-$ is only -0.13 V. On this basis, the mechanism is believed to involve a cyclic series of single-electron transferences

$$ArN_2^+ \bar{Cl} \xrightarrow{Cu^+} \begin{array}{c} Ar\text{---}\overset{+}{N_2} \\ | \\ \bar{Cl}\text{--}Cu^+ \end{array} \longrightarrow \begin{bmatrix} Ar & +N_2 \\ | & \\ \bar{Cl}\text{--}Cu^{2+} \end{bmatrix} \longrightarrow Ar\,Cl + N_2 + Cu^+$$

General reaction conditions

Copper powder is freshly prepared for the reaction by adding zinc to a copper sulphate solution and filtering off the resulting copper precipitate. Excess of the washed copper powder is added with the appropriate dilute acid to the diazotization mixture that has been maintained below 0°. It is then carefully heated to about 50° for 1 h, nitrogen being evolved during the process. When evolution ceases, the product is isolated either by steam distillation or extraction with a suitable solvent.

Bigelow's[6] detailed procedure for the preparation of *o*-bromotoluene is of wide application. Only a small quantity of copper powder is used, and a 45 per cent yield of pure product is obtained

o-H₂N-C₆H₄-CH₃ $\xrightarrow{NaNO_2/HBr}$ o-(N_2Br)-C₆H₄-CH₃ $\xrightarrow{Cu}$ o-Br-C₆H₄-CH₃ + N_2

Applications

Attempts to introduce a halogen different from that in the diazonium salt often lead to mixtures of products. Hodgson and co-workers[7] diazotized *p*-nitroaniline and isolated a mixture of

p-chloro- and p-bromonitrobenzene after refluxing the diazonium chloride with copper powder and 35 per cent hydrobromic acid

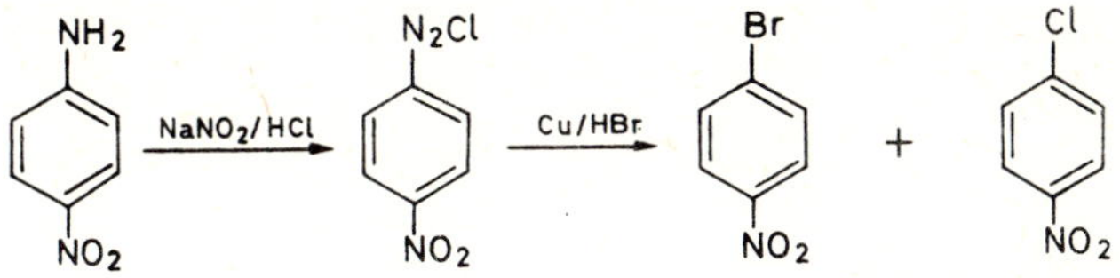

$$\underset{NO_2}{\underset{|}{C_6H_4}}-NH_2 \xrightarrow{NaNO_2/HCl} \underset{NO_2}{\underset{|}{C_6H_4}}-N_2Cl \xrightarrow{Cu/HBr} \underset{NO_2}{\underset{|}{C_6H_4}}-Br + \underset{NO_2}{\underset{|}{C_6H_4}}-Cl$$

Starkey[8] has described a similar procedure in which a nitro group is introduced by displacement of a diazofluoroborate group; the diazo compound was treated with sodium nitrite and a large quantity of copper powder

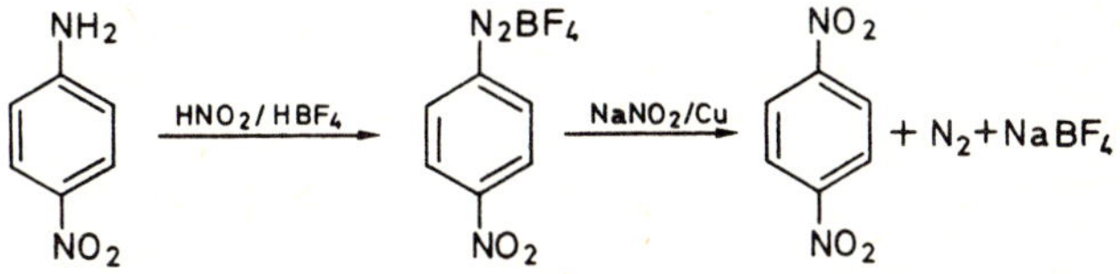

$$\underset{NO_2}{\underset{|}{C_6H_4}}-NH_2 \xrightarrow{HNO_2/HBF_4} \underset{NO_2}{\underset{|}{C_6H_4}}-N_2BF_4 \xrightarrow{NaNO_2/Cu} \underset{NO_2}{\underset{|}{C_6H_4}}-NO_2 + N_2 + NaBF_4$$

REFERENCES

[1] Gattermann, L. *et al.*, *Ber. dt. chem. Ges.* 27 (1890) 1218 [*J. chem. Soc.* (A) 50 (1890) 970]
[2] Cowdrey, W. A. and Davies, D. S. *Q. Rev. chem. Soc.* 6 (1952) 358
[3] Hodgson, H. H. *Chem. Rev.* 40 (1947) 259
[4] Saunders, K. H. *The Aromatic Diazo Compounds*, 2nd edn., p. 275, London (Arnold) 1949
[5] Waters, W. A. *J. chem. Soc.* (1942) 266
[6] Bigelow, L. A. *Org. Synth.*, *Coll. Vol.*, 2nd edn., 1 (1941) 135
[7] Hodgson, H. H., Birtwell, S. and Walker, J. *J. chem. Soc.* (1941) 770
[8] Starkey, E. B. *Org. Synth.*, *Coll. Vol.* 2 (1943) 225

50

SANDMEYER REACTION

Nature of the reaction

A diazonium group in an aromatic compound is replaced by halo, cyano or related groups in the presence of a corresponding cuprous salt

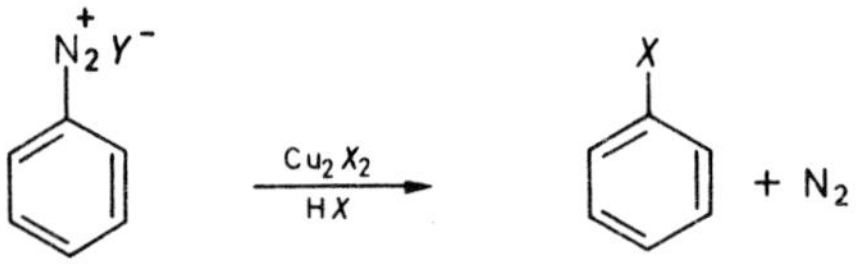

$$Ar{-}N_2^+Y^- \xrightarrow[HX]{Cu_2X_2} Ar{-}X + N_2$$

Historical development

Sandmeyer[1] observed that benzene diazonium chloride produced chlorobenzene when treated with cuprous acetylide and determined that in this case the active catalyst was, in fact, cuprous chloride

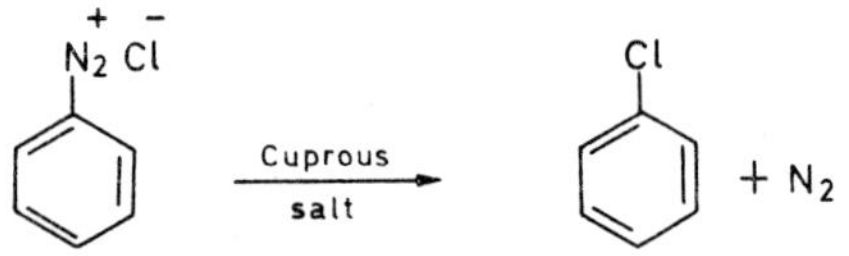

$$Ph{-}N_2^+Cl^- \xrightarrow{\text{Cuprous salt}} Ph{-}Cl + N_2$$

Further investigations by Sandmeyer[2] showed the extensive value of the reaction, as in the preparation of substituted benzoic acids (63 per cent yield)

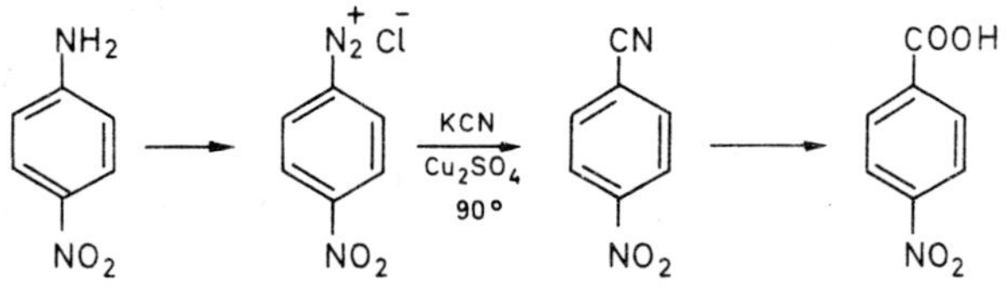

$$p\text{-}O_2N{-}C_6H_4{-}NH_2 \rightarrow p\text{-}O_2N{-}C_6H_4{-}N_2^+Cl^- \xrightarrow[Cu_2SO_4]{KCN, 90°} p\text{-}O_2N{-}C_6H_4{-}CN \rightarrow p\text{-}O_2N{-}C_6H_4{-}COOH$$

Yields from this reaction are between 40 and 90 per cent, with cuprous chloride and cuprous bromide as the best catalysts[3]; cupric and ferric salts have been found to be of no value.

The Sandmeyer reaction should be compared with the Gattermann reaction (No. 49) with which it is often reviewed[4].

SANDMEYER REACTION

Mechanism

Sandmeyer suggested that under normal reaction conditions the procedure involved the formation of an unstable intermediate between the catalyst and the diazonium chloride

$$ArN\!=\!NCl + Cu_2Cl_2 \longrightarrow \underset{\underset{CuCl\ \ CuCl}{|\ \ \ \ \ \ \ |}}{ArN\!-\!N\!-\!Cl}$$

Waentig and Thomas[5] have isolated compounds containing copper, at low temperatures, which decompose on raising the temperature.

Cowdrey and Davies[6] suggested a mechanism that involved formation of a complex which may follow several routes to account for the formation of by-products

$$ArN_2^+ + \bar{C}uCl_2 \longrightarrow [ArN_2CuCl_2] \longrightarrow ArCl + N_2 + CuCl$$

$$\downarrow ArN_2^+$$

$$[(ArN_2)_2CuCl_2]^+ \longrightarrow ArCl + N_2 + CuCl + ArN_2^+$$

$$\downarrow \bar{C}uCl_2$$

$$ArN\!=\!NAr + N_2 + Cu_2^{2+} + 4Cl^-$$

Investigations by Dickerman and co-workers[7] carried out in acetone have led to the suggestion of a radical mechanism

$$ArN_2^+ + \bar{C}uCl_2 \longrightarrow Ar^\cdot + N_2 + CuCl_2$$

$$Ar^\cdot + CuCl_2 \longrightarrow ArCl + CuCl$$

Zollinger[8] has discussed the implications of such a mechanism in the light of other work and emphasizes that the existence of free radicals in aqueous solution is still uncertain.

General reaction conditions

Cuprous chloride (or bromide) is stirred into concentrated acid and cooled to 0°. An aqueous solution of the diazonium salt at 0° to −5° is added slowly with stirring and cooling and, when addition is complete, the mixture is refluxed for about 30 min. Liquid products are often isolated by steam distillation, while solids usually crystallize out on cooling.

NAMED ORGANIC REACTIONS

Gunstone and Tucker[9] used a diazonium sulphate to prepare 1-chloro-2,6-dinitrobenzene, with a yield of over 70 per cent

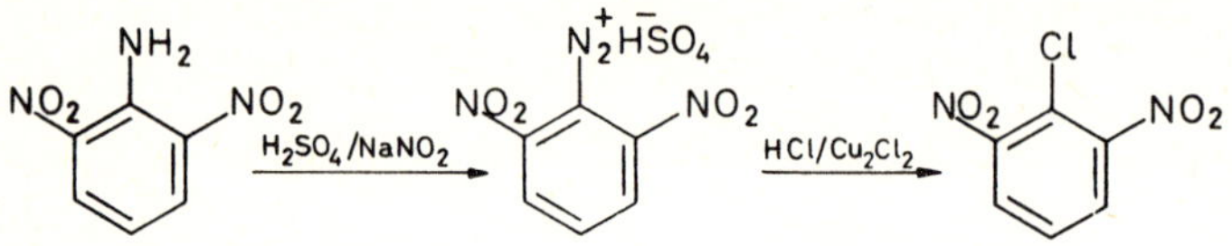

A 90–95 per cent yield of *o*-chlorobromobenzene may be obtained[10] from the diazonium bromide prepared from *o*-chloroaniline. In this case, the diazo compound is added directly to the boiling catalyst–acid mixture

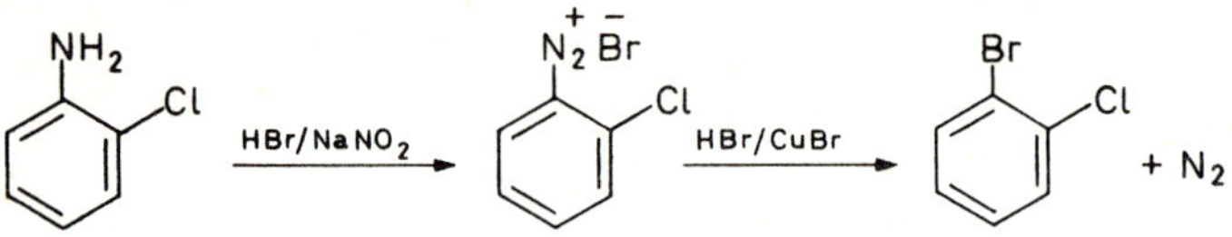

Applications

Kovar[11] has reported an interesting form of the Sandmeyer reaction in which diazonium groups are replaced by hydrogen, due to the action of acetic acid and ethanol in the presence of cuprous oxide

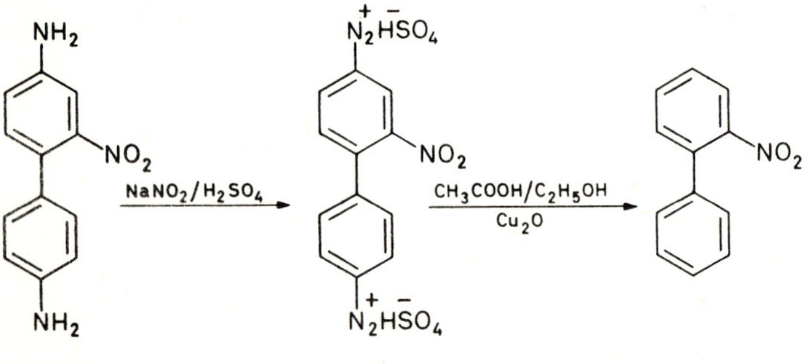

2—Nitrobenzidine   2—Nitrobiphenyl

REFERENCES

[1] Sandmeyer, T. *Ber. dt. chem. Ges.* 17 (1884) 1633 [*J. chem. Soc.* (A) 46 (1884) 1311]
[2] Sandmeyer, T. *Ber. dt. chem. Ges.* 18 (1885) 1492; 23 (1890) 1880 [*J. chem. Soc.* (A) 48 (1885) 981; 58 (1890) 1115]

3 Saunders, K. H. *The Aromatic Diazo Compounds*, 2nd edn., p. 277, London (Arnold) 1949
4 Cowdrey, W. A. and Davies, D. S. *Q. Rev. chem. Soc.* 6 (1952) 358; Hodgson, H. H. *Chem. Rev.* 40 (1947) 251
5 Waentig, P. and Thomas, J. *Ber. dt. chem. Ges.* 46 (1913) 3923 [*Chem. Abstr.* 8 (1914) 709]
6 Cowdrey, W. A. and Davies, D. S. *J. chem. Soc.* (1949) S48
7 Dickerman, S. C., Weiss, K. and Ingberman, A. K. *J. Am. chem. Soc.* 80 (1958) 1901
8 Zollinger, H. *Azo and Diazo Chemistry*, p. 165, New York (Interscience) 1961
9 Gunstone, F. D. and Tucker, S. H. *Org. Synth., Coll. Vol.* 4 (1963) 160
10 Hartwell, J. L. *Org. Synth., Coll. Vol.* 3 (1955) 185
11 Kovar, V., *Czech. Pat.* 105496 (1962) [*Chem. Abstr.* 60 (1964) 454]

51

BAEKELAND POLYMERIZATION PROCESS

Nature of the reaction
Thermosetting resins are produced by the polymerization of phenols and formaldehyde in the presence of a basic catalyst

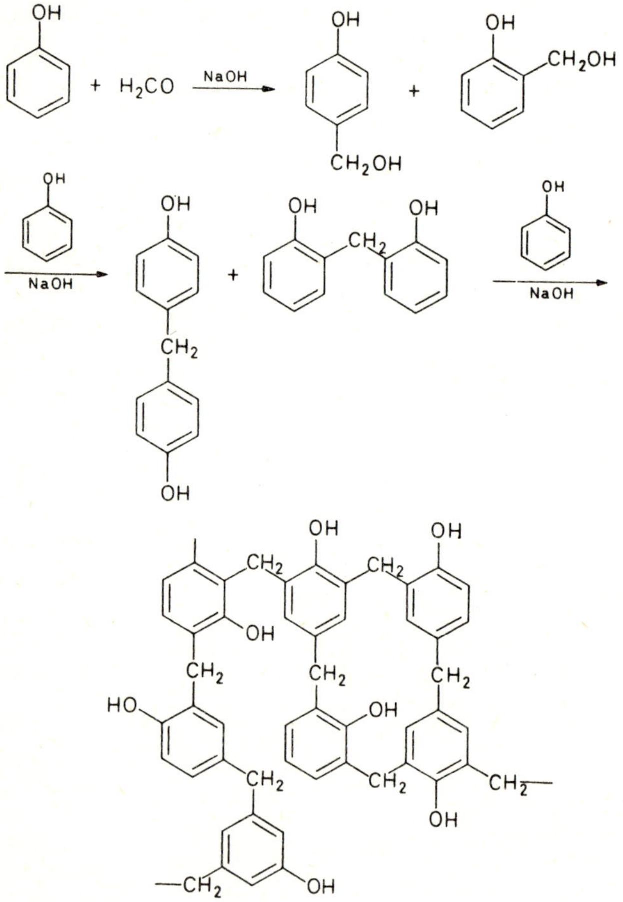

BAEKELAND POLYMERIZATION PROCESS

Historical development

The phenol formaldehyde resins formed by the Baekeland process were the first thermosetting polymers to be commercially produced. Conflicting results that had been reported on reactions between phenols and aldehydes attracted Baekeland's interest, and his investigations led eventually to the process for making Bakelite.

Baeyer[1] had reported that in acid media simple molecular condensations occurred to give hydroxyarylmethanes

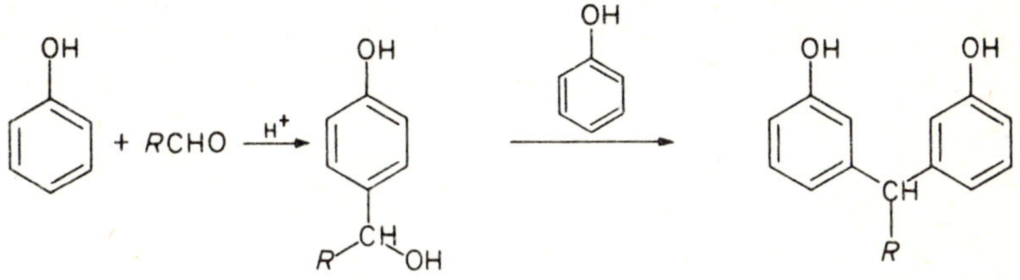

Lederer[2] and Manasse[3], working independently of each other, showed that under alkaline conditions only phenolic alcohols were obtained

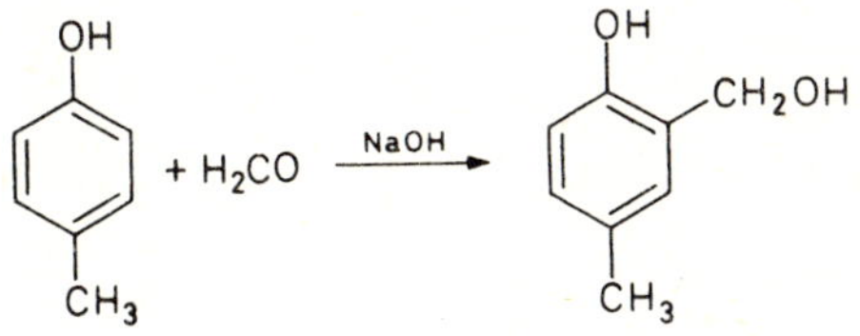

In 1909, Baekeland[4] described in detail the results of his investigations into the polymeric masses which often resulted from the condensation reactions previously reported. He showed that, by employing well-defined conditions, it was possible to control the degree of polymerization and obtain a resin that could be moulded and impregnated on to fabrics. His investigations into the properties of Bakelite extended over many years[5] and opened up a new realm of chemistry.

Bakelite is used extensively for insulation purposes and for giving structural rigidity to materials.

Phenol formaldehyde resins have been dealt with in considerable detail by several authors. Megson[6] has surveyed the growth of the subject from the days prior to Baekeland's results up to 1957.

Mechanism

The base-catalysed Lederer–Manasse reaction is considered to involve electrophilic attack by formaldehyde at the *ortho* or *para* position of the phenol molecule[7]

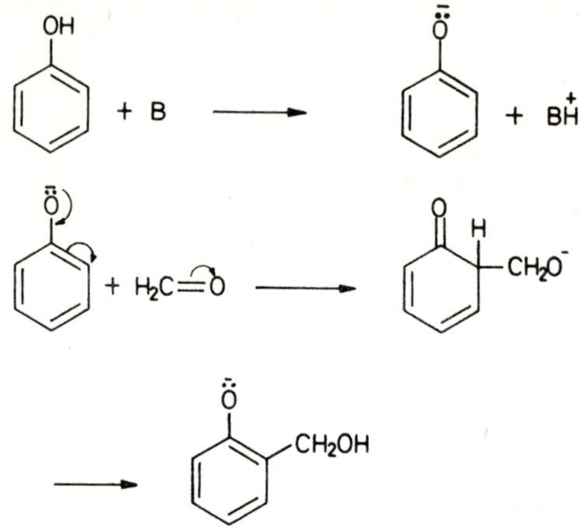

[reaction scheme: PhOH + B → PhO⁻ + BH⁺; PhO⁻ + H₂C=O → cyclohexadienone intermediate with CH₂O⁻ → o-hydroxymethylphenoxide with CH₂OH]

The acid-catalysed Baeyer reaction has been represented as a two-step process[8]

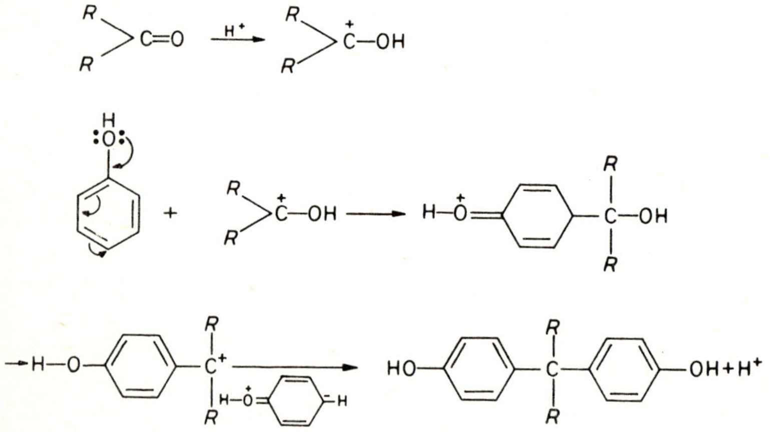

[reaction scheme: R₂C=O + H⁺ → R₂C⁺–OH; then attack by phenol giving R₂C(OH)(C₆H₄OH) and further condensation to HO–C₆H₄–CR₂–C₆H₄–OH + H⁺]

General reaction conditions

The complete Baekeland process involves three stages. An initial condensation product is prepared by heating equal amounts of phenol and formaldehyde with a small quantity of alkaline condensing agent. The mixture separates into an upper aqueous layer and a liquid condensation product as the lower layer (designated Bakelite A). This initial product may vary in its physical form from a mobile liquid to a solid, depending upon the amount of condensing agent employed.

Separation from the aqueous liquor and gradual heating at 70° converts the initial condensing agent into a secondary intermediate (Bakelite B), obtained as a brittle solid.

The process is completed by heating the secondary intermediate above 100° under a pressure of 50–100 lb/in^2 to give the final Bakelite C.

Applications

Bakelite A may be converted directly to Bakelite C by heating under pressure above 100°, so that solid forms of Bakelite A may be employed as moulding powders.

If Bakelite A is a liquid, it may be painted onto objects to give a solid Bakelite C coating after heat and pressure treatment.

For rapid moulding purposes, Bakelite A is moulded and heated to the brittle Bakelite B which retains the moulded shape and can be later processed in quantity to give objects in the Bakelite C form.

REFERENCES

[1] von Baeyer, A. *Ber. dt. chem. Ges.* 5 (1872) 25, 1094; [*J. chem. Soc. (A)* 25 (1872) 301; 26 (1873) 501]
[2] Lederer, L. *J. prakt. Chem.* 50 (1894) [2] 223 [*J. chem. Soc. (A)* 50 (1894) [1] 577]
[3] Manasse, O. *Ber. dt. chem. Ges.* 27 (1894) 2409 [*J. chem. Soc. (A)* 50 (1894) [1] 577]
[4] Baekeland, L. H. *Ind. Engng Chem.* 1 (1909) 149
[5] Baekeland, L. H. and Bender, H. L. *Ind. Engng Chem.* 17 (1925) 225; Megson, N. J. L. *Chemy Ind.* (1968) 632
[6] Megson, N. J. L. *Phenolic Resin Chemistry*, London (Butterworths) 1958
[7] Martin, R. W. *The Chemistry of Phenolic Resins*, p. 6, New York (Wiley) 1956
[8] Schnell, H. and Krimm, H. *Angew. Chem.* 75 (1963) 662; (*int. Edn*) 2 (1963) 373

52

BUCHERER REACTION

Nature of the reaction

Aromatic amines, particularly naphthylamine derivatives, are converted to the corresponding phenols in the presence of aqueous sodium bisulphite. The reverse reaction is achieved by the action of ammonia and ammonium sulphite on the phenol

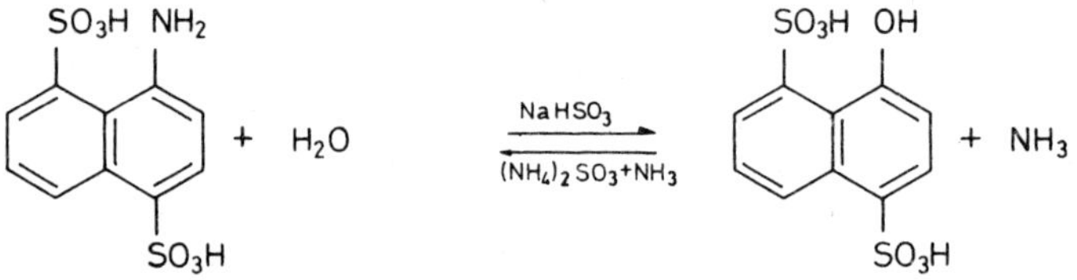

$$\underset{SO_3H}{\underset{SO_3H\;NH_2}{\text{[naphthalene]}}} + H_2O \xrightleftharpoons[(NH_4)_2SO_3+NH_3]{NaHSO_3} \underset{SO_3H}{\underset{SO_3H\;OH}{\text{[naphthalene]}}} + NH_3$$

Historical development

The first conversion of a naphthylamine to a naphthol was reported by Lepetit[1] who obtained 1-naphthol-4-sulphonic acid from naphthionic acid

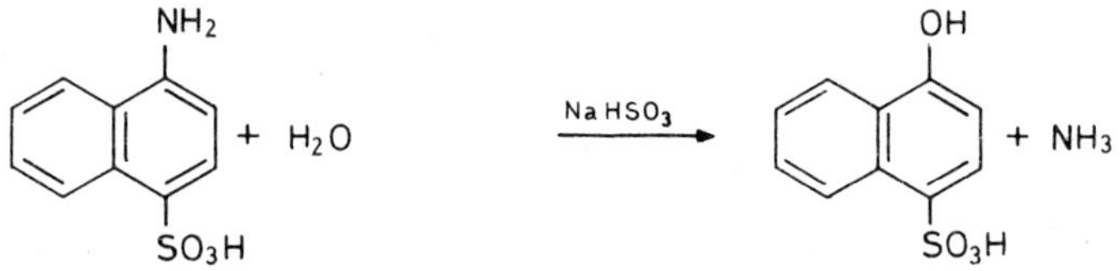

$$\underset{SO_3H}{\underset{NH_2}{\text{[naphthalene]}}} + H_2O \xrightarrow{NaHSO_3} \underset{SO_3H}{\underset{OH}{\text{[naphthalene]}}} + NH_3$$

During investigations on dyestuffs, Bucherer[2] discovered the reaction independently and showed it to be reversible and of wide application in organic syntheses.

With naphthalene compounds, yields are generally above 60 per cent but often much smaller with other aromatic systems. The many factors affecting the reaction have been discussed by Drake[3].

Mechanism

Fuchs and Stix[4] suggested that in the presence of bisulphite, tautomerism of naphthols to keto forms occurs. On this basis, the mechanism postulated was

BUCHERER REACTION

[reaction scheme: 2-naphthol ⇌ naphthalenone tautomer, + NaHSO₃ → tetralone-SO₃Na adduct with OH; then + NH₃/H₂O → amino-SO₃Na intermediate ⇌ 2-naphthylamine + (NH₄)NaSO₃]

The rate-determining step in the reaction was shown by Cowdrey and Hinshelwood[5,6] to be the bimolecular formation of the bisulphite addition compound. The process is first order with respect to both the naphthylamine derivative and bisulphite.

Rieche and Seeboth[7] have demonstrated the existence of other intermediates during the course of the Bucherer reaction. They isolated tetrahydrosulphonic acids, as stable compounds, from reactions in which naphthols are converted to naphthylamines. The mechanism has been given as

[reaction scheme: 1-naphthol ⇌ dihydronaphthalenone, + NaHSO₃ → tetrahydro ketone-SO₃Na adduct]

This work and earlier mechanistic studies have been reviewed by Seeboth[8].

General reaction conditions

Amination reactions are carried out in an autoclave fitted with a high-speed stirrer, as vigorous agitation is of prime importance. The naphthol is placed in the autoclave with freshly prepared ammonium sulphite and the sealed vessel heated at 150° for several hours. On cooling, the required product is filtered off and may be purified by forming a water-soluble hydrochloride, pure amine being reprecipitated by sodium hydroxide.

In some cases, metal halide catalysts have been used to facilitate the naphthol–naphthylamine conversion. Allan and Bell[9] used zinc chloride in their preparation of 3-amino-2-naphthoic acid (70 per cent yield)

[reaction scheme: 3-hydroxy-2-naphthoic acid + ZnCl₂/NH₃ → 3-amino-2-naphthoic acid]

Conversion of naphthylamines to naphthols is carried out by refluxing the amine with 40 per cent sodium bisulphite for several days. The product is in the form of an addition compound, and sodium hydroxide is added and the solution boiled until the ammonia has been expelled. The product is isolated by acidifying the aqueous solution.

Applications

Werbel and co-workers[10] have used the Bucherer reaction to prepare secondary amines containing a naphthyl group. 1-Naphthol was reacted with primary amines, in place of ammonia, in the presence of bisulphite

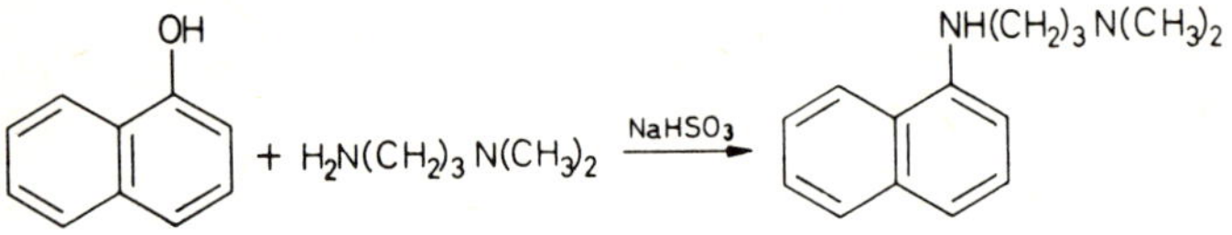

$$\text{1-Naphthol} + H_2N(CH_2)_3N(CH_3)_2 \xrightarrow{NaHSO_3} \text{1-[NH(CH_2)_3N(CH_3)_2]-naphthalene}$$

REFERENCES

[1] Lepetit, R., *Sealed Communication* 888 (1896); *Bull. Soc. ind. Mulhouse* (1903) 326
[2] Bucherer, H. T. *Z. Farb.-u. TextChem.* 2 (1903) 193; *J. prakt. Chem.* 69 (1904) [2] 49; with Seyde, F. 77 (1908) [2] 403 [*J. chem. Soc. (A)* 84 (1903) [1]627; 86 (1904) [1] 309; 94 (1908) [1] 455]
[3] Drake, N. L. *Org. React.* 1 (1942) 105
[4] Fuchs, W. and Stix, W. *Ber. dt. chem. Ges.* 55B (1922) 658 [*Chem. Abstr.* 16 (1922) 2860]
[5] Cowdrey, W. A. and Hinshelwood, C. N. *J. chem. Soc.* (1946) 1036
[6] Cowdrey, W. A. *J. chem. Soc.* (1946) 1041, 1046
[7] Reiche, A. and Seeboth, H. *Angew. Chem.* 70 (1958) 52; *Justus Liebigs Annln Chem.* 671 (1964) 70 [*Chem. Abstr.* 52 (1958) 11798]; 60 (1964) 10612]
[8] Seeboth, H. *Angew. Chem.* 79 (1967) 329; (*int. Edn*) 6 (1967) 307
[9] Allen, C. F. H. and Bell, A. *Org. Synth., Coll. Vol.* 3 (1955) 78
[10] Werbel, L. M. *et al., J. mednl pharm. Chem.* 6 (1963) 637

53

DAKIN REACTION

Nature of the reaction
The carbonyl function in phenolic aldehydes and ketones is replaced by an hydroxyl group through the action of hydrogen peroxide under alkaline conditions

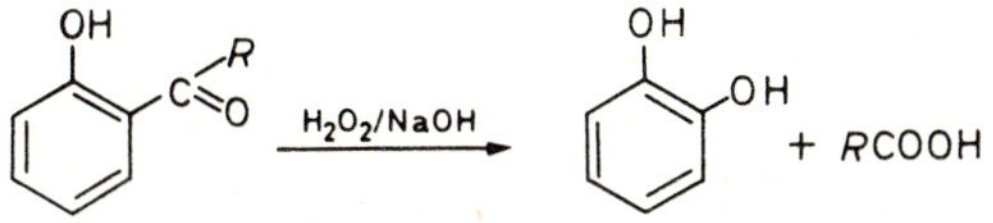

$$\underset{}{\text{o-HOC}_6\text{H}_4\text{C(=O)R}} \xrightarrow{\text{H}_2\text{O}_2/\text{NaOH}} \text{o-HOC}_6\text{H}_4\text{OH} + R\text{COOH}$$

Historical development
Dakin[1] found a large number of substituted benzaldehydes and acetophenones to undergo this reaction when an hydroxyl group was *ortho* or *para* to the carbonyl group

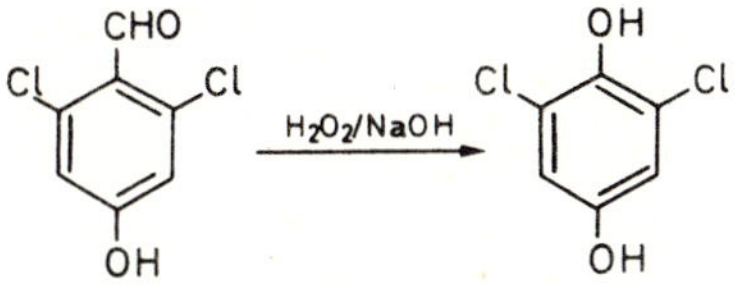

$$\underset{}{2,6\text{-Cl}_2\text{-4-HO-C}_6\text{H}_2\text{CHO}} \xrightarrow{\text{H}_2\text{O}_2/\text{NaOH}} 2,6\text{-Cl}_2\text{-4-HO-C}_6\text{H}_2\text{OH}$$

The scope of the reaction was extended to higher ketones, such as hydroxyphenyl benzyl ketones, by Baker and co-workers[2] using elevated temperatures and inert atmospheres.

Yields from Dakin reactions are high, often over 90 per cent.

Mechanism
The mechanism of this reaction has not received detailed study, but by analogy with related processes[3] it is believed that the peroxide ion attacks the carbonyl group and gives rise to a rearrangement

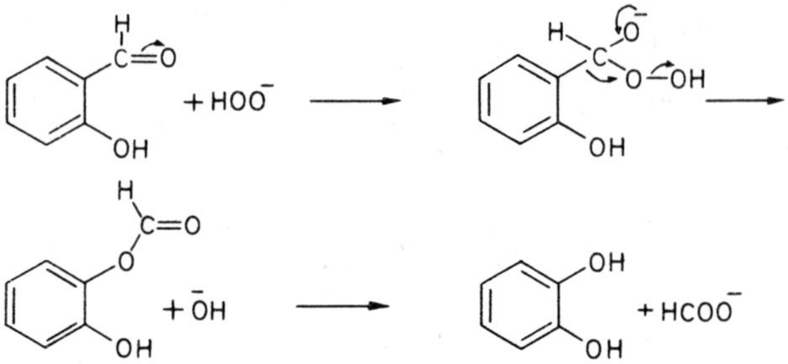

Lee and Uff[4] have indicated that, in the case of salicylaldehyde, the mechanism may include a cyclic intermediate formed between the aldehyde group, the phenolic group and a peroxide ion.

General reaction conditions

Reactions are carried out under inert atmospheres, since phenols in alkaline solution are readily oxidized. The phenolic aldehyde or ketone is dissolved in aqueous sodium hydroxide and an excess of 3–6 per cent hydrogen peroxide added portionwise during 1 h. During addition, the temperature is kept below 50°, then the mixture is left to stand for several hours at room temperature. The solution is made just acid and the product isolated either by extraction or evaporation.

Complete experimental details for the preparation of catechol from salicylaldehyde have been given by Dakin[5]. A similar procedure has been reported by Surrey[6] for the preparation of pyrogallol 1-methyl ether from 2-hydroxy-3-methoxybenzaldehyde (80 per cent yield)

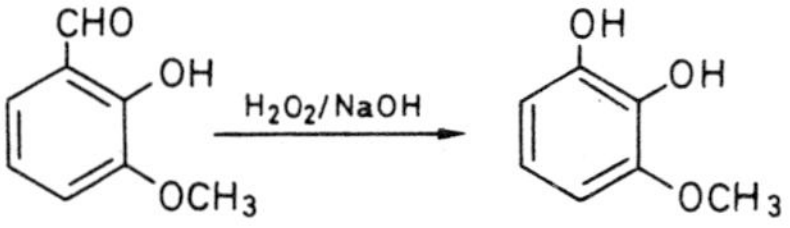

Applications

Nikforov and Ershov[7] employed the Dakin reaction to prepare 2,6-dialkylhydroquinones in high yields from the corresponding 4-hydroxydialkyl benzaldehydes

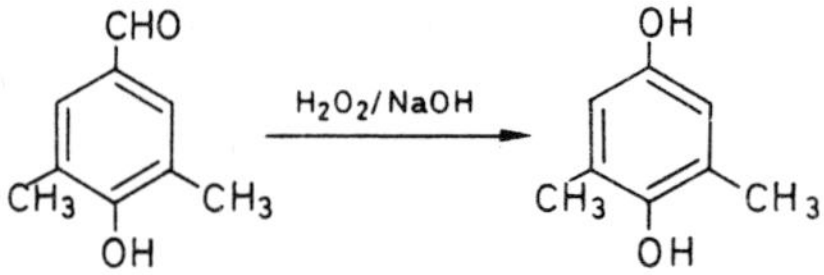

DAKIN REACTION

Baker et al.[8] found that, owing to lack of solubility, sodium hydroxide was unsuitable for a Dakin reaction applied to 5-acetyl-indan-4-ol. This difficulty was overcome by using tetramethyl ammonium hydroxide, and the required indane-4,5-diol successfully obtained

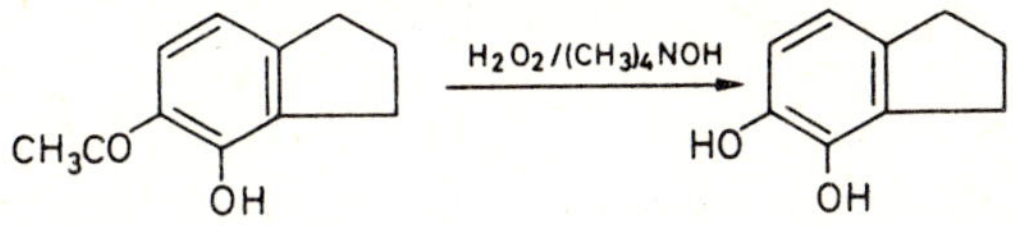

$$\text{CH}_3\text{CO}\text{-aryl-OH} \xrightarrow{H_2O_2/(CH_3)_4NOH} \text{HO-aryl-OH}$$

This procedure was later used[9] to give a 25 per cent yield of 3,4-dimethylcatechol from 2-hydroxy-3,4-dimethyl acetophenone, as compared with a 2·5 per cent yield by the conventional method.

REFERENCES

[1] Dakin, H. D. *Am. Chem. J.* 42 (1909) 477
[2] Baker, W., Jukes, E. H. T. and Sabrahmanyam, C. A. *J. chem. Soc.* (1934) 1681
[3] Hine, J. *Physical Organic Chemistry*, 2nd edn., p. 341, New York (McGraw-Hill) 1962
[4] Lee, J. B. and Uff, B. C. *Q. Rev. chem. Soc.* 21 (1967) 454
[5] Dakin, H. D. *Org. Synth., Coll. Vol.* 1 (1941) 149
[6] Surrey, A. R. *Org. Synth., Coll. Vol.* 3 (1955) 759
[7] Nikoforov, G. A. and Ershov, V. V. *Izv. Akad. Nauk SSSR, Ser. Khim.* 1 (1964) 176 [*Chem. Abstr.* 60 (1964) 9188]
[8] Baker, W., McOmie, J. F. W. and Ulbricht, T. L. V. *J. chem. Soc.* (1952) 1825
[9] Baker, W. et al., *J. chem. Soc.* (1953) 1615

54

DARZENS' PROCEDURE

Nature of the reaction
Alkyl chlorides are prepared by treating primary and secondary alcohols with thionyl chloride in the presence of tertiary amines

$$ROH + SOCl_2 \xrightarrow{Pyridine} RCl + SO_2 + HCl$$

Historical development
In reporting this approach for the formation of alkyl chlorides, Darzens[1] stated that suitable bases for enabling high yields to be obtained were pyridine, quinoline and diethyl aniline. The method was applied to a wide range of alcohols, and yields were in most cases of the order of 90 per cent. Among the first compounds treated in this manner were β-phenyl ethanol and ethylene chlorohydrin

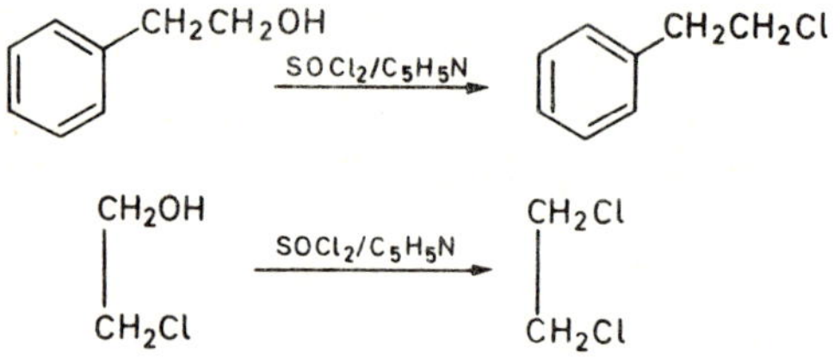

$$PhCH_2CH_2OH \xrightarrow{SOCl_2/C_5H_5N} PhCH_2CH_2Cl$$

$$\underset{CH_2Cl}{\overset{CH_2OH}{|}} \xrightarrow{SOCl_2/C_5H_5N} \underset{CH_2Cl}{\overset{CH_2Cl}{|}}$$

Alkyl bromides may be prepared in a similar manner by using thionyl bromide. Darzens' procedure is unsuitable, however, for replacing the hydroxy group in phenols.

Mechanism
Hughes and Ingold[2] found retention of configuration at the active centre in optically active compounds and proposed an S_{Ni} (substitution, nucleophilic, internal) mechanism for the process.

It was later shown, by Gerrard[3], that an intermediate sulphite is formed when only 0·5 mol of thionyl chloride has reacted. On the basis of this work a three-step mechanism was proposed

DARZENS' PROCEDURE

$ROH:NC_5H_5$

$$\underset{\underset{Cl}{\overset{Cl}{\diagup}}}{\overset{+}{\bar{O}-S}} \longrightarrow \underset{\underset{R-O}{\overset{R-O}{\diagup}}}{\overset{+}{\bar{O}-S^+}} + 2C_5H_5\overset{+}{N}H\bar{C}l$$

$ROH:NC_5H_5$

$$\underset{\underset{R-O}{\overset{R-O}{\diagup}}}{\overset{+}{\bar{O}-S^+}} \underset{\underset{Cl}{\overset{Cl}{\diagup}}}{\overset{+}{\bar{O}-S}} \longrightarrow 2 \underset{\underset{Cl}{\overset{R-O}{\diagup}}}{\overset{+}{\bar{O}-S^+}} \xrightarrow{C_5H_5\overset{+}{N}H\bar{C}l} 2RCl + 2SO_2$$

Further work by Gerrard and Hudson[4] involving vapour-phase chromatography has confirmed that the configuration of optically active compounds is retained, as would be the case with the above mechanism.

General reaction conditions

The alcohol and pyridine (1 mol each) are mixed together and cooled to about 0°. Thionyl chloride is then added over 1–2 h to the stirred mixture. Sulphur dioxide and hydrogen chloride are evolved and the exothermic reaction is kept below 60° before finally being left to cool for several hours. The product, if a solid, is filtered off and water-washed to remove the pyridine hydrochloride. Liquid products are isolated by ether extraction.

A detailed preparation of α-chloro-α-phenyl acetophenone has been given by Ward[5]. Brooks and Snyder[6] have reported conditions for that of tetrahydrofurfuryl chloride, based upon Kirner's[7] method

$$\text{(tetrahydrofurfuryl alcohol)} \xrightarrow{SOCl_2/C_5H_5N} \text{(tetrahydrofurfuryl chloride)} + SO_2 + HCl$$

Applications

In some cases, successful halogenation reactions have been carried out employing only a small quantity of base. Newman[8] used only a few drops of pyridine in the reaction with *o*-methyl benzyl alcohol and obtained an 89 per cent yield of the halide

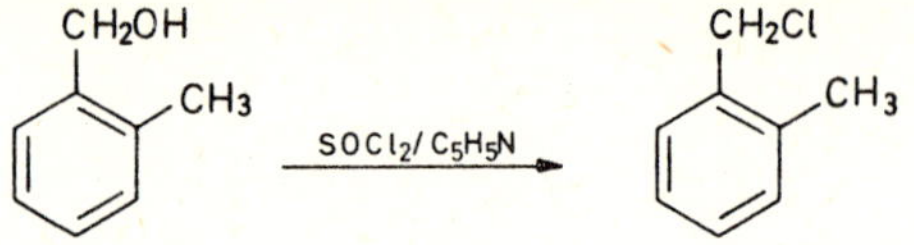

NAMED ORGANIC REACTIONS

[o-methylbenzyl alcohol] —SOCl$_2$/C$_5$H$_5$N→ [o-methylbenzyl chloride]

Darzens' procedure has been widely used as an industrial method for preparing alkyl halides. Ludsteck and co-workers[9] have worked with dialkyl amides at moderate temperatures in place of tertiary amines to prepare alkyl halides (94 per cent yield)

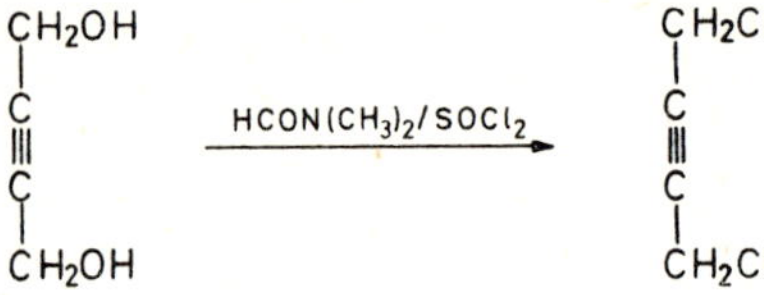

HOCH$_2$–C≡C–CH$_2$OH —HCON(CH$_3$)$_2$/SOCl$_2$→ ClCH$_2$–C≡C–CH$_2$Cl

REFERENCES

[1] Darzens, G. *C. r. hebd. Séanc. Acad. Sci., Paris* 152 (1911) 1314, 1601 [*Chem. Abstr.* 5 (1911) 3410]
[2] Hughes, E. D., Ingold, C. K. and Scott, A. D. *J. chem. Soc.* (1937) 1201
[3] Gerrard, W. *J. chem. Soc.* (1939) 99
[4] Gerrard, W. and Hudson, H. R. H. *J. chem. Soc.* (1963) 1059
[5] Ward, A. M. *Org. Synth., Coll. Vol.* 2 (1943) 159
[6] Brooks, L. A. and Snyder, H. R. *Org. Synth., Coll. Vol.* 3 (1955) 698
[7] Kirner, W. R. *J. Am. chem. Soc.* 52 (1930) 3251
[8] Newman, M. S. *J. Am. chem. Soc.* 62 (1940) 2295
[9] Ludsteck, D. *et al., Germ. Pat.* 1133716 (1962) [*Chem. Abstr.* 57 (1962) 16396]

55

DIELS–ALDER REACTION

Nature of the reaction
Six-membered ring compounds are formed by the 1,4 addition of compounds possessing double or triple bonds (dienophiles) to conjugated dienes

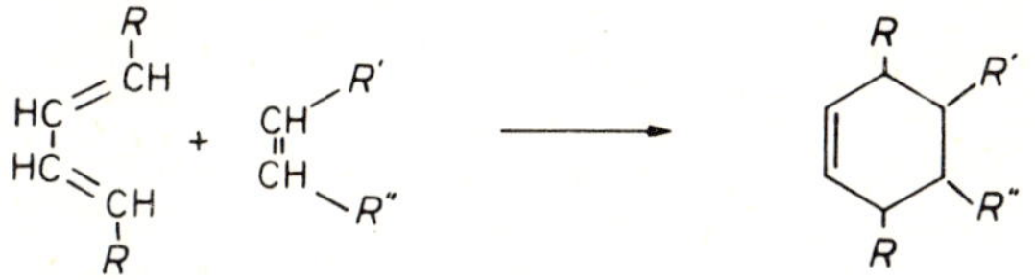

Historical development
   The reaction, for which Diels and Alder[1] were granted patent rights, was found to be of very wide application and has the advantage of not normally requiring the use of a catalyst. Much of the early work was involved with the formation of ring systems related to camphor, the reaction of maleic anhydride with cyclopentadiene being one example

   High yields are obtained, usually above 60 per cent and often quantitative. Most of the early work on the Diels–Alder method was dealt with in extensive review articles[2]. Other reviews[3] have covered more recent aspects, that of Needleman and Chang Kuo[4] especially reactions in which one or more atoms of the reacting system are not carbon. A Diels–Alder reaction with cyclopentadiene and a –N=N– system is the basis of an assay[5] for benzothiadiole 1,1-dioxide

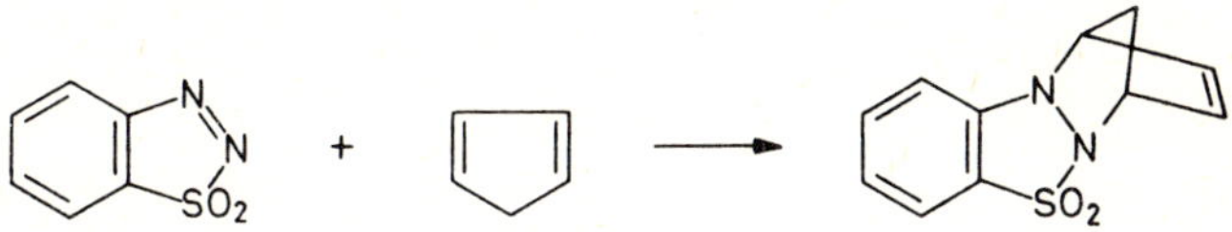

Numerous examples of retro-Diels–Alder reactions have been reported; in some cases, elevation of temperature is enough to cause reversion to adducts, in others catalytic conditions have been employed[6]. Rosenblum[7] found that a mixture of cyclopentadiene and cyclopentenone was obtained when 1-dicyclopentadiene was heated at 140°–150°. The cyclopentenone was formed owing to isomerization of 2,4-cyclopentadienol

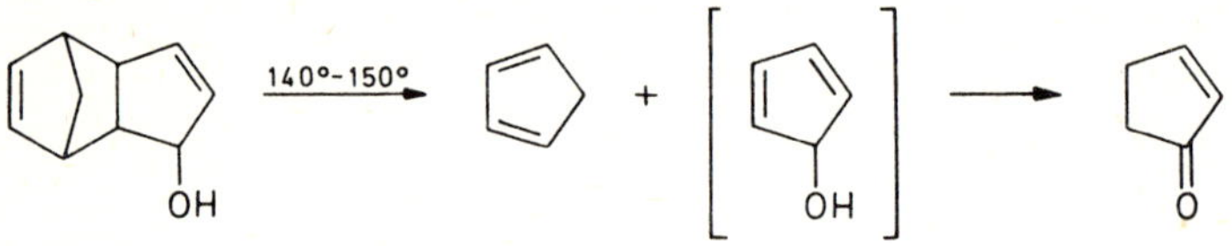

Mechanism

Uncertainty over the mechanism of the Diels–Alder reaction still exists[8] and the usual simple representation does not reveal the complete story

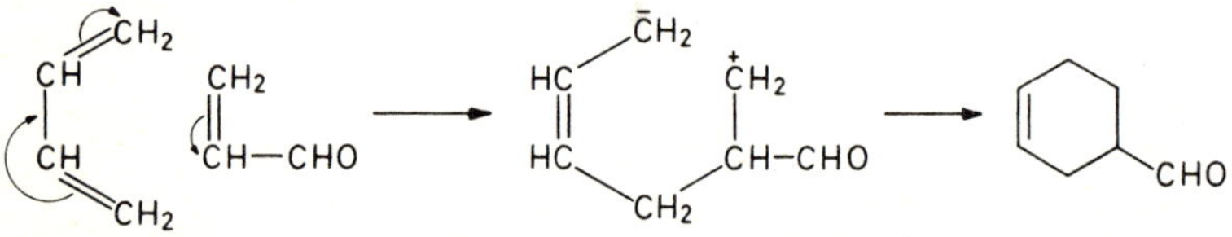

The controversy revolves around whether it is a one-step concerted mechanism[9] or a two-step reaction proceeding by ionic or radical attack[10]. The studies have necessitated a great deal of work on the stereospecificity of the reaction.

Dewar[11] represented the one-step mechanism as proceeding through a cyclic pseudoaromatic transition state

In the Woodward and Katz[10] mechanism, the rate-controlling step is considered to be the formation of a single bond between one end of the diene and one of the unsaturated carbon atoms in the dienophile

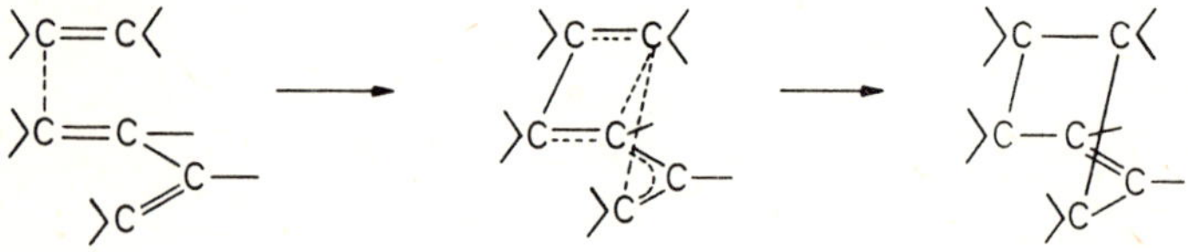

DIELS–ALDER REACTION

Support of the one-step concerted mechanism has been given by Lambert and Roberts[12] who examined systems which were expected to favour the formation of diradical intermediates. By reacting hexachlorocyclopentadiene with *cis-* and *trans-*α-methyl styrene labelled with deuterium, they showed that the resulting products had *endo/exo* ratios identical with the original *cis/trans* ratios

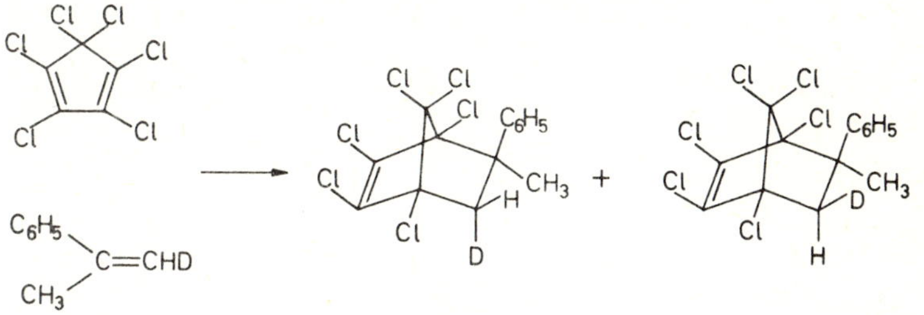

If a diradical had been an intermediate, interconversion between the two forms would have occurred.

The various aspects of the Diels–Alder mechanism have been summarized by Sauer[13].

General reaction conditions

The Diels–Alder reaction is rapid and easy, and although catalysts have been employed, they are rarely necessary. Many reactions involving uncomplicated dienes and dienophiles have been carried out by mixing molar proportions together. Under these circumstances, an exothermic reaction occurs.

More generally, the process involves mixing the reactants in an inert solvent; this is essential when the diene is a gas such as butadiene. In these cases, the gas is slowly bubbled into a stirred solution of the dienophile over a period of several hours[14]. The exothermic reaction slowly subsides, and after standing for several hours the product either crystallizes from solution or is isolated by evaporation.

Allen and Bell[15] used a fairly standard procedure in refluxing 1,4-naphthoquinone with 2,3-dimethyl butadiene in alcohol for 5 h to obtain a 96 per cent yield of dimethyl tetrahydroanthroquinone

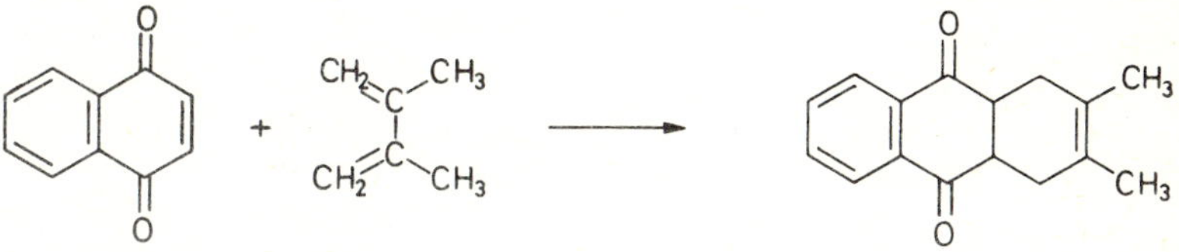

Modification

It has been established[16] that the use of Lewis acids not only accelerates a Diels–Alder reaction but also affects the isomerism of the products. Thus, methyl vinyl ketone and isoprene can give two products

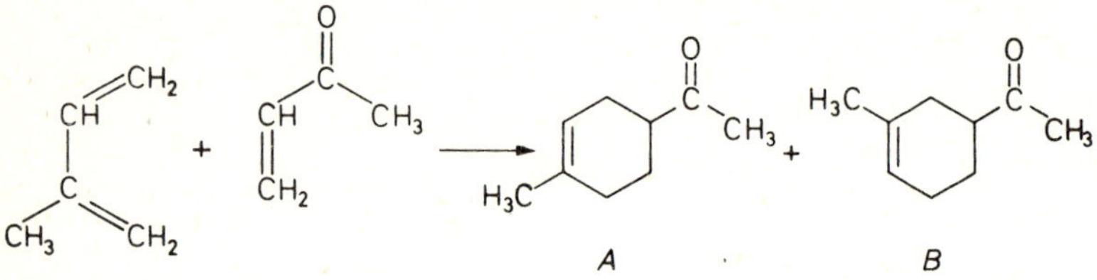

If no catalyst is used, 71 per cent *A* is obtained compared to 29 per cent *B*. With stannic chloride catalyst, the percentages are 93 and 7, respectively.

Applications

Investigations into intramolecular Diels–Alder reactions have aroused a great deal of interest[17]. Klemm and Gopinath[18] used the idea in their synthesis of γ-apopicropodophyllin

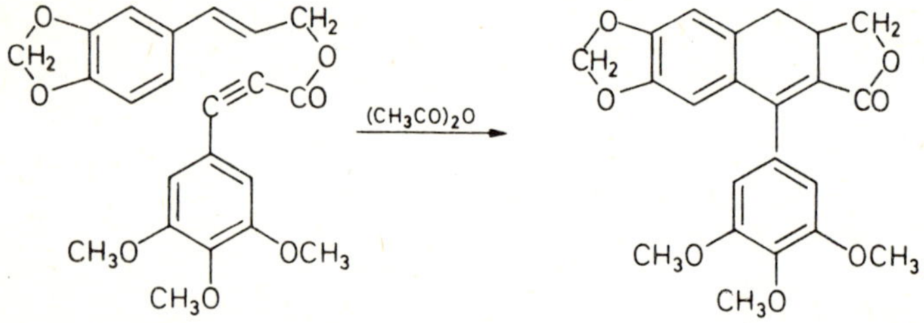

Reactions between alkynes and 2,3,4,5-tetraphenylpentadienone have been shown to form aromatic systems with loss of carbon monoxide[19]

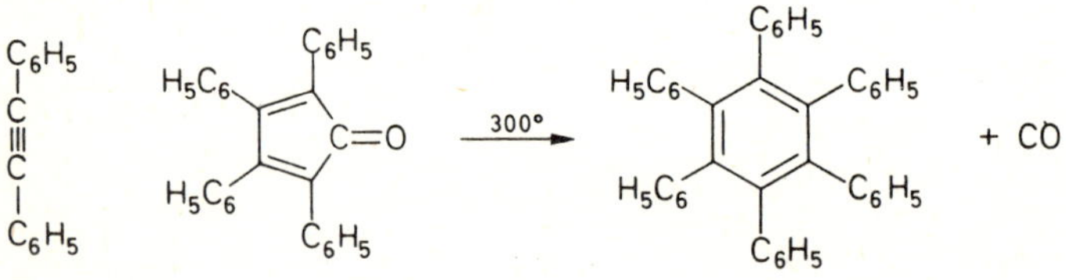

REFERENCES

[1] Diels, O. and Alder, K. *Br. Pat.* 300130 (1927); *Justus Liebigs Annln Chem.* 460 (1928) 98; *Ber. dt. chem. Ges.* 62 (1929) 2081 [*Chem. Abstr.* 22 (1928) 1144; 24 (1930) 96]
[2] Kloetzel, M. C. *Org. React.* 4 (1948) 1; Holmes, H. L. *ibid.* 60; Butz, L. W. and Rytina, A. W. 5 (1949) 136
[3] Martin, J. G. and Hill, R. K. *Chem. Rev.* 61 (1961) 537
[4] Needleman, S. and Chang Kuo, M. C. *Chem. Rev.* 62 (1962) 405
[5] Wittig, G. and Hoffmann, R. W. *Org. Synth.* 47 (1967) 4
[6] Wasserman, A. *Diels Alder Reactions*, p. 61, London (Elsevier) 1965
[7] Rosenblum, M. *J. Am. chem. Soc.* 79 (1957) 3179
[8] Huisgen, R., Grashey, R. and Sauer, J. *The Chemistry of Alkenes* (ed. Patai, S.), p. 878, New York (Interscience) 1964
[9] Berson, J. A. and Remanick, A. *J. Am. chem. Soc.* 83 (1961) 4947
[10] Woodward, R. B. and Katz, T. J. *Tetrahedron* 5 (1959) 70
[11] Dewar, M. J. S. *Tetrahedron Lett.* 5 (1959) 19
[12] Lambert, J. B. and Roberts, J. D. *Tetrahedron Lett.* 20 (1965) 1457
[13] Sauer, J. *Angew. Chem. (int. Edn)* 5 (1966) 211; 6 (1967) 16
[14] Cope, A. C. and Herrick, E. C. *Org. Synth., Coll. Vol.* 4 (1963) 890
[15] Allen, C. F. H. and Bell, A. *Org. Synth., Coll. Vol.* 3 (1955) 310
[16] Lutz, E. F. and Bailey, G. M. *J. Am. chem. Soc.* 86 (1964) 3899
[17] House, H. O. and Cronin, T. H. *J. org. Chem.* 30 (1965) 1061
[18] Clemm, L. H. and Gopinath, K. W. *Tetrahedron Lett.* 19 (1963) 1243
[19] Fieser, L. F. *Org. Synth.* 46 (1966) 44; with Haddadin, M. J. *ibid.* 107

56
FISCHER–SPEIER ESTERIFICATION

Nature of the reaction

The formation of esters by refluxing an alcohol and a carboxylic acid together is greatly facilitated by the addition of a small quantity of mineral acid to the mixture

$$RCOOH + R'OH \xrightarrow{H^+} RCOOR' + H_2O$$

Historical development

In the absence of any catalyst, an alcohol and a carboxylic acid will slowly interact to form an equilibrium mixture with the ester. Fischer and Speier[1] found that the addition of a small quantity of either dry hydrogen chloride gas or concentrated sulphuric acid produced yields of esters as high as 90 per cent within periods of 4 h. They applied the procedure to aromatic and aliphatic monobasic and dibasic acids.

Cinnamic acid, under various catalytic conditions, gave the ethyl ester in yields ranging from 75 to 90 per cent

$$\text{Ph-CH=CH-COOH} + C_2H_5OH \xrightarrow{H^+} \text{Ph-CH=CH-CO}_2C_2H_5 + H_2O$$

The structures of the reactants affect the ease or difficulty of the esterification; α-substituents on aliphatic acids or on the carbinol carbon of alcohols can lead to steric hindrance and retard the reaction.

Mechanism

Two mechanisms for esterification are possible[2]:

(a) Acyl–oxygen fission $RCO|OH + H|OR' \longrightarrow RCOOR' + H_2O$

(b) Alkyl–oxygen fission $RCOO|H + HO|R' \longrightarrow RCOOR' + H_2O$

Under acidic conditions, the former is the more common, involving formation of the conjugate acid of the carboxylic acid, and represented by the reversible route[3]

$$RCO_2H + R'OH \xrightleftharpoons{H^+} \underset{\text{Conjugate acid}}{RC(OH)_2^+} + R'OH \rightleftharpoons R-\underset{OH}{\overset{OH}{\underset{|}{C}}}-\overset{+}{O}HR'$$

$$\rightleftharpoons R-\underset{OH}{\overset{OH}{\underset{|}{C}}}-OR' \xrightleftharpoons{H^+} R-\underset{^+OH_2}{\overset{OH}{\underset{|}{C}}}-OR' \xrightarrow{-H_2O} \left[R-\overset{OH}{\underset{}{C}}\text{↔}OR' \right]^+$$

$$\xrightleftharpoons{-H^+} R-\overset{O}{\underset{}{\overset{\|}{C}}}-OR'$$

Much of the confirmation of this mechanism was provided by Bender[4] using isotopic procedures with ^{18}O. The whole concept of esterification and the reverse procedure, hydrolysis, has since been reviewed in detail[5].

FISCHER–SPEIER ESTERIFICATION

General reaction conditions

On the basis of the law of mass action, the equilibrium of any esterification may be altered by employing different ratios of starting materials. Because of this, a large excess of one of the reactants, usually the alcohol, is employed.

The acid is dissolved in an excess of the pure alcohol to which either a small quantity of concentrated sulphuric acid has been added or in which dry hydrogen chloride has been passed to give a 3 per cent increase in weight. The mixture is then refluxed for 2–4 h and left to stand for a few hours. Most of the excess alcohol is distilled off and water and sodium bicarbonate added to the organic material. The ester is then isolated by extraction and distillation.

Applications

Natelson and Gottfried[6] used benzene as a solvent in the preparation of ethyl bromoacetate and were able to distil off and measure the water formed in the reaction by means of a Dean Starke trap

$$BrCH_2COOH + C_2H_5OH \xrightarrow{H_2SO_4} BrCH_2CO_2C_2H_5 + H_2O$$

A method of wide applicability for the formation of diethyl esters utilizing dry hydrogen chloride catalyst has been described in the preparation of ethyl acetone dicarboxylate[7]

$$\underset{\text{Acetone dicarboxylic acid}}{CO\begin{cases}CH_2\text{—}COOH\\CH_2\text{—}COOH\end{cases}} + 2C_2H_5OH \xrightarrow{HCl} \underset{\text{Ethyl acetone dicarboxylate}}{CO\begin{cases}CH_2COOC_2H_5\\CH_2COOC_2H_5\end{cases}}$$

REFERENCES

[1] Fischer, E. and Speier, A. *Ber. dt. chem. Ges.* 28 (1895) 3252 [*J. chem. Soc.* (*A*) 70 (1896) [1] 201]
[2] Ingold, C. K. *Structure and Mechanism in Organic Chemistry*, p. 753, Ithaca, N.Y. (Cornell Univ. Press) 1953
[3] Hine, J. *Physical Organic Chemistry*, 2nd edn., p. 277, New York (McGraw-Hill) 1962
[4] Bender, M. L. *J. Am. chem. Soc.* 73 (1951) 1626; 75 (1953) 5986
[5] Bender, M. L. *Chem. Rev.* 60 (1960) 53
[6] Natelson, S. and Gottfried, S. *Org. Synth., Coll. Vol.* 3 (1955) 382
[7] Adams, R. and Chiles, H. M. *Org. Synth., Coll. Vol.* 1 (1932) 237

57
GABRIEL SYNTHESIS

Nature of the reaction

Pure primary amines are prepared by formation and subsequent cleavage of N-alkyl phthalimides

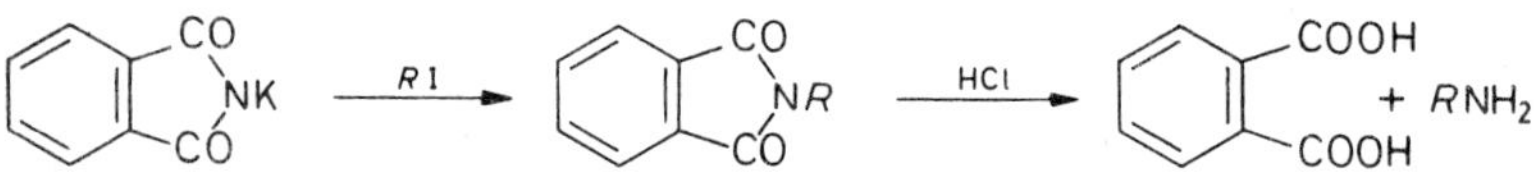

$$\text{Phth-NK} \xrightarrow{RI} \text{Phth-NR} \xrightarrow{HCl} \text{phthalic acid (COOH, COOH)} + RNH_2$$

Historical development

The formation of a number of alkyl phthalimides by reacting potassium phthalimide and alkyl iodides was reported by Graebe and Pictet[1] in 1884, though they attributed incorrect structures to the compounds obtained. Gabriel[2] found that hydrolysis of the alkyl phthalimides gave phthalic acid and a primary amine. The first reaction given was the preparation of ethylene diamine from ethylene dibromide

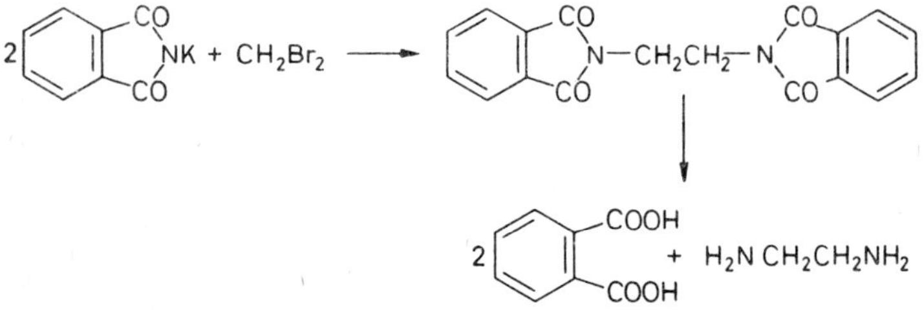

$$2\,\text{Phth-NK} + CH_2Br_2 \longrightarrow \text{Phth-N}-CH_2CH_2-\text{N-Phth}$$
$$\downarrow$$
$$2\,\text{phthalic acid} + H_2N\,CH_2CH_2NH_2$$

Amino acids are prepared by this route by reacting potassium phthalimide with the appropriate α-halogeno ester

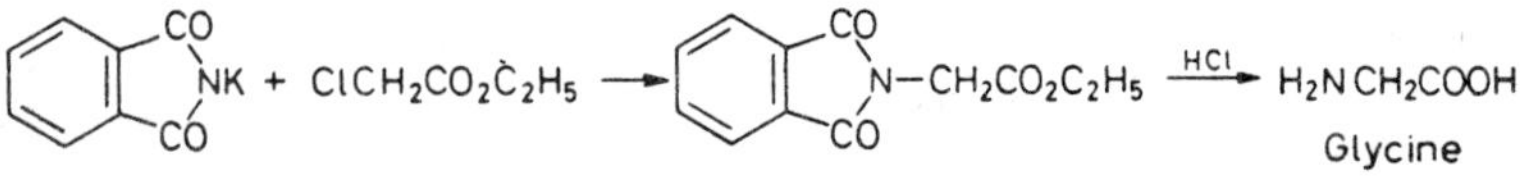

$$\text{Phth-NK} + ClCH_2CO_2C_2H_5 \longrightarrow \text{Phth-N}-CH_2CO_2C_2H_5 \xrightarrow{HCl} H_2N\,CH_2COOH$$
Glycine

GABRIEL SYNTHESIS

The method is of particular value since the primary amines formed are not contaminated with secondary or tertiary ones. Yields are usually very good, regularly above 70 and often >90 per cent.

General reaction conditions

The first stage is carried out by directly heating a mixture of potassium phthalimide and the alkyl halide above 150° for up to 2 h. The cooled mixture is treated with 20 per cent hydrochloric acid and refluxed for several hours. Phthalic acid crystallizes out on cooling and is filtered off. The soluble primary amines are isolated as the hydrochlorides by evaporation of the bulk of the solution.

In his procedure for γ-aminobutyric acid, based upon earlier work by Gabriel[3], De Witt[4] used sulphuric acid for the hydrolysis and precipitated the sulphate as the barium salt, prior to evaporating the aqueous solution

$$\text{C}_6\text{H}_4(\text{CO})_2\text{NK} + \text{Cl}(\text{CH}_2)_3\text{CN} \longrightarrow \text{C}_6\text{H}_4(\text{CO})_2\text{N}(\text{CH}_2)_3\text{CN} + \text{KCl}$$

$$\xrightarrow{\text{H}_2\text{SO}_4} \text{C}_6\text{H}_4(\text{COOH})_2 + \text{H}_2\text{N}(\text{CH}_2)_3\text{COOH}$$

γ-Aminobutyric acid

Modifications

Formation of *N*-substituted phthalimides can often be accomplished under mild conditions by heating the alkyl halide with potassium phthalimide using dimethyl formamide as the solvent to improve solution. This method, introduced by Sheehan and Bolhofer[5], was extended by Müller and Rieck[6] who prepared potassium phthalimide *in situ* by adding methanolic KOH to phthalimide in the dimethyl formamide solution and heated the mixture between 50° and 70°.

As the hydrolysis step is often difficult, Ing and Manske[7] developed a procedure involving the formation of an intermediate that could be more readily hydrolysed. The *N*-substituted phthalimide was treated with hydrazine hydrate and the intermediate hydrolysed with 10 per cent hydrochloric acid

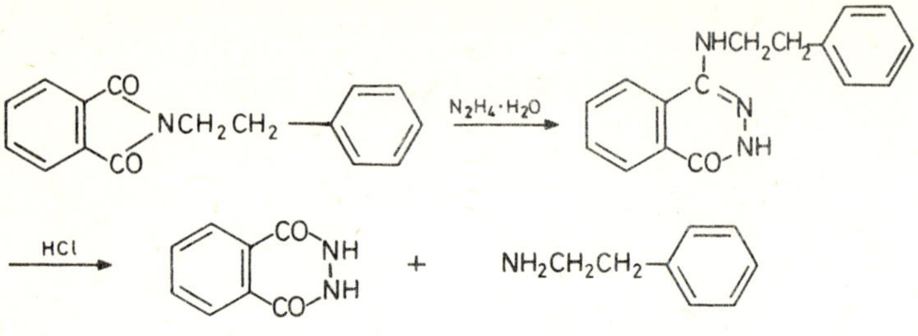

[reaction scheme showing phthalimide derivative + N2H4·H2O, then HCl, yielding phthalhydrazide and β-phenylethylamine]

β-Phenylethylamine

Yields of substituted benzylamines > 90 per cent were obtained by this procedure.

Applications for amino acids

One of the most important features of Gabriel's synthesis has been its use to prepare amino acids in a pure state[8] and labelled with ^{15}N by using labelled potassium phthalimide[9]

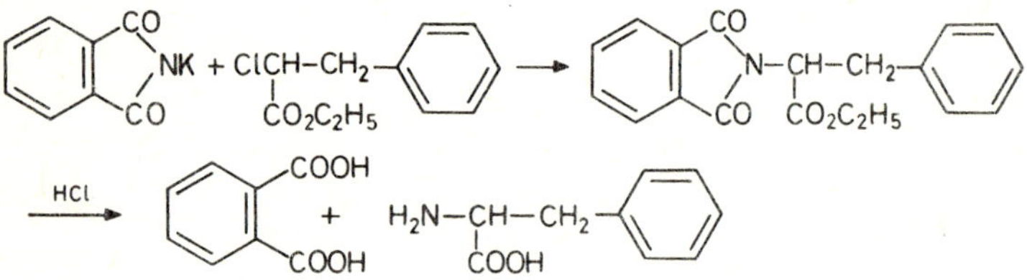

[reaction scheme showing potassium phthalimide + ClCH(CO$_2$C$_2$H$_5$)CH$_2$Ph, then HCl, yielding phthalic acid and phenylalanine]

Phenyl alanine

Kosolapoff[10] employed Gabriel's synthesis to prepare 2-aminoethylphosphonic acid which has since been found to be naturally occurring in certain ciliated protozoa[11] and marine organisms[12]

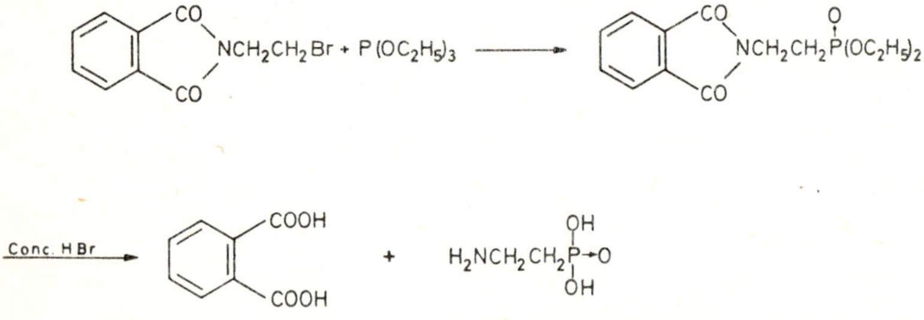

[reaction scheme showing N-(2-bromoethyl)phthalimide + P(OC$_2$H$_5$)$_3$ → phthalimide phosphonate ester, then Conc. HBr → phthalic acid + H$_2$NCH$_2$CH$_2$P(OH)$_2$→O]

2-Aminoethyl phosphonic acid

REFERENCES

[1] Graebe, C. and Pictet, A. *Ber. dt. chem. Ges.* 17 (1884) 1173; *Justus Liebigs Annln Chem.* 247 (1888) 302 [*J .chem. Soc. (A)* 46 (1884) 1019; 56 (1889) 141]
[2] Gabriel, S. *Ber. dt. chem. Ges.* 20 (1887) 2224 [*J. chem. Soc. (A)* 52 (1887) 1037]
[3] Gabriel, S. *Ber. dt. chem. Ges.* 23 (1890) 1771 [*J. chem. Soc. (A)* 58 (1890) 1129]
[4] De Witt, C. C. *Org. Synth., Coll. Vol.* 2 (1955) 25
[5] Sheehan, J. C. and Bolhofer, W. A. *J. Am. chem. Soc.* 72 (1950) 2786
[6] Müller, V. H. K. and Rieck, G. *J. prakt. Chem.* 9 (1959) [4] 30 [*Chem. Abstr.* 54 (1960) 2339]
[7] Ing, H. R. and Manske, R. H. F. *J. chem. Soc.* (1926) 2348
[8] Gabriel, S. and Kroseberg, K. *Ber. dt. chem. Ges.* 22 (1889) 426 [*J. chem. Soc. (A)* 56 (1889) 590]
[9] Greenstein, J. P. and Winitz, M. *Chemistry of the Amino Acids*, 1, 701, New York (Wiley) 1961
[10] Kosolapoff, G. M. *J. Am. chem. Soc.* 69 (1947) 2112
[11] Horiguchi, M. and Kandatsu, M. *Nature, Lond.* 184 (1959) [12] 901
[12] Kittredge, J. S. and Hughes, R. R. *Biochemistry, N.Y.* 3 (1964) 991

58

HOFMANN EXHAUSTIVE METHYLATION

(The name 'Hofmann Degradation' should be avoided because of ambiguity with the Hofmann rearrangement (No. 4) which is also often called the 'Hofmann Degradation'.)

Nature of the reaction

The thermal decomposition of quaternary ammonium hydroxides leads to the formation of a tertiary amine and an olefin

$$[(CH_3)_2 N(C_2H_5)_2]^+ \; \bar{O}H \xrightarrow{100°-200°} (CH_3)_2 N C_2H_5 + CH_2{=}CH_2 + H_2O$$

Where any of the original groups are methyl, these are retained and one of the other groups eliminated. By progressively eliminating groups and forming quaternary salts with methyl iodide, the nature of the groups originally attached to the nitrogen atom may be established

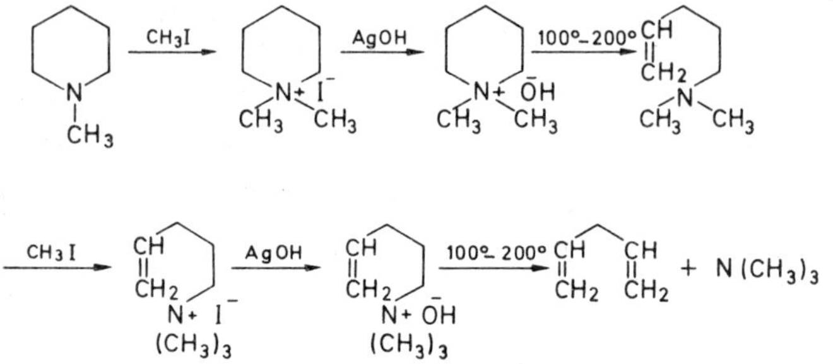

Historical development

The procedure was first reported by Hofmann[1] in 1851, but its value was not fully appreciated until he published[2a] his work on the constitution of piperidine. Since then, the method has been extensively employed, particularly in the determination of alkaloid structures.[2b] Cope and Trumbull[3] have reviewed the reaction in the context of the formation of olefins from amines. Yields vary but generally are above 50 per cent.

Mechanism

The Hofmann exhaustive methylation has been investigated in detail over many years. Ingold *et al.*[4] designated the decomposition of quaternary ammonium compounds as predominantly bimolecular elimination (E2) processes. More completely, the evidence suggests that the E2 mechanism is only a limiting form of a very variable process which ranges from stepwise (E1) processes[5].

The E2 mechanism is represented as involving a *trans* intermediate

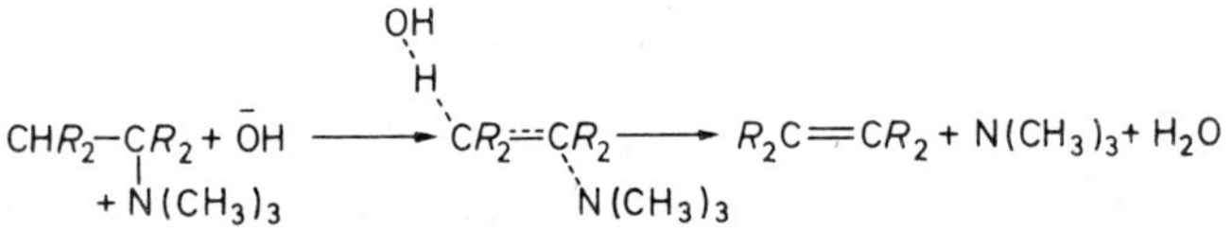

$$CHR_2\text{-}CR_2 + \bar{O}H \longrightarrow CR_2 \overset{\cdot\cdot}{=} CR_2 \longrightarrow R_2C=CR_2 + N(CH_3)_3 + H_2O$$
$$+N(CH_3)_3 \qquad\qquad N(CH_3)_3$$

Bourns and Smith[6] have studied the mechanism in connection with 2-arylethylammonium ions by employing trimethyl-β,β-dideuterophenethylammonium bromide, $C_6H_5CD_2CH_2N^+(CH_3)_3$ Br^-, in ethanol. They showed that no exchange of deuterium occurred and concluded that a simple one-step E2 process occurred in this case rather than a zwitterion necessary for a stepwise reaction.

HOFMANN EXHAUSTIVE METHYLATION

General reaction conditions

(i) Formation of quaternary salt—The quaternary salt in most cases is prepared from methyl iodide, although dimethyl sulphate and related compounds have been employed. The amine, in alcoholic solution, is treated with methyl iodide and alcoholic potassium hydroxide at ice temperature for several hours followed by moderate warming. The solution is made neutral, alcohol distilled off and the product precipitated by addition of sodium hydroxide.

(ii) Formation of quaternary hydroxide—The quaternary halide in aqueous solution is shaken with freshly prepared silver oxide and the precipitate filtered off. The solution of the quaternary hydroxide is then employed directly.

(iii) Pyrolysis—The aqueous solution is stirred and heated above 100° until evolution of water vapour and gases ceases. When trimethylamine is one of the products, the progress of the reaction may be followed by absorption in dilute hydrochloric acid.

Longone and Simanyi[7] carried out the pyrolysis by adding the aqueous solution dropwise to a flask of hot toluene fitted with a Dean Starke trap. The reaction was complete after 12 h. By this means, pentamethylbenzyltrimethylammonium hydroxide was converted to tetramethyl-*p*-xylylene which underwent a number of dimerization reactions

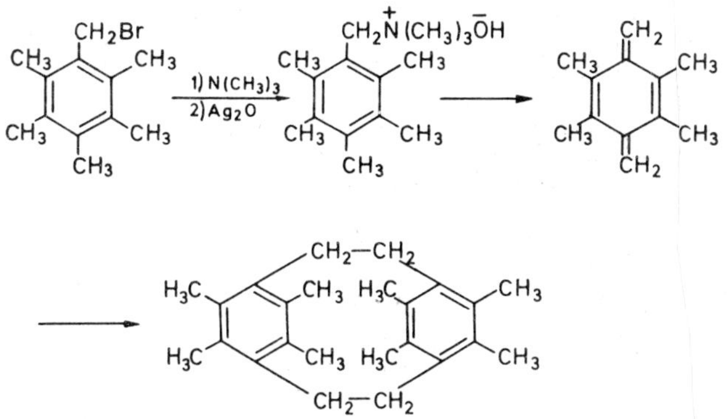

Applications

Willstätter and Waser[8] employed the exhaustive methylation approach to convert the alkaloid pseudopelletierine to cyclo-octadiene

$$\begin{array}{c}CH_2\text{—}CH\text{—}CH_2\\|\quad\;\;|\quad\;\;|\\CH_2\;\;NCH_3\;\;CO\\|\quad\;\;|\quad\;\;|\\CH_2\text{—}CH\text{—}CH_2\end{array} \longrightarrow \begin{array}{c}CH_2\text{—}CH\text{—}CH_2\\|\quad\;\;|\quad\;\;|\\CH_2\;\;NCH_3\;\;CH_2\\|\quad\;\;|\quad\;\;|\\CH_2\text{—}CH\text{—}CH_2\end{array} \xrightarrow[2)\,Ag_2O]{1)\,CH_3I} \begin{array}{c}CH_2\text{—}CH\text{———}CH_2\\|\quad\;\;|\quad\;\;\;\;\;|\\CH_2\;\;N(CH_3)_2OH\;\;CH_2\\|\quad\;\;|\quad\;\;\;\;\;|\\CH_2\text{—}CH\text{———}CH_2\end{array}$$

$$\xrightarrow{\text{Pyrolysis}} \begin{array}{c}CH_2\text{—}CH\text{—}CH_2\\|\quad\;\;|\quad\;\;|\\CH_2\;\;N(CH_3)_2CH_2\\|\quad\;\;|\quad\;\;|\\CH_2\text{—}CH\!=\!CH\end{array} \xrightarrow[\substack{2)\,Ag_2O\\3)\,\text{Pyrolysis}}]{1)\,CH_3I} \begin{array}{c}CH\!=\!CH\text{—}CH_2\\|\quad\quad\quad|\\CH_2\quad\quad CH_2\\|\quad\quad\quad|\\CH_2\text{—}CH\!=\!CH\end{array}$$

Cyclo-octadiene

The application of the Hofmann procedure to the corresponding quaternary phosphorus compounds is of current interest. Bestman and co-workers[9] have reviewed the work in this field and have carried out mechanistic studies on tritium-labelled triphenyl phosphonium salts

$$\left[\begin{array}{c}C_6H_5\;CH\;CHT\;CO_2CH_3\\|\\P(C_6H_5)_3\end{array}\right]^+ Br^-$$

REFERENCES

[1] Hofmann, A. W. *Justus Liebigs Annln Chem.* 78 (1851) 253; 79 (1851) 11
[2] (a) Hofmann, A. W. *Ber. dt. chem. Ges.* 14 (1881) 494, 659, 705 [*J. chem. Soc.* (A) 40 (1881) 570, 621, 745]; (b) Battersby, A. R. et al., *J. chem. Soc. C* (1968) 2163
[3] Cope, A. C. and Trumbull, E. R. *Org. React.* 11 (1960) 317
[4] Hanhart, W. H. and Ingold, C. K. *J. chem. Soc.* (1927) 997; Ingold, C. K. and Vass, C. C. N. (1928) 3125; Hughes, E. D. and Ingold, C. K. (1933) 526
[5] Simon, H. and Muellhofer, G. *Ber. dt. chem. Ges.* 96 (1963) 3167 [*Chem. Abstr.* 40 (1964) 3973]
[6] Bourns, A. N. and Smith, P. J. *Proc. chem. Soc.* (1964) 366
[7] Longone, D. T. and Simanyi, L. H. *J. org. Chem.* 29 (1964) 3245
[8] Willstätter, R. and Waser, E. *Ber. dt. chem. Ges.* 43 (1910) 1176; 44 (1911) 3423 [*J. chem. Soc.* (A) 98 (1910) 366; 102 (1912) 17]
[9] Bestmann, H. J., Haberlein, H. and Pils, I. *Tetrahedron* 20 (1964) 2079 [*Chem. Abstr.* 61 (1964) 14496]

59

HOFMANN–MARTIUS REARRANGEMENT

Nature of the reaction
C-alkylaniline hydrohalides are formed by heating *N*-alkyl aniline hydrohalides

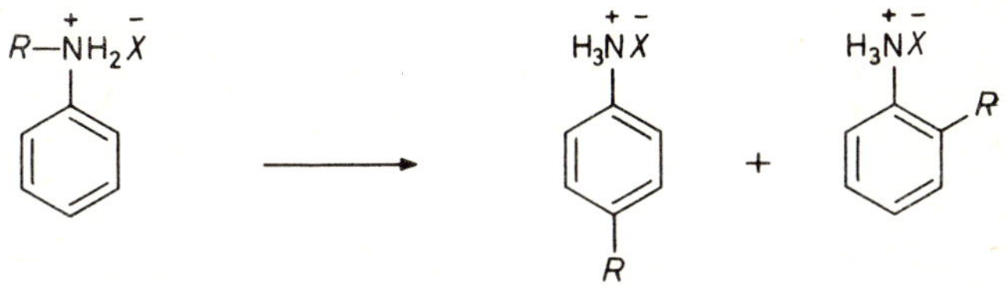

Historical development
The rearrangement was first discovered by Hofmann and Martius[1] during work on the methylation of aniline, but it was not fully explained at the time. Hofmann[2] clarified the situation later when he showed that the hydrochloride of methyl aniline rearranges at elevated temperatures to toluidine hydrochloride

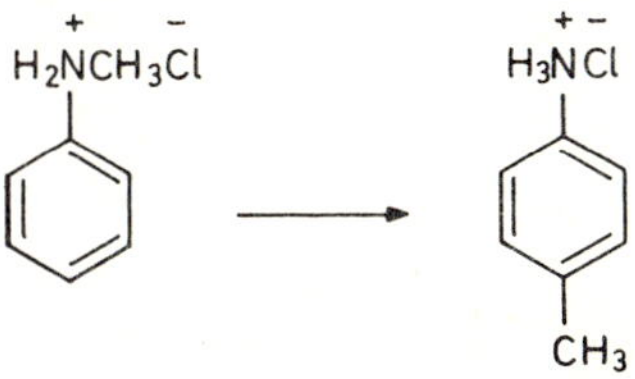

The migration occurs preferentially to the *para* position, unless occupied, in which case migration to the *ortho* position is the only possibility. Temperatures up to 300° are used to rearrange primary alkyl groups, but secondary and tertiary groups rearrange at lower temperatures. The rearrangement may be carried out employing anhydrous metallic halide catalysts[3].

Yields are very variable. Monoalkyl anilines rearrange to give total yields of 50–70 per cent of the corresponding C-substituted anilines.

Mechanism

The Hofmann–Martius rearrangement was originally considered to be intramolecular[4]. However, Hughes and Ingold[5] have discussed this in the light of the extensive work carried out by Hickinbottom[6] which indicated the formation of alkyl carbonium ions. They have, therefore, proposed a mechanism which involves an $S_N 2$ mechanism

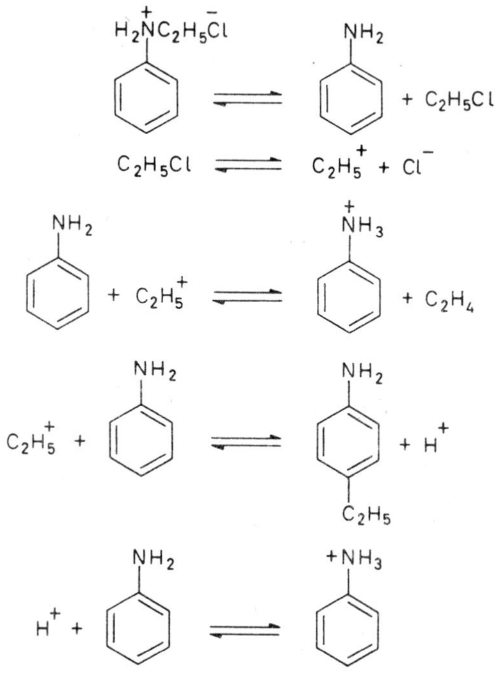

$$C_6H_5\overset{+}{N}H_2C_2H_5 \, \overset{-}{Cl} \rightleftharpoons C_6H_5NH_2 + C_2H_5Cl$$

$$C_2H_5Cl \rightleftharpoons C_2\overset{+}{H_5} + Cl^-$$

$$C_6H_5NH_2 + C_2\overset{+}{H_5} \rightleftharpoons C_6H_5\overset{+}{N}H_3 + C_2H_4$$

$$C_2\overset{+}{H_5} + C_6H_5NH_2 \rightleftharpoons (4\text{-}C_2H_5)C_6H_4NH_2 + H^+$$

$$H^+ + C_6H_5NH_2 \rightleftharpoons C_6H_5\overset{+}{N}H_3$$

This also justifies the isolation of olefins which Hickinbottom and co-workers[7] identified in various reactions.

General reaction conditions

The pure dry salt may be rearranged by simply heating at about 300° in an open vessel for 1 to 2 h. On cooling and solidifying, the resulting mixture is treated with alkali. Primary amines that have been formed may be precipitated as zinc chlorides. Hickinbottom[8] employed this procedure to prepare *o*- and *p*-toluidines from methylaniline hydriodide.

The separation of mixtures of products is the major problem. It

HOFMANN–MARTIUS REARRANGEMENT

may be accomplished initially by distillation and, if necessary, by resorting to a further procedure such as that of Hinsberg.

The modified form of the reaction[9] involves heating the N-alkyl aniline with half its weight of anhydrous metallic salt at 200°–250° for 20 h during which time dry nitrogen is passed through. On cooling, the mixture is acidified with hydrochloric acid, diluted and treated with excess ammonia. By this means ethyl aniline gave a 70 per cent yield of p-aminoethyl benzene

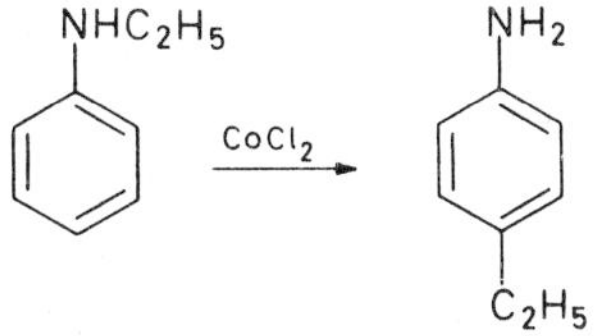

Application

Fischer and co-workers[10] found that the monohydrobromide of N,N'-diphenyl trimethylenediamine undergoes a Hofmann–Martius rearrangement when heated above 230°. They obtained 1,2,3,4-tetrahydroquinoline in 50 per cent yields

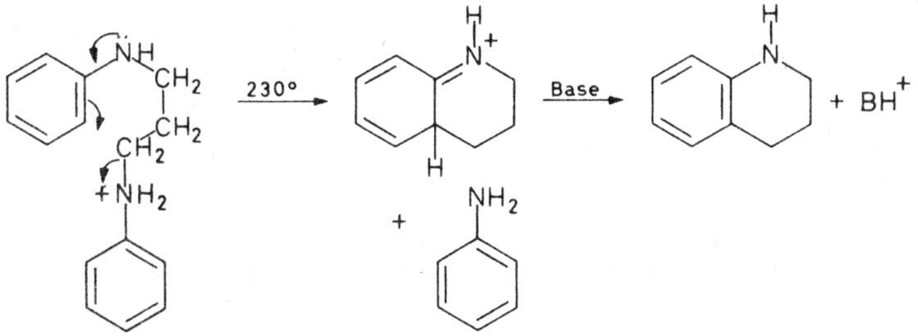

The corresponding o-tolyl compound also undergoes this rearrangement to give the 8-methyl-1,2,3,4-tetrahydroquinoline.

REFERENCES

[1] Hofmann, A. W. and Martius, C. A. *Ber. dt. chem. Ges.* 4 (1871) 742 [*J. chem. Soc.* (A) 24 (1871) 1060]
[2] Hofmann, A. W. *Ber. dt. chem. Ges.* 5 (1872) 704, 720; 7 (1874) 526 [*J. chem. Soc.* (A) 25 (1872) 1021, 1022; 27 (1874) 807]
[3] Reilly, J. and Hickinbottom, W. J. *J. chem. Soc.* 117 (1920) 103
[4] Dewar, M. J. S. *Electronic Theory of Organic Chemistry*, p. 227, Oxford (Univ. Press) 1949

[5] Hughes, E. D. and Ingold, C. K. *Q. Rev. chem. Soc.* 6 (1952) 43; Ingold, C. K. *Structures and Mechanism in Organic Chemistry*, p. 615, Ithaca, N.Y. (Cornell Univ. Press) 1953
[6] Hickinbottom, W. J. *J. chem. Soc.* (1932) 2396
[7] Hickinbottom, W. J., with Ryder, S. E. A. *J. chem. Soc.* (1931) 1281; with Preston, G. H. (1930) 1566; (1937) 404
[8] Hickinbottom, W. J. *J. chem. Soc.* (1934) 1700
[9] Hickinbottom, W. J. and Waine, A. C. *J. chem. Soc.* (1930) 1558
[10] Fischer, A., Topson, R. D. and Vaughan, J. *J. org. Chem.* 25 (1960) 463

60

HUNSDIECKER (BORODINE–HUNSDIECKER) REACTION

Nature of the reaction

Halogens in anhydrous media are used to degrade the silver salts of carboxylic acids to alkyl halides with one less carbon atom

$$RCO_2Ag + X_2 \longrightarrow RX + CO_2 + AgX$$

Historical development

The reaction was reported by Borodine[1] to give methylbromide from silver acetate

$$CH_3CO_2Ag + Br_2 \longrightarrow CH_3Br + CO_2 + AgBr$$

This was studied in great detail by H. and C. Hunsdiecker[2] who extended it to a large number of other carboxylic acids. They showed that mercury and potassium salts could also be used and were able to obtain dibromides from the silver salts of dicarboxylic acids

$$\begin{array}{l} CH_2CO_2Ag \\ | \\ CH_2CO_2Ag \end{array} + 2\,Br_2 \longrightarrow \begin{array}{l} CH_2Br \\ | \\ CH_2Br \end{array} + 2\,CO_2 + 2\,AgBr$$

HUNSDIECKER (BORODINE–HUNSDIECKER) REACTION

The reaction proceeds much more easily with aliphatic compounds than with the aromatic series. Although silver benzoate does give bromobenzene, the yields are not always reproducible[3]

$$\text{C}_6\text{H}_5\text{CO}_2\text{Ag} \xrightarrow{\text{Br}_2/\text{CCl}_4} \text{C}_6\text{H}_5\text{Br}$$

The reaction has been reviewed[4, 5] in considerable detail for the period of extensive investigation between 1936 and 1956.

Yields are usually very good, within the range of 65 to 80 per cent.

Mechanism

The initial reaction product under anhydrous conditions between the silver salt of a carboxylic acid and a halogen is an acyl hypohalite[6]

$$R\,CO_2\,Ag + X_2 \longrightarrow R\,CO_2\,X + AgX$$

Experimental evidence suggests that a free-radical mechanism is involved, proposed by Wilson[3] to be

$$R\,CO_2\,X \longrightarrow R\,CO_2^{\bullet} + Br^{\bullet}$$
$$R\,CO_2^{\bullet} \longrightarrow R^{\bullet} + CO_2$$
$$R^{\bullet} + R\,CO_2\,X \longrightarrow RX + R\,CO_2^{\bullet}$$
$$2\,R^{\bullet} \longrightarrow R\text{-}R \text{ or } RH + \text{olefin}$$
$$R\,CO_2^{\bullet} + R^{\bullet} \longrightarrow R\,CO_2\,R$$

This explains the existence of some of the by-products that have been isolated. Further support for the radical mechanism has come from the work of Cristol *et al.*[7]

General reaction conditions

The metallic salt of the carboxylic acid is suspended in a dry organic solvent (carbon tetrachloride, carbon disulphide), and bromine or other halogen in organic solution is added dropwise. No initial heating is generally necessary, carbon dioxide being steadily evolved as the reaction proceeds. When addition is complete, the mixture is warmed for a short time, then cooled. The metallic halide is filtered off and solvent distilled off.

Allen and Wilson[8] have reported a detailed procedure for the preparation of methyl 5-bromo valerate with a 65 per cent yield from methyl silver adipate. For this they employed carbon tetrachloride and bromine

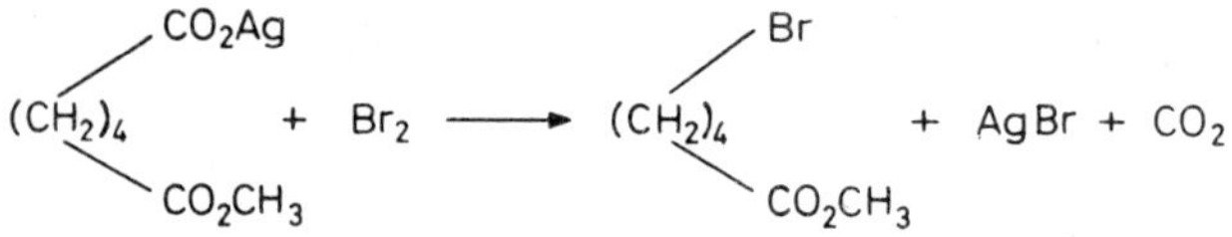

$$(CH_2)_4\begin{smallmatrix}CO_2Ag\\ \\CO_2CH_3\end{smallmatrix} + Br_2 \longrightarrow (CH_2)_4\begin{smallmatrix}Br\\ \\CO_2CH_3\end{smallmatrix} + AgBr + CO_2$$

Modification

Cristol and Frith[9] showed that the Hunsdiecker–Borodine reaction can be carried out without isolation of a metallic salt of the carboxylic acid. The aliphatic carboxylic acid is slurried with mercuric oxide in carbon tetrachloride to which bromine is added and the mixture refluxed.

Meek and Osuga[10] have used the procedure to prepare bromo-cyclopropane (46 per cent yield)

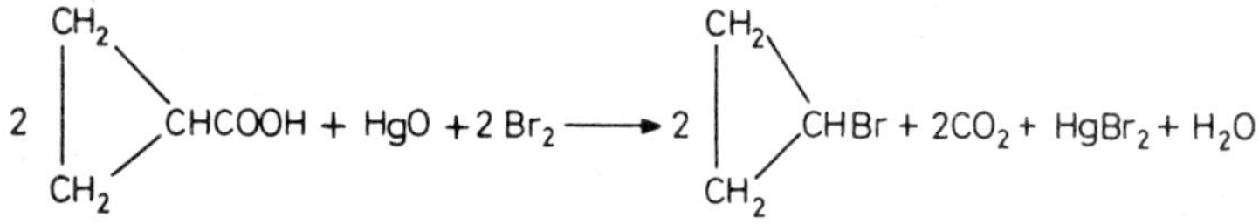

$$2\;\begin{smallmatrix}CH_2\\ \\CH_2\end{smallmatrix}\!\!>\!\!CHCOOH + HgO + 2\,Br_2 \longrightarrow 2\;\begin{smallmatrix}CH_2\\ \\CH_2\end{smallmatrix}\!\!>\!\!CHBr + 2CO_2 + HgBr_2 + H_2O$$

The value of other metallic oxides in place of mercuric oxide has been studied by Davis and co-workers[11]. They found that the reaction could be carried out if lead, silver or cadmium oxides were used.

Applications

Dyatkin and co-workers[12] have used the reaction to prepare fluorocarbons possessing additional halogen atoms

			Per cent yield
$(CF_3)_2CHCO_2K$	$\xrightarrow{Cl_2}$	$(CF_3)_2CHCl$	56·8
	$\xrightarrow{Br_2}$	$(CF_3)_2CHBr$	78·0
	$\xrightarrow{KI/I_2}$	$(CF_3)_2CHI$	76·6

With bromine and silver acetate, Grewe and Vangermain[13] obtained an abnormal reaction with the tetra-acetate of guinic acid

HUNSDIECKER (BORODINE–HUNSDIECKER) REACTION

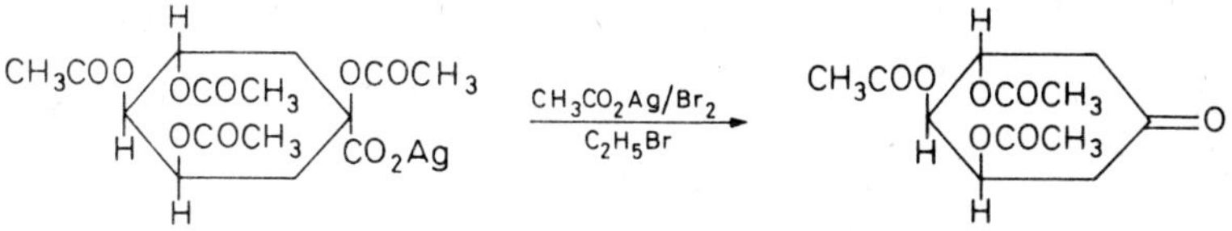

[Reaction scheme: pentaacetoxy cyclohexane carboxylic acid silver salt + CH_3CO_2Ag/Br_2, C_2H_5Br → 3,4,5-triacetoxy-cyclohexanone]

3,4,5 — Triacetoxy-cyclohexanone

REFERENCES

[1] Borodine, A. *Justus Liebigs Annln Chem.* 119 (1861) 121
[2] Hunsdiecker, H. and C. *Ber. dt. chem. Ges.* 75B (1942) 291; with Vogt, E. U.S. Pat. 2176181 (1939); Brit. Pat. 456565 (1936) [*Chem. Abstr.* 37 (1943) 3734; 34 (1940) 1685; 31 (1937) 2233]
[3] Oldam, J. W. H. *J. chem. Soc.* (1950) 100
[4] Johnson, R. G. and Ingham, R. K. *Chem. Rev.* 56 (1956) 219
[5] Wilson, C. V. *Org. React.* 9 (1957) 332
[6] Uschakov, M. I. and Chistov, W. O. *Ber. dt. chem. Ges.* 68 (1935) 824 [*Chem. Abstr.* 29 (1935) 4686]
[7] Cristol, S. J. et al., *J. Am. chem. Soc.* 82 (1960) 1829
[8] Allen, C. F. H. and Wilson, C. V. *Org. Synth., Coll. Vol.* 3 (1955) 578
[9] Cristol, S. J. and Frith, W. C. *J. org. Chem.* 26 (1961) 280
[10] Meek, J. S. and Osuga, D. T. *Org. Synth.* 43 (1963) 9
[11] Davis, J. A. et al., *J. org. Chem.* 30 (1965) 415
[12] Dyatkin, B. L. et al., *Zh. vses. khim. Obshch.* 10 (1965) 469 [*Chem. Abstr.* 63 (1965) 14691]
[13] Crewe, R. and Vangermain, E. *Ber. dt. chem. Ges.* 98 (1965) 104 [*Chem. Abstr.* 62 (1965) 11894]

61

KOLBE–SCHMITT REACTION

Nature of the reaction
Phenolic carboxylic acids are prepared by the action of carbon dioxide on metal phenolates under pressure

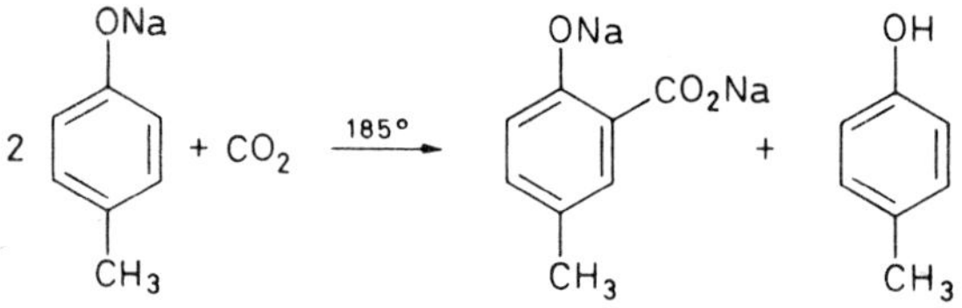

$$2 \underset{CH_3}{\underset{|}{C_6H_3}}\text{-ONa} + CO_2 \xrightarrow{185°} \underset{CH_3}{\underset{|}{C_6H_3}}(ONa)(CO_2Na) + \underset{CH_3}{\underset{|}{C_6H_4}}\text{-OH}$$

Historical development
　The reaction, originally reported by Kolbe[1], was for a method of preparing salicylic acid by heating sodium phenoxide at 180° with a continuous stream of carbon dioxide. As shown above, half the phenol is regenerated. Schmitt[2] improved upon Kolbe's procedure by preparing the sodium phenyl carbonate (*A*) at ordinary temperatures and then heating the salt at 120°–140° under pressure

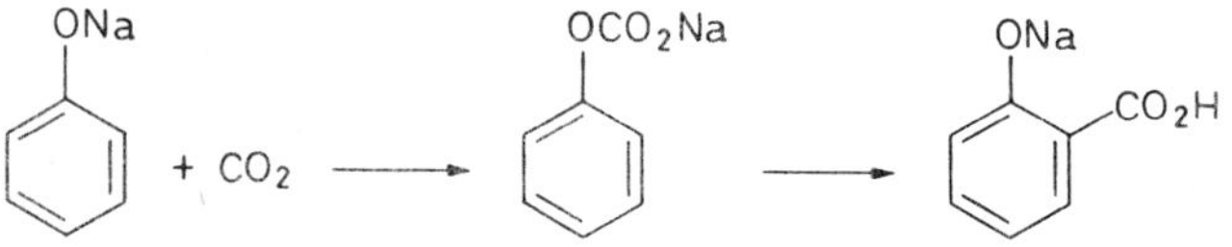

$$C_6H_5\text{-ONa} + CO_2 \longrightarrow \underset{(A)}{C_6H_5\text{-OCO}_2Na} \longrightarrow C_6H_4(ONa)(CO_2H)$$

This method gives greater yields without regeneration of the phenol and is the standard approach for carrying out the Kolbe–Schmitt reaction. Lindsey and Jeskey[3], who have reviewed the subject up to 1956, listed a wide variety of carboxylic acids prepared by this method. The main side reaction arises from substitution in the *para* position when this is vacant. The proportion of *para*-substituted material can usually be increased by carrying out the reaction at a higher temperature. Yields of the main product are between 50 and 90 per cent.

KOLBE–SCHMIDT REACTION

Mechanism

Schmitt believed the reaction to proceed by rearrangement of the sodium phenyl carbonate. However, Jones and co-workers[4] showed on the basis of infra-red spectra that this is actually a weak chelate of sodium phenoxide and carbon dioxide. They suggested that the complex undergoes an intramolecular rearrangement with displacement of hydrogen from the ring due to electrophilic attack. In the case of sodium phenoxide, this may proceed via a π complex[5]

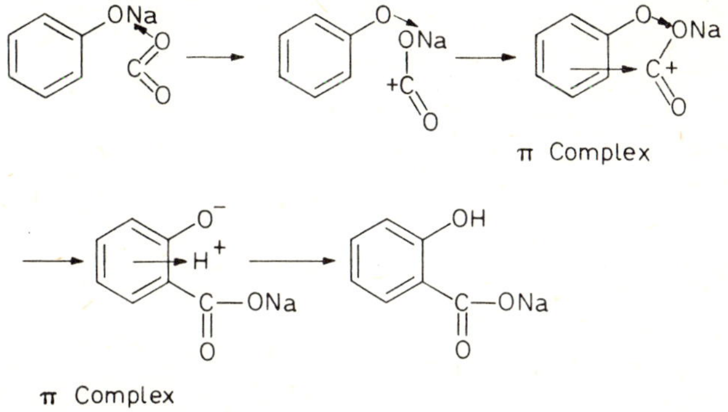

General reaction conditions

The conditions for the Kolbe–Schmitt reaction necessitate the preparation of the dry sodium or potassium salt of the phenol, since the presence of water leads to lower yields. The salt is then heated at 120°–130° in an autoclave with carbon dioxide at a pressure of up to 90 atm for several hours. The acid is obtained by acidification of the product.

The procedure has been extensively employed for preparing aromatic carboxylic acids and is used industrially.

Modifications

In 1893, Marasse[6] prepared the potassium salts of aromatic carboxylic acids by heating the free phenol with excess anhydrous potassium carbonate under pressure. This procedure in many cases gives superior yields and has been much used for laboratory preparations. Industrially it is expensive, as the cheaper sodium carbonate does not undergo the reaction.

Hill and co-workers[7] used this procedure to prepare 1-hydroxy-5-methoxy-2-naphthoic acid, with 90 per cent yield, by heating 5-methoxy-1-naphthol with potassium bicarbonate at 220° for 3 h in a sealed tube

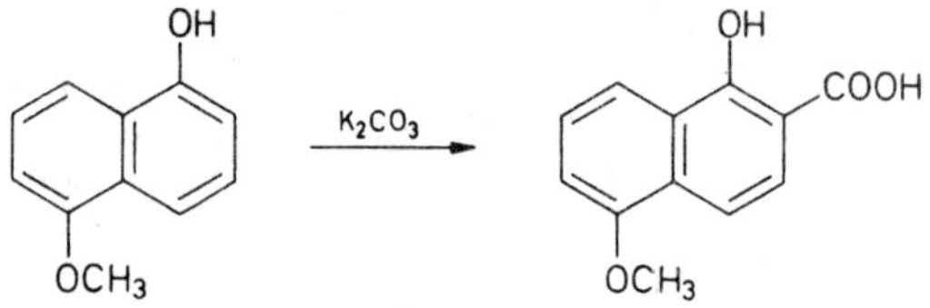

[Reaction scheme: 1-hydroxy-5-methoxynaphthalene + K_2CO_3 → 1-hydroxy-5-methoxy-2-naphthoic acid]

Applications

A method of wide application to dihydroxyphenols has been given by Nierenstein and Clibbens[8]. Resorcinol is dissolved in aqueous potassium hydrogen carbonate solution and refluxed for several hours while carbon dioxide is bubbled through. Acidification of the cooled solution gives a 60 per cent yield of β-resorcylic acid

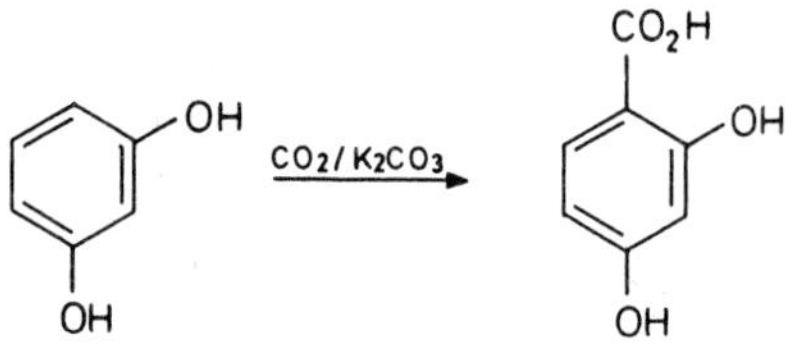

[Reaction scheme: resorcinol + CO_2/K_2CO_3 → β-resorcylic acid]

The reaction has been applied to heterocyclic systems. Ciamician and Silber[9] reported a 50 per cent yield of pyrrole-3-carboxylic acid from potassio-pyrrole when heated in a sealed tube with aqueous ammonium carbonate

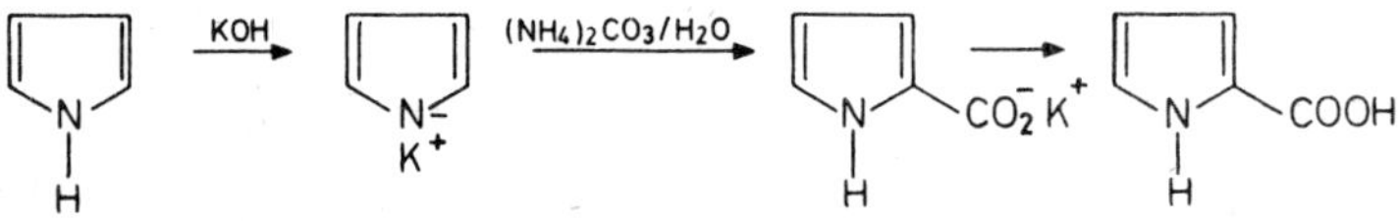

[Reaction scheme: pyrrole + KOH → potassio-pyrrole + $(NH_4)_2CO_3/H_2O$ → pyrrole-2-$CO_2^-K^+$ → pyrrole-2-COOH]

REFERENCES

[1] Kolbe, H. *Justus Liebigs Annln Chem.* 113 (1860) 125; *J. prakt. Chem.* 8 (1874) 336; 10 (1874) 89 [*J. chem. Soc.* (A) 27 (1874) 477; 28 (1875) 260]
[2] Schmitt, R. *Dinglers polytech. J.* 255 (1885) 259; *J. prakt. Chem.* 31 (1885) 397 [*J. chem. Soc.* (A) 48 (1885) 709, 982]
[3] Lindsey, A. S. and Jeskey, H. *Chem. Rev.* 57 (1957) 583
[4] Hales, J. L., Jones, J. I. and Lindsey, A. S. *Chemy Ind.* (1954) 49; *J. chem. Soc.* (1954) 3145
[5] Dewar, M. J. S. *The Electronic Theory of Organic Chemistry*, p. 228, London (Oxford Univ. Press) 1949
[6] Marasse, S. *Germ. Pat.* 73279 (1893); 78708 (1894)

[7] Hill, P., Short, W. F. and Stromberg, H. *J. chem. Soc.* (1937) 937
[8] Nierenstein, M. and Clibbens, D. A. *Org. Synth., Coll. Vol.* 2 (1943) 557
[9] Ciamician, G. L. and Silber, P. *Gazz. chim. ital.* 14 (1884) 162 [*J. chem. Soc.* (A) 48 (1885) 246]

62

REIMER–TIEMANN SYNTHESIS

Nature of the reaction

Phenolic aldehydes are prepared by reactions between chloroform and phenols under alkaline conditions

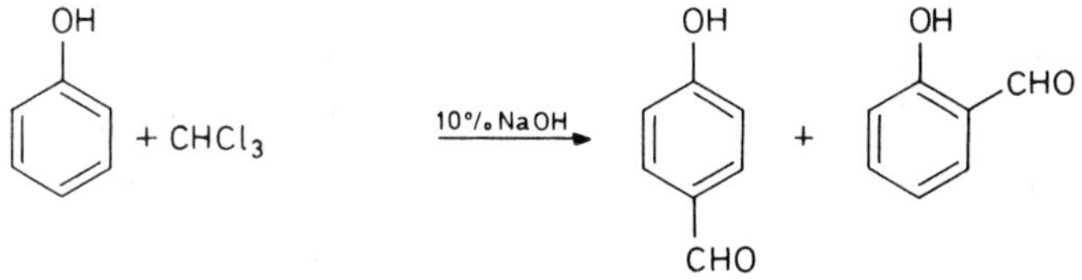

Historical development

In 1876, Reimer and Tiemann[1] reported their method for preparing aromatic aldehydes, based upon Schiff's[2] earlier observation that a coloured complex is formed when chloroform is heated with sodium phenoxide. Under similar conditions, with carbon tetrachloride in place of chloroform, carboxylic acids are formed[3]

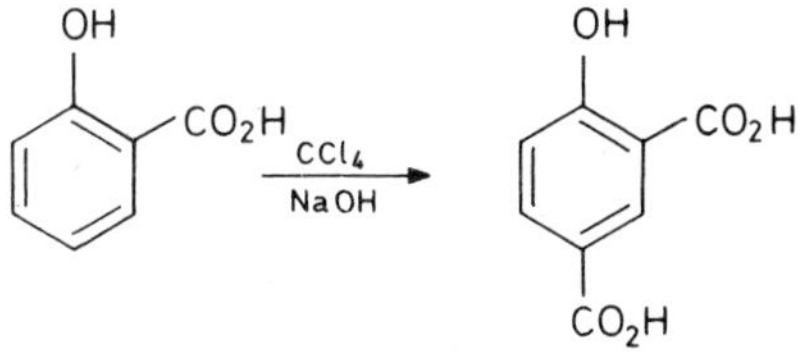

Of the direct methods available for preparing aromatic aldehydes, the Reimer–Tiemann synthesis is the only one carried out under alkaline conditions. In many cases it works when other procedures

fail, although yields are usually below 50 per cent and frequently consist of a mixture of isomers. The synthesis has been discussed[4] with other general procedures for preparing aldehydes. The only comprehensive review of it as a synthetic method with its own virtues is by Wynberg[5].

Mechanism

Alkaline hydrolysis of chloroform leads to the formation of dichloromethylene[6], and the work of Wynberg[7] and of Hine and co-workers[8] has shown this to be the active intermediate in the Reimer–Tiemann reaction

$$CHCl_3 + \bar{O}H \rightleftharpoons \bar{C}Cl_3 + H_2O$$

$$\bar{C}Cl_3 \longrightarrow CCl_2 + Cl^-$$

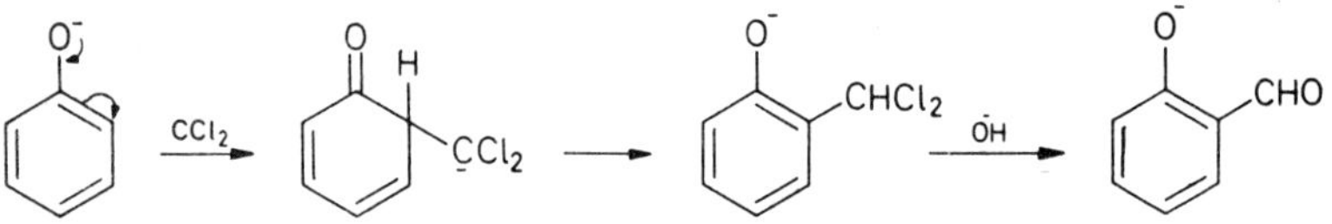

The results of kinetic studies by Robinson[9] are in accordance with this mechanism.

General reaction conditions

The reaction is carried out by dissolving the phenol and a small quantity of ethanol in aqueous sodium hydroxide. Chloroform is added to the heated solution at a rate which maintains a steady reflux. After 3–4 h, excess ethanol and chloroform are distilled off and the solution acidified, liberating the aldehyde. The product may be purified by preparing a derivative or by recrystallization.

Russell and Lockhart[10] have described a detailed procedure for the preparation of 2-hydroxy-1-naphthaldehyde, with a yield of 38–48 per cent

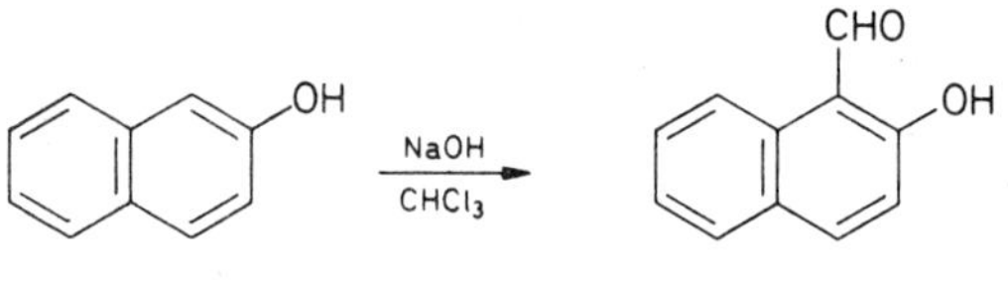

Applications

Substituted pyrimidine-5-carboxaldehydes were prepared by Wiley and Yamamoto[11], the yields increasing from 13 to 42 per cent with the electron-releasing characteristics of the substituents

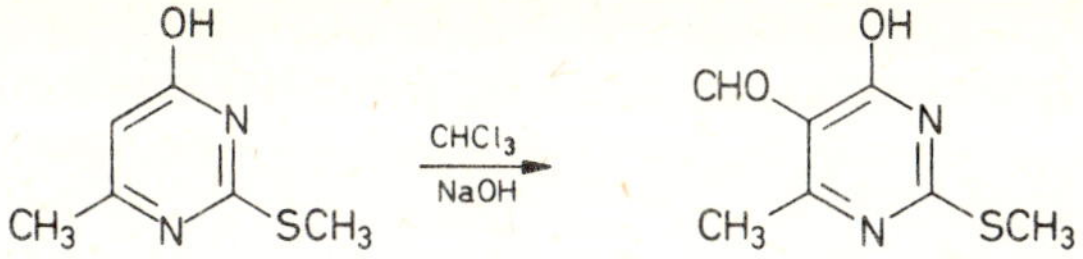

REIMER–TIEMANN SYNTHESIS

Cohen and co-workers[12] have found an unexpected Reimer–Tiemann reaction to occur between phenol and 6-trichloromethyl purine under basic conditions

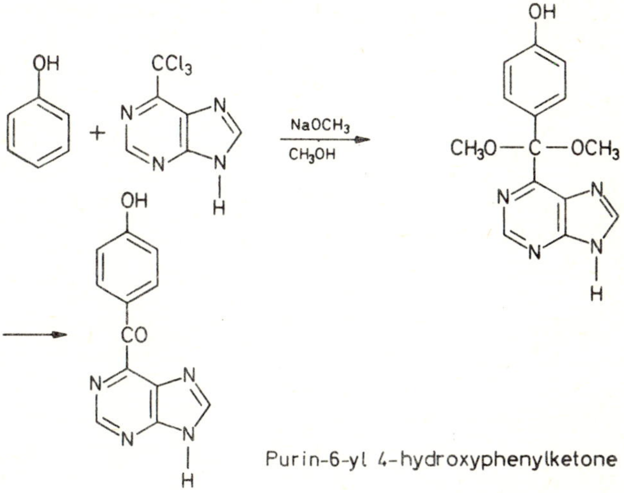

Purin-6-yl 4-hydroxyphenylketone

REFERENCES

[1] Reimer, K. *Ber. dt. chem. Ges.* 9 (1876) 423; with Tiemann, F. *ibid.* 824; 10 (1877) 1562) [*J. chem. Soc.* (*A*) 30 (1876) 82, 632; 34 (1878) 225]
[2] Schiff, H. *Ber. dt. chem. Ges.* 5 (1872) 1055
[3] Reimer, K. and Tiemann, F. *Ber. dt. chem. Ges.* 9 (1876) 1285 [*J. chem. Soc.* (*A*) 31 (1877) 77]
[4] Ferguson, L. N. *Chem. Rev.* 38 (1946) 227
[5] Wynberg, H. *Chem. Rev.* 60 (1960) 169
[6] Finar, I. L. *Organic Chemistry*, 4th edn., 1, 117, London (Longmans) 1963
[7] Wynberg, H. *J. Am. chem. Soc.* 76 (1954) 4998
[8] Hine, J., with Dowell, A. M. *J. Am. chem. Soc.* 76 (1954) 2688; with van der Veen, J. M. *J. org. Chem.* 26 (1961) 1406
[9] Robinson, E. A. *J. chem. Soc.* (1961) 1663
[10] Russell, A. and Lockhart, L. B. *Org. Synth.*, *Coll. Vol.* 3 (1955) 463
[11] Wiley, R. H. and Yamamoto, Y. *J. org. Chem.* 25 (1960) 1906
[12] Cohen, S., Thom, E. and Bendich, A. *J. org. Chem.* 28 (1963) 1379

63

SOMMELET REACTION

Nature of the reaction

Aldehydes are formed from organic halides by the action of hexamethylenetetramine in aqueous alcohol

$$RCH_2X + C_6H_{12}N_4 \longrightarrow [RCH_2 \cdot C_6H_{12}N_4]^+ X^- \longrightarrow RCHO$$

Historical development

In 1913, Sommelet[1] showed that benzyl chloride formed an addition compound with hexamethylenetetramine that could be decomposed by boiling water to give benzaldehyde; the method was of general application to substituted benzyl chlorides

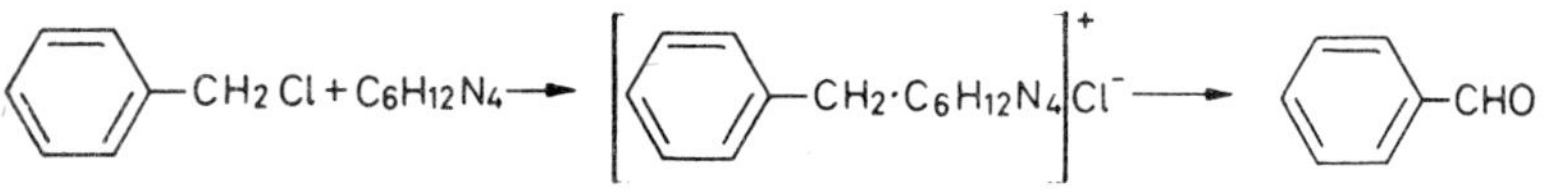

$$PhCH_2Cl + C_6H_{12}N_4 \longrightarrow [PhCH_2 \cdot C_6H_{12}N_4]^+ Cl^- \longrightarrow PhCHO$$

The procedure has been employed to prepare a wide variety of alkyl and aryl aldehydes. It is also successful when more than one methylene halide group exists in the molecule. Wood *et al.*[2] showed that aromatic dialdehydes could readily be formed by this route

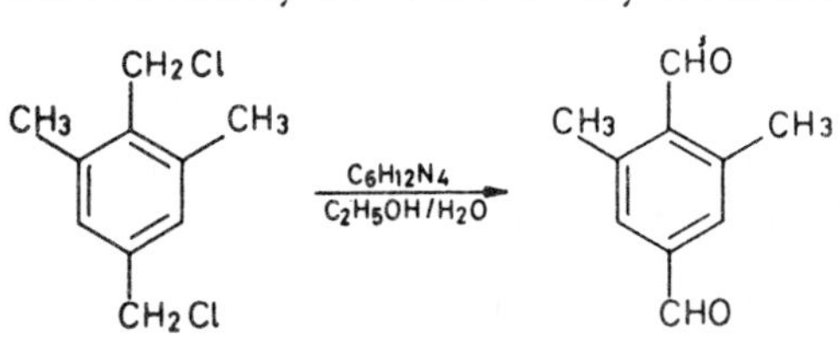

2,5 Dimethylterephthalaldehyde

A detailed review by Angyal[3] has covered the work on this reaction up to 1954. Yields of the hexaminium salt when isolated are normally above 80 per cent. Those of the corresponding aldehydes, however, are 50–70 per cent, based upon the starting materials.

SOMMELET REACTION

Mechanism

Angyal and co-workers[4], who carried out much of the work on the mechanism, believe that the reaction is essentially an hydrogenation–dehydrogenation process involving transfer of an hydride ion.

Initially the organic halide forms a complex salt with hexamethylenetetramine which then gives a primary amine; this in turn reacts with hexamethylenetetramine by means of an hydride ion transfer

$$RCH_2X + C_6H_{12}N_4 \longrightarrow \left[RCH_2 \cdot C_6H_{12}N_4\right]^+ X^- \xrightarrow{H_2O} RCH_2NH_2$$

$$R-\underset{H}{\overset{}{C}H}-NH_2 \rightleftharpoons \left[RCH\cdots NH_2\right]^+ \xrightarrow{H_2O} RCHO + NH_3$$

Le Henaff[5] has suggested that the mechanism involves the formation of a carbonium ion followed by a rearrangement

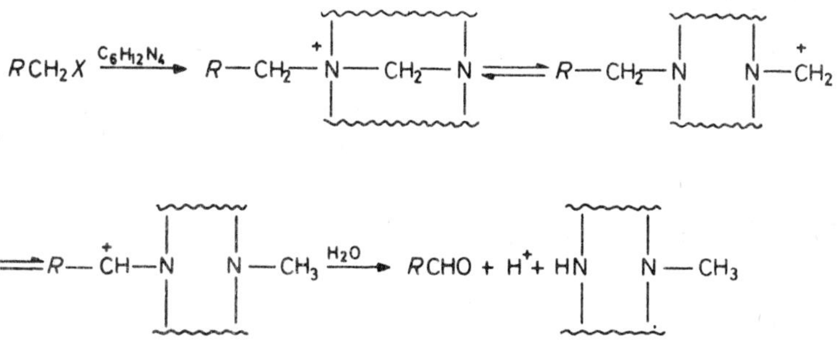

$$RCH_2X \xrightarrow{C_6H_{12}N_4} R-CH_2-\overset{+}{N}-CH_2-N \rightleftharpoons R-CH_2-N \quad \overset{+}{N}-CH_2$$

$$\rightleftharpoons R-\overset{+}{C}H-N \quad N-CH_3 \xrightarrow{H_2O} RCHO + H^+ + HN \quad \overset{+}{N}-CH_3$$

General reaction conditions

Reactions are usually carried out without isolating the intermediate salt. Although water may be used as a solvent, solubility factors more often necessitate the addition of ethanol or glacial acetic acid.

When an hexamethylenetetrammonium chloride salt is required, the alkyl halide is refluxed for 30–60 min with hexamethylenetetramine in chloroform. The salt formed generally precipitates from solution and is filtered off ready for the next process.

Straight through reactions, and those of the isolated salts, are carried out by refluxing the materials for several hours in aqueous ethanol or dilute acetic acid. Volatile aldehydes are isolated from the reaction mixture by steam distillation. Alternatively, formation of a sodium bisulphite addition compound enables the aldehyde to be obtained.

Wiberg[6] isolated the intermediate salt in his preparation of 2-thiophenealdehyde (50 per cent yield)

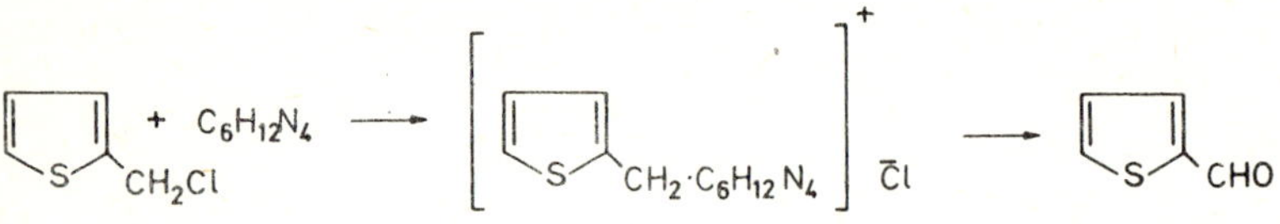

$$\underset{S}{\square}-CH_2Cl + C_6H_{12}N_4 \longrightarrow \left[\underset{S}{\square}-CH_2 \cdot C_6H_{12}N_4 \right]^+ \bar{C}l \longrightarrow \underset{S}{\square}-CHO$$

Similarly, Campaigne and co-workers[7] prepared the corresponding 3-aldehyde after aqueous extraction of the salt formed from mixing chloroform solutions of 3-chloromethyl thiophene and hexamethylenetetramine. In contrast, the straight through method for preparing 1-naphthaldehyde[8] from 1-chloromethylnaphthalene was carried out (with 80 per cent yield) by refluxing the reagents in 50 per cent acetic acid

Applications

The method has been employed for the preparation of aldehydes from alkyl halides possessing labile double and triple bonds that may be affected by using other synthetic routes. Schulte and Goes[9] obtained aldehydes from the corresponding alkynyl bromides in yields of 17–22 per cent

$$CH_3 CH_2 CH_2 CH_2-C\equiv C-CH_2 Br \longrightarrow CH_3 CH_2 CH_2 CH_2-C\equiv C-CHO$$

2-Heptynal

Cherkasova and co-workers[10] used the Sommelet method to confirm the structures of a number of brominated alkyl dienes by preparing the corresponding known aldehydes

$$CH_3-CH=CH-\underset{CH_3}{\overset{|}{C}H}-CH_2-CH=CH-CH_2Br \longrightarrow CH_3-CH=CH-\underset{CH_3}{\overset{|}{C}H}-CH_2-CH=CH-CHO$$

SOMMELET REACTION

The first application of the Sommelet reaction to convert a diamine to a dialdehyde has been reported by Ackerman and Surrey[11]; a 60 per cent yield of isophthalaldehyde was obtained from α,α'-diamino-m-xylene

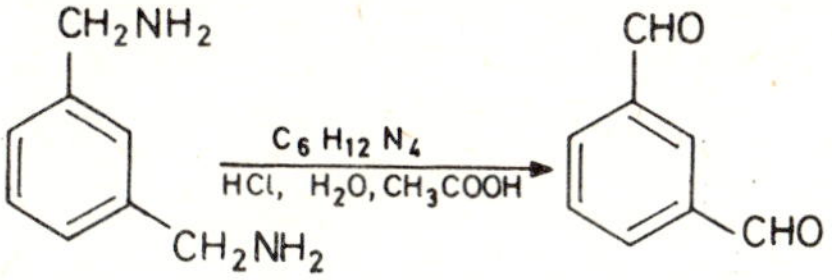

REFERENCES

[1] Sommelet, M. *C. r. hebd. Séanc. Acad. Sci.*, *Paris* 157 (1913) 852 [*Chem. Abstr.* 8 (1914) 660]; *Bull. Soc. chim. Fr.* 13 (1918) [4] 1085; 23 (1918) [4] 95
[2] Wood, J. H. et al., *J. Am. chem. Soc.* 72 (1950) 2992
[3] Angyal, S. J. *Org. React.* 8 (1954) 197
[4] Angyal, S. J., Penman, D. R. and Warwick, G. P. *J. chem. Soc.* (1953) 1742
[5] Le Henaff, P. *Annls Chim. Phys.* 7 (1962) [13] 367 [*Chem. Abstr.* 58 (1963) 2331]
[6] Wiberg, K. B. *Org. Synth., Coll. Vol.* 3 (1955) 811
[7] Campaigne, E., Bourgeois, R. C. and McCarthy, W. C. *Org. Synth., Coll. Vol.* 4 (1963) 918
[8] Angyal, S. J., Tetaz, J. R. and Wilson, J. G. *Org. Synth., Coll. Vol.* 4 (1963) 690
[9] Schulte, K. E. and Goes, M. *Arch. Pharm., Berl.* 290 (1957) 157 [*Chem. Abstr.* 52 (1958) 248]
[10] Cherkasova, L. A., Bal'yan, K. V. and Zubritskii, L. M. *Zh. obshch. Khim.* 34 (1964) 1917 [*Chem. Abstr.* 61 (1964) 8172]
[11] Ackerman, J. H. and Surrey, A. R. *Org. Synth.* 47 (1967) 76

64

STEPHEN REACTION

Nature of the reaction

Nitriles are reduced in ether solution, by hydrogen chloride gas and stannous chloride, to give aldimines which are hydrolysed to aldehydes

$$R-C\equiv N + SnCl_2 + 3 HCl \longrightarrow RCH=NH \cdot HCl + SnCl_4$$

$$RCH=NH \cdot HCl + H_2O \longrightarrow RCHO + NH_4Cl$$

Historical development

The method of preparing aldehydes introduced by Stephen[1] in 1925 has been applied to a variety of compounds of both aromatic and non-aromatic systems

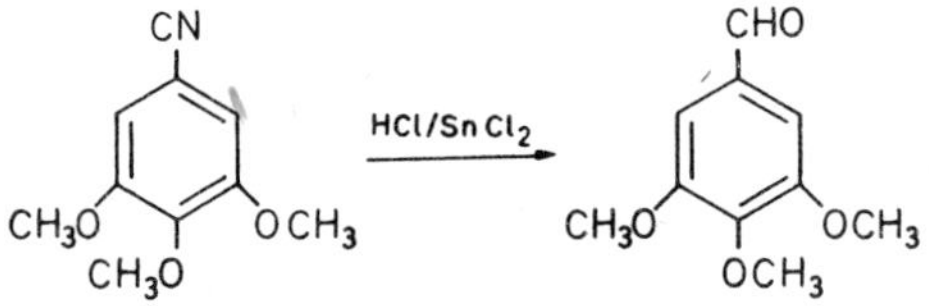

Mosettig[2] has reviewed Stephen's method with a number of other procedures for the preparation of aldehydes. In many cases yields are very good, above 80 per cent, but aromatic systems with groups *ortho* to the nitrile cause steric hindrance, and compounds such as *o*-toluonitrile give only small yields (<10 per cent) of the corresponding aldehyde.

Mechanism

Stephen assumed the route to involve the formation of an imidoyl chloride, $R-CCl=NH$, from the addition of hydrogen chloride to the nitrile. Hantzsch[3], however, showed that a nitrilium salt, $[R-C\equiv NH]^+Cl^-$, was more probable, and this idea is supported by kinetic studies carried out by Turner[4]. A mechanism proposed by Meerwein[5] has been generally accepted

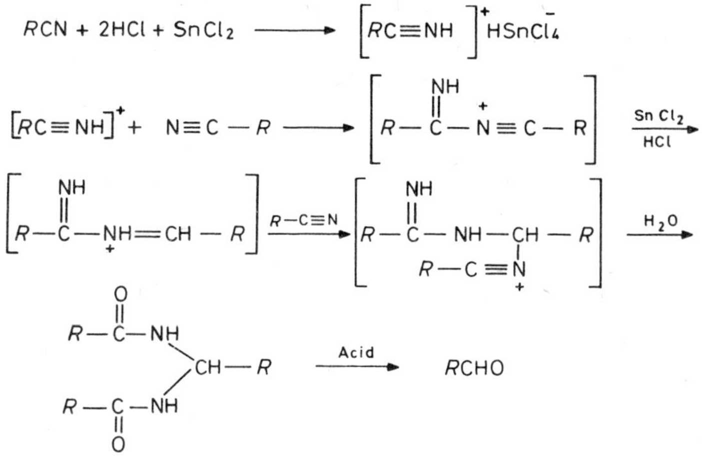

STEPHEN REACTION

$$RCN + 2HCl + SnCl_2 \longrightarrow [RC\equiv NH]^+ HSnCl_4^-$$

$$[RC\equiv NH]^+ + N\equiv C-R \longrightarrow \left[R-\underset{\underset{+}{\overset{\overset{NH}{\|}}{C}}}{}-N\equiv C-R\right] \xrightarrow[HCl]{SnCl_2}$$

$$\left[R-\underset{\overset{\|}{\underset{+}{}}}{\overset{NH}{C}}-NH=CH-R\right] \xrightarrow{R-C\equiv N} \left[R-\overset{\overset{NH}{\|}}{C}-NH-\underset{\underset{+}{R-C\equiv N}}{CH}-R\right] \xrightarrow{H_2O}$$

$$\begin{array}{c} R-\overset{O}{\overset{\|}{C}}-NH \\ \diagdown \\ CH-R \\ \diagup \\ R-\underset{\|}{\underset{O}{C}}-NH \end{array} \xrightarrow{Acid} RCHO$$

A number of the intermediates postulated in this mechanism have been isolated by Zil'berman and Pyralova[6].

General reaction conditions

Anhydrous stannous chloride is suspended in dry ether and saturated with gaseous hydrogen chloride until two layers are obtained. The nitrile is then added with vigorous agitation, which leads to separation of the nitrilium salt as a solid after a few minutes. The salt is then hydrolysed in the reaction mixture by warm water and volatile aldehydes are isolated by steam distillation.

Williams[7] has given a detailed method for the preparation of β-naphthaldehyde employing this type of procedure (73–80 per cent yield)

Modifications

The formation of two layers that occurs with ether was later avoided by Stephen[8] who recommended the use of either ethyl formate or ethyl acetate.

The isolation of pure aldimines, RCH=NH, was accomplished by Tolbert and Houston[9] by decomposing the intermediates under anhydrous conditions in ether with excess triethylamine (72 per cent yield)

Application

By reacting the stannous chloride–nitrile complex of β-cyanopyridine with thiosemicarbazone, Gardner and co-workers[10] prepared derivatives of nicotinaldehyde (83 per cent yield) suitable for the chemotherapy of tuberculosis

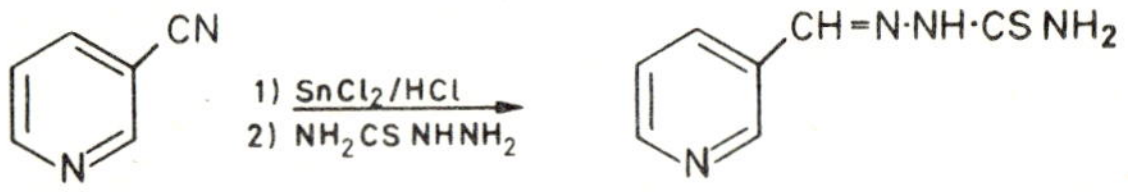

β – Cyanopyridine   Nicotinaldehyde thiosemicarbazone

REFERENCES

[1] Stephen, H. *J. chem. Soc.* 127 (1925) 1874
[2] Mosettig, E. *Org. React.* 8 (1954) 246
[3] Hantzsch, A. and Geidel, W. *Ber. dt. chem. Ges.* 64B (1931) 667 [*Chem. Abstr.* 25 (1931) 3960]
[4] Turner, L. *J. chem. Soc.* (1956) 1687
[5] Meerwein, H., private communication to O. Bayer, *Methoden der Organischen Chemie* (*Houben–Weyl*) 7 (1954) [1] 301, Stuttgart (Thieme)
[6] Zil'berman, E. N. and Pyryalova, P. S. *Zh. obshch. Khim.* 33 (1963) 3420 [*Chem. Abstr.* 60 (1964) 4005]
[7] Williams, J. W. *Org. Synth., Coll. Vol.* 3 (1955) 626
[8] Stephen, T. and H. *J. chem. Soc* (1956) 4695
[9] Tolbert, T. L. and Houston, B. *J. org. Chem.* 28 (1963) 695
[10] Gardner, T. S. et al., *J. org. Chem.* 16 (1951) 1121

65

STRECKER SYNTHESIS

Nature of the reaction

α-Amino acids are synthesized by reacting aldehydes with ammonia in the presence of hydrogen cyanide, followed by acid hydrolysis of the product

$$R\text{CHO} + \text{NH}_3 + \text{HCN} \longrightarrow R-\text{CH}\!\!\begin{array}{c}\nearrow \text{NH}_2 \\ \searrow \text{CN}\end{array} + \text{H}_2\text{O}$$

$$R-\text{CH}\!\!\begin{array}{c}\nearrow \text{NH}_2 \\ \searrow \text{CN}\end{array} \xrightarrow{\text{Hydrolysis}} R-\text{CH}\!\!\begin{array}{c}\nearrow \text{NH}_2 \\ \searrow \text{COOH}\end{array}$$

Historical development

In 1850, Strecker[1] reported that he had prepared alanine from acetaldehyde by hydrolysis of the intermediate condensation product of acetaldehyde, ammonia and hydrogen cyanide. Erlenmeyer and Sigel[2] used the procedure for the preparation of a number of compounds and concluded that the intermediate was an α-amino nitrile

$$\text{CH}_3\text{CHO} + \text{NH}_3 \longrightarrow \text{CH}_3\text{CH}\!\!\begin{array}{c}\nearrow \text{OH} \\ \searrow \text{NH}_2\end{array} \xrightarrow{\text{HCN}} \text{CH}_3\text{CH}\!\!\begin{array}{c}\nearrow \text{CN} \\ \searrow \text{NH}_2\end{array}$$

α-Amino nitrile

$$\xrightarrow{\text{Hydrolysis}} \text{CH}_3\text{CH}\!\!\begin{array}{c}\nearrow \text{COOH} \\ \searrow \text{NH}_2\cdot\text{HCl}\end{array} \xrightarrow{\text{Pb(OH)}_2} \text{CH}_3\text{CH}\!\!\begin{array}{c}\nearrow \text{COOH} \\ \searrow \text{NH}_2\end{array}$$

Alanine

Strecker's was the first such synthesis of a naturally occurring amino acid. The method he developed is still used, with little alteration, to prepare a number of amino acids, usually with yields >50 per cent.

The applications of the Strecker synthesis have been discussed by Block[3] in his review of preparative methods for the naturally occurring amino acids.

Mechanism

The mechanism had appeared to be straightforward, with an initial addition to the carbonyl group by ammonia followed by hydrogen cyanide. Tiemann[4], however, showed that better yields were obtained by forming a cyanohydrin first and then reacting with ammonia

$$RCHO + HCN \longrightarrow RCH\begin{matrix}CN\\OH\end{matrix} \xrightarrow{NH_3} RCH\begin{matrix}CN\\NH_2\end{matrix} \xrightarrow{Hydrolysis} RCH\begin{matrix}COOH\\NH_2\end{matrix}$$

Mowry[5] has discussed the confused state of the mechanistic studies. Despite doubts concerning the stability of the aldehyde–ammonia intermediate, the general view is that this *is* formed in the Strecker synthesis. The work of Stewart and Li[6] suggested that in Tiemann's procedure the cyanohydrin is dissociated by the action of amines to regenerate the carbonyl compound which then forms the aldehyde–ammonia intermediate

$$R-\underset{CN}{\overset{OH}{CH}} \rightleftharpoons RCHO + HCN \underset{}{\overset{R'_2NH}{\rightleftharpoons}} \underset{NR'_2}{\overset{OH}{RCH}} + HCN$$

$$\rightleftharpoons \underset{NR'_2}{\overset{CN}{RCH}} + H_2O \longrightarrow \underset{NR'_2}{\overset{COOH}{RCH}}$$

General reaction conditions

Although earlier reactions were carried out on aldehydes, the reaction has been extended to include ketones. Instead of a series of separate addition reactions, the preparation is now usually carried out in a mixture of ammonia (or amine) and the nitrile.

The carbonyl compound and an aqueous solution of sodium or potassium cyanide are added gradually to a cold (0–10°) aqueous solution of ammonium chloride and ammonium hydroxide. The cold solution is left to stand for an extended period (6–48 h) or

STRECKER SYNTHESIS

warmed at 50° for a few hours. After acidification with hydrochloric acid, any hydrogen cyanide is driven off and the bulk of the solution reduced in volume. The amino-acid hydrochloride is taken up in ethanol, the alcohol removed and the free amino acid obtained by treating with excess lead hydroxide.

Detailed procedures for the preparation of three amino acids have been given by Steiger[7]. α-Phenylglycine was obtained with a yield of 37 per cent

$$\text{Ph-CHO} + \text{NaCN} + \text{NH}_4\text{Cl} \longrightarrow \text{Ph-CH(CN)(NH}_2\text{)} \longrightarrow$$

$$\text{Ph-CH(COOH)(NH}_2 \cdot \text{HCl)} \xrightarrow{\text{NH}_4\text{OH}} \text{Ph-CH(COOH)(NH}_2\text{)}$$

Various attempts have been made to avoid the use of dangerous cyanides, but no general alternative is available[8].

Applications

The Strecker synthesis was used for the first successful preparation of six of the naturally occurring α-amino acids, including leucine, methionine and phenylalanine. In addition, many other synthetic amino acids have been prepared by this method.

Apart from conventional applications, Geipel and co-workers[9] used Strecker conditions to prepare β-hydroxy-α-amino acids from 1-alkoxy-1,2-diacetoxy alkanes

$$\underset{\text{OCOCH}_3\ \ \text{OCOCH}_3}{\text{CH}_2-\text{CH}-\text{OC}_2\text{H}_5} \xrightarrow[\text{KCN}]{\text{NH}_4\text{Cl/NH}_4\text{OH}} \underset{\text{OH}\ \ \ \text{NH}_2}{\text{CH}_2-\text{CHCO}_2\text{H}}$$

REFERENCES

[1] Strecker, A. *Justus Liebigs Annln Chem.* 75 (1850) 27
[2] Erlenmeyer, E. and Sigel, O. *Justus Liebigs Annln Chem.* 176 (1875) 341; 177 (1875) 111 [*J. chem. Soc.* (A) 28 (1875) 1007, 1012]
[3] Block, R. J. *Chem. Rev.* 38 (1946) 523
[4] Tiemann, F., with Friedländer, L. *Ber. dt. chem. Ges.* 13 (1880) 381; with Piest, K. 15 (1882) 2029 [*J. chem. Soc.* (A) 38 (1880) 473; 44 (1883) 198]
[5] Mowry, D. T. *Chem. Rev.* 42 (1948) 236
[6] Stewart, T. D. and Li, C. *J. Am. chem. Soc.* 60 (1938) 2782

[7] Steiger, R. E. *Org. Synth., Coll. Vol.* 3 (1955) 66, 84, 88
[8] Greenstein, J. P. and Winitz, M. *Chemistry of the Amino Acids* 1, 698, New York (Wiley) 1961
[9] Geipel, H. *et al., Ber. dt. chem. Ges.* 98 (1965) 1677 [*Chem. Abstr.* 63 (1965) 3030]

66

ULLMANN REACTION

Nature of the reaction

Biaryls are prepared by condensing aryl halides at elevated temperatures in the presence of copper

$$2 \; C_6H_5{-}X + Cu \longrightarrow C_6H_5{-}C_6H_5 + CuX_2$$

Historical development

This reaction was the first of several associated with the name of Ullmann and involving the use of copper. It was reported[1] in 1901 in connection with the formation of several nitrodiphenyls

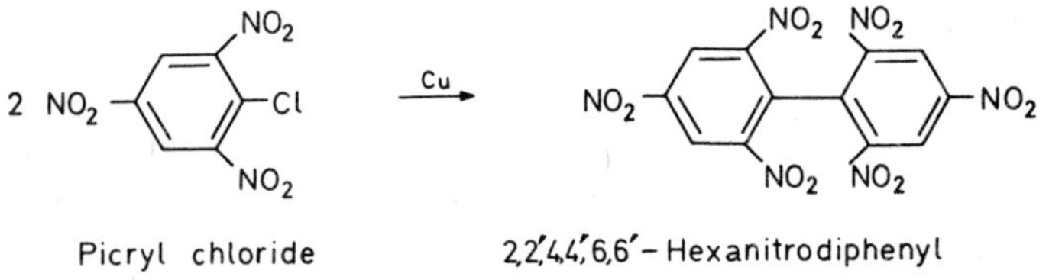

Picryl chloride   2,2′,4,4′,6,6′-Hexanitrodiphenyl

Further investigation[2] showed the reaction to be of wide applicability, and a large number of diphenyls were prepared. The order of reactivity of halogens was found to be I > Br > Cl; aryl bromides and chlorides in general give satisfactory reactions only when activated by electron-withdrawing groups in *ortho* and *para* positions. Yields are low if large groups occupy the *ortho* positions of the aryl halide. Simple aryl halides react to give high yields of the corresponding diaryl, usually above 60 per cent.

A review of the reaction by Fanta[3] includes a survey of compounds prepared to 1944. The extensive use of the Ullmann reaction necessitated a supplementary review[4] which covers the literature to 1963.

Mechanism

A mechanism has not been clearly established. The reaction had been assumed[5,6] to arise from the formation of neutral radicals by the aryl halide on the catalyst surface. This is not supported by the work of Forrest[7] on the formation of unsymmetrical diaryls from mixtures of two different aryl halides. He obtained them from mixtures of aryl halides in which one component formed intermediates with the copper in preference to the other.

This suggests that a molecule attached to the catalyst reacts with one that is not attached. Fanta[4] has expressed this in terms of an activated complex in which the electronegative group on the more reactive aryl halide complexes with the catalyst

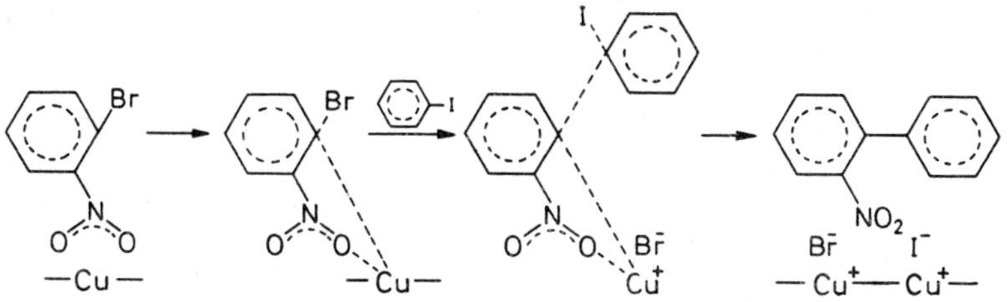

General reaction conditions

Reactions are carried out by heating the aryl halide between 100° and 300° in an open container and gradually adding the copper powder until a two- or three-fold excess has been used. The heating is continued for several hours, then the mixture is left to cool before extracting the product.

A detailed method for preparation of 2,2′-dinitrodiphenyl with 50–60 per cent yields has been given by Fuson and Cleveland[8]

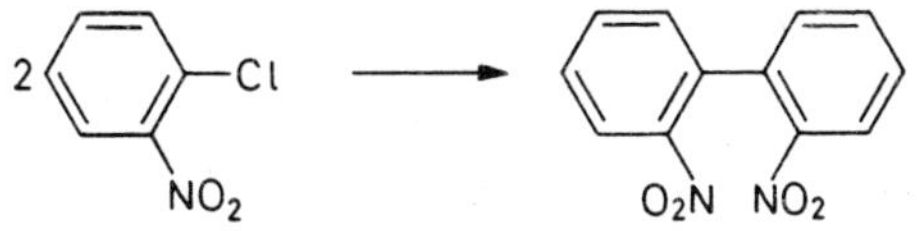

High-boiling organic solvents have often been employed as diluents in the Ullmann reaction. In the case of nitrobenzene, this often leads to side reactions between the aryl halide and the solvent[9]. Increased yields have been obtained by using dimethylformamide as solvent, believed to be due to the maintenance of an uncontaminated surface on the copper[10].

It has also been claimed[11] that the nature of the copper affects the yields from the reaction. High yields (72 per cent) of 2,2'-bipyridine were obtained from 2-bromopyridine by employing electrolytic copper rather than copper bronze or alloys.

Applications

Wirth and co-workers[12], using mixtures of monofunctional and bifunctional aryl halides, have produced mixtures of *p*-oligophenylenes

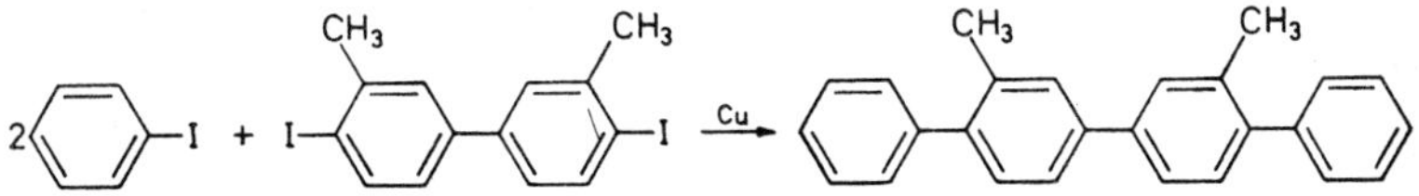

The method has also been used to prepare condensed-ring systems such as perylene[13]

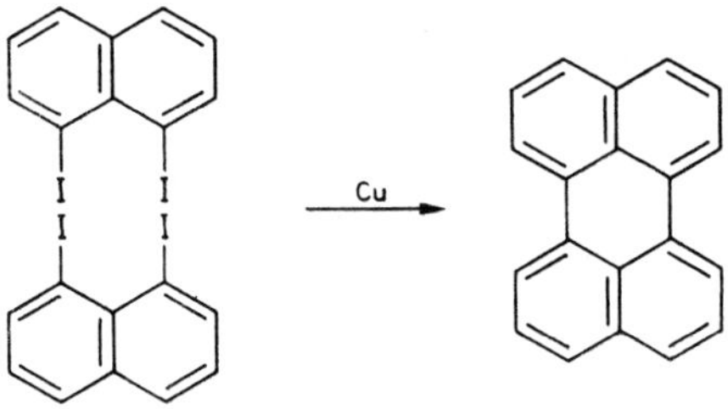

REFERENCES

[1] Ullmann, F. and Bielecki, J. *Ber. dt. chem. Ges.* 34 (1901) 2174 [*J. chem. Soc.* (A) 80 (1901) [1] 586]
[2] Ullmann, F. et al., *Justus Liebigs Annln Chem.* 332 (1904) 38 [*J. chem. Soc.* (A) 86 (1904) [1] 725]
[3] Fanta, P. E. *Chem. Rev.* 38 (1946) 139
[4] Fanta, P. E. *Chem. Rev.* 64 (1964) 613
[5] Waters, W. A. *The Chemistry of Free Radicals*, 2nd edn., p. 210, Oxford (Univ. Press) 1948
[6] Rapson, W. S. and Shuttleworth, R. G. *Nature, Lond.* 147 (1941) 675

[7] Forrest, J. *J. chem. Soc.* (1960) 594
[8] Fuson, R. C. and Cleveland, E. A. *Org. Synth., Coll. Vol.* 3 (1955) 339
[9] Forrest, J. *J. chem. Soc.* (1960) 581
[10] Braithwaite, R. S. W. and Holt, P. F. *J. chem. Soc.* (1959) 3025
[11] Kulicki, Z. and Jarminski, W. *Zesz. nauk. Politech. śląsk.* 16 (1963) [85] 11 [*Chem. Abstr.* 62 (1965) 4001]
[12] Wirth, H. O. *et al., Makromolec. Chem.* 63 (1963) 30 [*Chem. Abstr.* 60 (1964) 5363]
[13] Scholl, R., Seer, C. and Weitzenböck, R. *Ber. dt. chem. Ges.* 43 (1910) 2202 [*Chem. Abstr.* 4 (1910) 2929]

67

VON RICHTER REACTION

Nature of the reaction

Substituted benzoic acids are prepared by treating substituted aromatic nitro compounds with ethanolic potassium cyanide. The carboxyl group enters a position *ortho* to that of the original nitro group

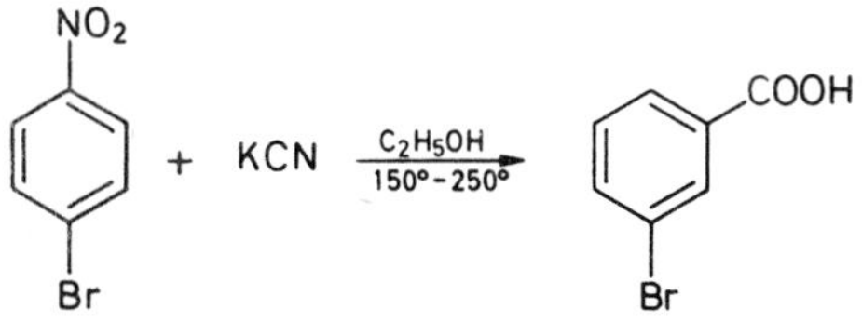

$$\underset{Br}{\underset{|}{C_6H_4}}-NO_2 + KCN \xrightarrow[150°-250°]{C_2H_5OH} \underset{Br}{\underset{|}{C_6H_4}}-COOH$$

Historical development

In an attempt to replace the bromine atom in *p*-bromonitrobenzene by a cyano group, von Richter[1] found that the nitro group was lost under the conditions employed. His further researches[2] showed that the reaction took place with other halogen-substituted nitrobenzenes, including those possessing more than one halogen atom

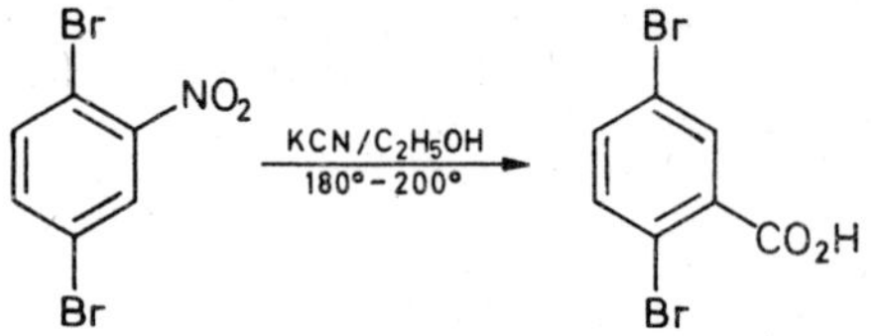

Due to the large number of by-products formed, yields of the benzoic acids tend to be low, usually between 10 and 50 per cent.

Mechanism

Virtually no work was carried out on this reaction until 1950 when Bunnett and his collaborators[3] investigated the mechanism. A short review of this work has been given[4].

Controversy has arisen between the work of Bunnett *et al.* and that of Rosenblum[5] over the possible intermediates formed in the process. The most likely route appears to include the formation of substituted *o*-nitrosobenzamides which cyclize to the corresponding substituted 3-indazolones. This is supported by the observation[6] that using such compounds as starting materials under von Richter conditions does lead to the formation of benzoic acids. This mechanism is represented as

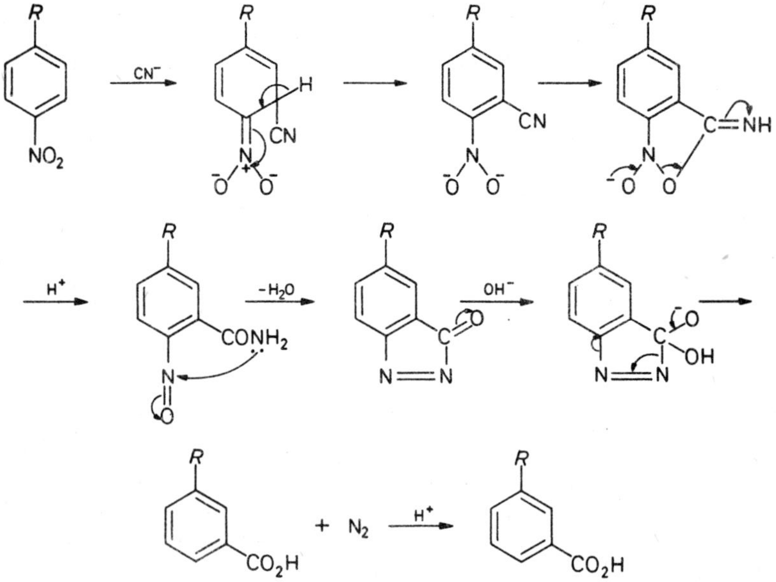

VON RICHTER REACTION

General reaction conditions

The early reactions were carried out by heating an aromatic nitro compound with ethanolic potassium cyanide at 180°–200° in sealed tubes. The mixture from the tubes was boiled with ethanolic potassium hydroxide, and the free acid obtained by neutralizing with hydrochloric acid.

Bunnett and co-workers[3] modified this procedure by carrying out preparations in a flask of refluxing aqueous ethanolic potassium cyanide. After cooling and treating with potassium hydroxide, steam distillation was used to remove solvent and unreacted starting material. The products were isolated by acidifying the residue left after the distillation.

A detailed procedure for the preparation of 2-bromo-3-methyl benzoic acid from 2-bromo-4-nitrotoluene has been given[7]

$$\text{4-NO}_2\text{-2-Br-C}_6\text{H}_3\text{-CH}_3 + \text{KCN} + 2\text{H}_2\text{O} \longrightarrow \text{2-Br-3-CH}_3\text{-C}_6\text{H}_3\text{-CO}_2\text{H} + \text{KNO}_2 + \text{NH}_3$$

Applications

Cullen and L'Ecuyer[8] have carried out an extensive study on all the products obtained in the reaction with *p*-chloronitrobenzene. They showed that, apart from a 46 per cent yield of *m*-chlorobenzoic acid, the mixture contained a wide variety of other compounds such as 4,4'-dichloro-2,2'-dicarbamoylazobenzene.

Holleman[9] demonstrated that the reaction also occurs with nitrobenzenesulphonic acids. *m*-Nitrobenzenesulphonic acid gave a mixture of *o*- and *p*-sulphobenzoic acids

$$\text{3-NO}_2\text{-C}_6\text{H}_4\text{-SO}_3\text{H} \xrightarrow{\text{KCN/C}_2\text{H}_5\text{OH}} \text{2-COOH-C}_6\text{H}_4\text{-SO}_3\text{H} + \text{4-COOH-C}_6\text{H}_4\text{-SO}_3\text{H}$$

REFERENCES

[1] von Richter, V. *Ber. dt. chem. Ges.* 4 (1871) 21, 459, 553 [*J. chem. Soc.* (*A*) 24 (1871) 220, 686, 824]

[2] von Richter, V. *Ber. dt. chem. Ges.* 7 (1874) 1145; 8 (1875) 1418 [*J. chem. Soc.* (*A*) 28 (1875) 73; 29 (1876) 387]

[3] Bunnett, J. F., with Cormack, J. F. and McKay, F. C. *J. org. Chem.* 15 (1950) 481; with Rauhut, M. M. 21 (1956) 944
[4] Bunnett, J. F. *Q. Rev. chem. Soc.* 12 (1958) 15
[5] Rosenblum, M. *J. Am. chem. Soc.* 82 (1960) 3796
[6] Ibne-Rasa, K. M. and Koubek, E. *J. org. Chem.* 28 (1963) 3240
[7] Bunnett, J. F. and Rauhut, M. M. *Org. Synth., Coll. Col.* 4 (1963) 114
[8] Cullen, E. and L'Ecuyer, P. *Can. J. Chem.* 39 (1961) 144, 382, 862 [*Chem. Abstr.* 62 (1965) 2726, 2745]
[9] Holleman, M. M. *Recl Trav. chim. Pays-Bas Belg.* 24 (1905) 194 [*J. chem. Soc. (A)* 88 (1905) [1] 595]

68

WAGNER–MEERWEIN REARRANGEMENT

Nature of the reaction

The rearrangement of a carbonium ion intermediate, formed from a monoalcohol or halide, occurs under acidic conditions. In the terpene series, it usually leads to ring expansion and/or contraction

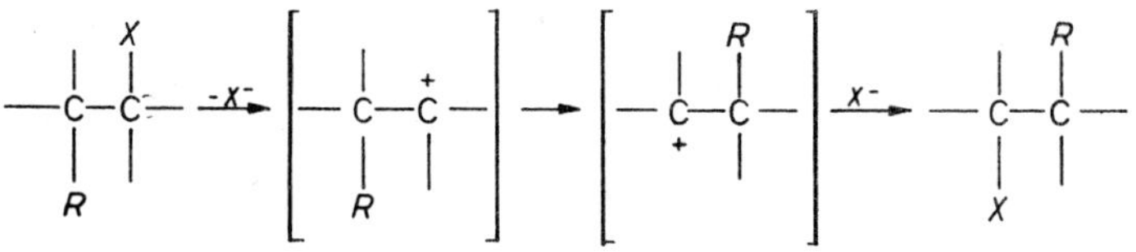

$$\underset{R}{\overset{X}{-\overset{|}{\underset{|}{C}}-\overset{|}{\underset{|}{C}}-}} \xrightarrow{-X^-} \left[\underset{R}{-\overset{|}{\underset{|}{C}}-\overset{+}{\underset{|}{C}}-} \right] \longrightarrow \left[-\overset{R}{\underset{+}{\overset{|}{C}}}-\overset{|}{\underset{|}{C}}- \right] \xrightarrow{X^-} -\overset{R}{\underset{X}{\overset{|}{C}}}-\overset{|}{\underset{|}{C}}-$$

Historical development

Blanc[1] was the first to point out that results of experiments carried out in acid solution upon compounds related to camphor were unreliable, as molecular rearrangements could occur. No structures were suggested for these rearrangements.

A short time later, Wagner[2] suggested that the formation of bornyl chloride involved an initial addition, followed by a rearrangement

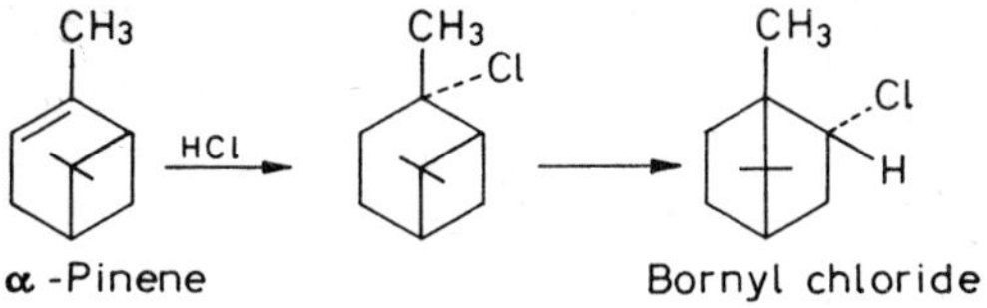

WAGNER–MEERWEIN REARRANGEMENT

α-Pinene → Bornyl chloride (via HCl)

As a result of work upon the pinacol–pinacolone rearrangement, Meerwein and his collaborators[3] became interested in the rearrangements in the terpene series. Their investigations, extended over a number of years, showed that in the terpenes the rearrangement involves a simultaneous ring expansion and contraction, usually resulting in a change of a 5-membered to a 6-membered ring and vice versa. The acid dehydration of borneol to camphene was explained on this basis

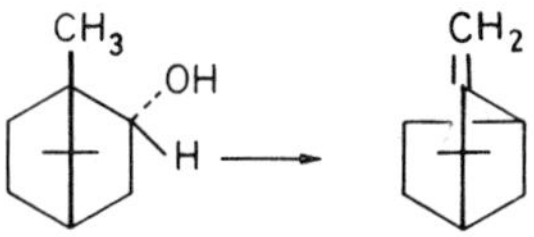

Rearrangements related to pinane derivatives have been reviewed by Banthorpe and Whittaker[4].

The rearrangement is not restricted to terpenes but can occur in any suitable organic system. Thus Collins and his co-workers[5] made chrysene by a Wagner–Meerwein rearrangement

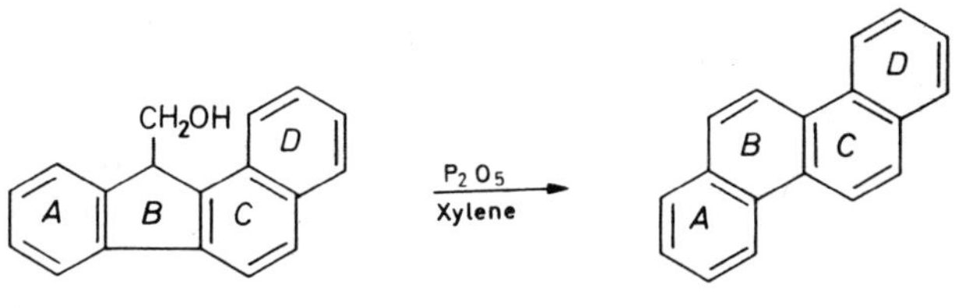

Mechanism

Mechanistic studies have been carried out almost continuously since this rearrangement was recognized as a reaction of wide application. Meerwein and van Emster[6] believed that it involved ionization of the alcohol or halide to give a carbonium ion which then rearranged.

On the basis of work by Whitmore *et al.*[7] in addition to their own investigations, Hughes and Ingold[8] supported the idea of a carbonium ion associated with a unimolecular nucleophilic substitution (S_N1) and elimination (E1).

The mechanistic work has been reviewed by Streitwieser[9] and discussed in connection with related reactions by Pocker[10]. In terms of the rearrangement of camphene hydrochloride, the mechanism may be shown as

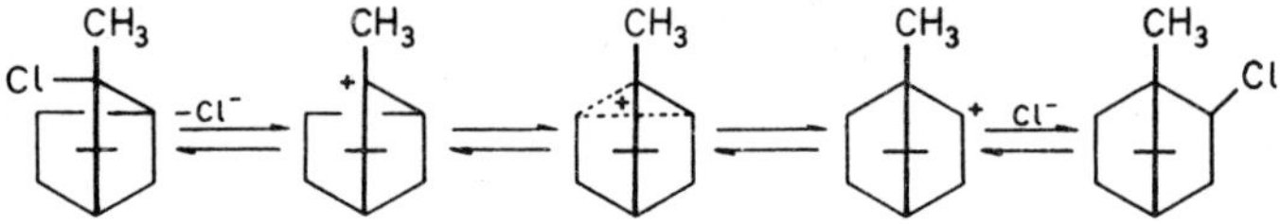

General reaction conditions

Many different systems have been used to obtain successful rearrangements. In general terms, procedures are carried out in polar solvents—nitrobenzene, ethyl bromide—and are often catalysed by zinc chloride, ferric chloride or similar Lewis acids. Hydrogen chloride gas is passed into a solution of the organic compound which is then heated between 50° and 150° for as long as several days. This type of procedure was used by Nevell and co-workers[11] to prepare isobornyl chloride from camphene

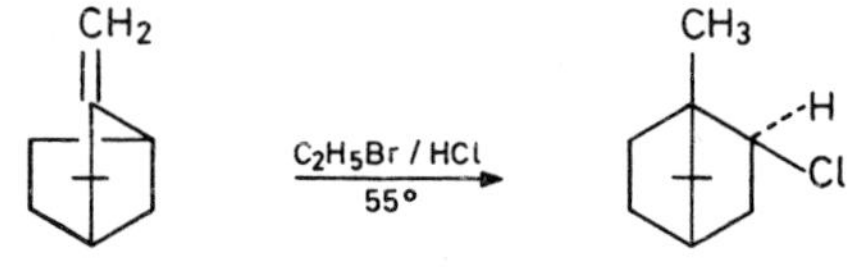

Camphene → (C$_2$H$_5$Br / HCl, 55°) → Isobornyl chloride

Modifications

Other rearrangements have been carried out by refluxing in formic acid[12] and by using phosphorus pentoxide in xylene[5] as reaction media. Polyphosphoric acid has been of particular value for a number of Wagner–Meerwein rearrangements[13], for example, to obtain phenanthrenes from 9-fluorenylmethylacetates[14]

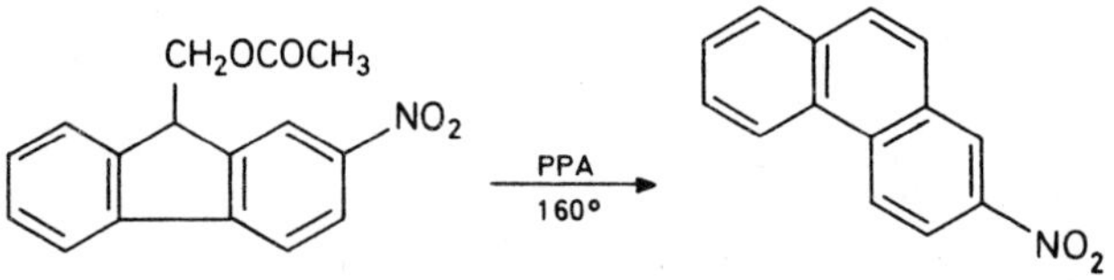

2-Nitro-9-fluorenylmethylacetate → (PPA, 160°) → 2-Nitrophenanthrene

WAGNER–MEERWEIN REARRANGEMENT

The compound is stirred into the hot PPA for about 30 min, then the mixture is poured into a large quantity of iced water. The product separates as a solid which is filtered off or extracted by a suitable solvent.

Applications

Tortorella and Romeo[15] have found that the rearrangement occurs with steroids possessing an hydroxy group in the 17 position

They have used this to prepare a number of $\triangle^{12}$ and $\triangle^{13}$ steroids.

REFERENCES

[1] Blanc, G. *Bull. Soc. chim. Fr.* 19 (1898) [3] 214 [*J. chem. Soc.* (A) 76 (1899) [1] 442, 444]
[2] Wagner, G. and Brykner, W. *Ber. dt. chem. Ges.* 32 (1899) 2302; 33 (1900) 2121 [*J. chem. Soc.* (A) 78 (1900) [1] 46, 554]
[3] Meerwein, H., with Probst, H. and Muhlendyk, W. *Justus Liebigs Annln Chem.* 405 (1914) 129; *et al.*, 453 (1927) 16 [*Chem. Abstr.* 8 (1914) 3183; 21 (1927) 2890]
[4] Banthorpe, D. V. and Whittaker, D. *Q. Rev. chem. Soc.* 20 (1966) 373
[5] Collins, C. J. *et al.*, *J. Am. chem. Soc.* 75 (1953) 397
[6] Meerwein, H. and van Emster, K. *Ber. dt. chem. Ges.* 55 (1922) 2500 [*Chem. Abstr.* 17 (1923) 747]
[7] Whitmore, F. C., with Rothrock, H. S. *J. Am. chem. Soc.* 54 (1932) 3431; with Fleming, G. H. 55 (1933) 4161
[8] Dostrovsky, I., Hughes, E. D. and Ingold, C. K. *J. chem. Soc.* (1946) 192
[9] Streitwieser, A. *Chem. Rev.* 56 (1956) 698
[10] Pocker, Y. *Molec. Rearrangements* 1 (1963) 1
[11] Nevell, T. P., de Salas, E. and Wilson, C. L. *J. chem. Soc.* (1939) 1188
[12] Anet, F. A. L. and Bavin, P. M. G. *Can. J. Chem.* 34 (1956) 991
[13] Popp, F. D. and McEwen, W. E. *Chem. Rev.* 58 (1958) 375
[14] Bavin, P. M. G. and Dewar, M. J. S. *J. chem. Soc.* (1955) 4477
[15] Tortorella, V. and Romeo, A. *Gazz. chim. ital.* 92 (1962) 1118 [*Chem. Abstr.* 58 (1963) 12627]

69

WALDEN INVERSION

Nature of the reaction

The inversion of configuration that occurs in many reactions when one substituent on an asymmetric centre is replaced by another is known as the Walden inversion

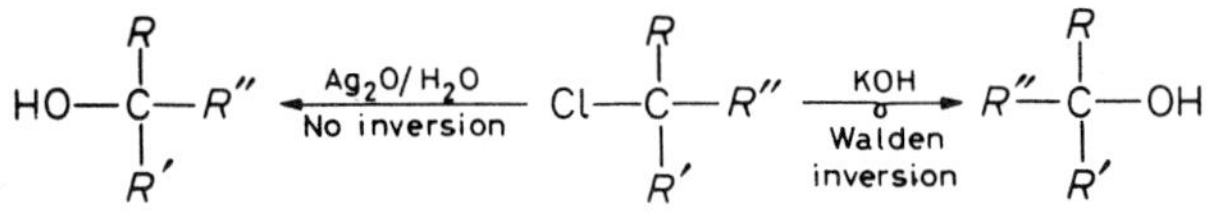

$$\text{HO}-\underset{R'}{\overset{R}{\underset{|}{\overset{|}{C}}}}-R'' \xleftarrow[\text{No inversion}]{\text{Ag}_2\text{O}/\text{H}_2\text{O}} \text{Cl}-\underset{R'}{\overset{R}{\underset{|}{\overset{|}{C}}}}-R'' \xrightarrow[\text{Walden inversion}]{\text{KOH}} R''-\underset{R'}{\overset{R}{\underset{|}{\overset{|}{C}}}}-\text{OH}$$

where R, R' and R'' are different substituent groups.

Historical development

During studies on optically active compounds, Walden[1] found that natural optically active aspartic acid could be converted by two different routes into the two optical isomers of dimethyl bromosuccinate

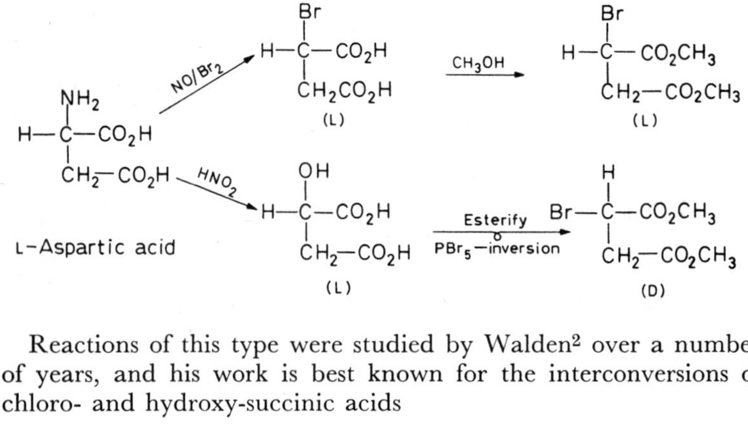

Reactions of this type were studied by Walden[2] over a number of years, and his work is best known for the interconversions of chloro- and hydroxy-succinic acids

WALDEN INVERSION

$$\underset{(D)}{\overset{Cl}{\underset{H}{>}}C\overset{CO_2H}{\underset{CH_2CO_2H}{<}}} \xrightarrow{Ag_2O} \underset{(D)}{\overset{HO}{\underset{H}{>}}C\overset{CO_2H}{\underset{CH_2CO_2H}{<}}}$$

$$\uparrow PCl_5 \qquad\qquad \downarrow PCl_5$$

$$\underset{(L)}{\overset{H}{\underset{HO}{>}}C\overset{CO_2H}{\underset{CH_2CO_2H}{<}}} \xleftarrow{Ag_2O} \underset{(L)}{\overset{H}{\underset{Cl}{>}}C\overset{CO_2H}{\underset{CH_2CO_2H}{<}}}$$

The study of the inversions was further extended by Fischer[3] who chose to name them after their discoverer. The growth and development of the subject were summarized by Walden[4] in a series of lectures at Cornell University.

A change in configuration is not necessarily accompanied by one in the direction of the observed rotation, as this is dependent upon the nature of the groups attached to the asymmetric carbon atom.

Many reactions are known in which a 100 per cent inversion to the new species is obtained. The majority of substances undergoing inversion, however, give a mixture of the two possible optical isomers. Whether inversion, retention or racemization occurs depends upon the reaction conditions and the mechanism of the process. It is now common practice to apply the term 'Walden inversion' to the actual step in which this occurs rather than to a complete cycle.

Mechanism

The problem of why some reactions led to inversion while others did not, confused researchers for many years. In 1925, Lowry[5] made the suggestion that under experimental conditions which produced carbonium ions the carbon atom could pass from its normal tetrahedral configuration to an intermediate trigonal planar configuration. If no carbonium ion was formed, the intermediate was believed to be a 5-membered trigonal bipyramide.

Extensive studies into the mechanism, first by Kenyon and Phillips[6], a little later by Hughes and Ingold[7], led to support for Lowry's main concept and to the development of the ideas of bimolecular (S_N2) and unimolecular (S_N1) nucleophilic modes of substitution. This work has been discussed in considerable detail by Ingold[8].

In the case of the S_N2 mechanism, inversion always occurs, as the attacking nucleophile does so from the side of the molecule opposite to that occupied by the leaving group. This leads to a planar intermediate in which neither the entering nor the departing group is completely attached to the central carbon atom. When the entering group is completely attached, an inversion of the molecular configuration has taken place

$$X^- \quad \underset{R'''}{\overset{R'\,R''}{\underset{|}{C}}\!\!-\!Y} \quad \longrightarrow \quad \underset{R'''}{\overset{R'\,R''}{\underset{|}{X\!-\!-\!-\!C\!-\!-\!-\!Y}}}^{\delta-\quad\delta-} \quad \longrightarrow \quad X\!-\!\underset{R'''}{\overset{R''\,R'}{\underset{|}{C}}} \quad Y^-$$

Where the S_N1 mechanism comes into play, the final result may be retention, inversion or racemization. In this mechanism, the molecule undergoing reaction is believed to ionize before being attacked by the nucleophile. In instances where the carbonium ion can assume a planar configuration, racemization is the normal result. This is not always the case, as steric factors may lead to one configuration being preferred over another. The nature of the attacking species also affects the proportions of isomers formed

$$\underset{R'''}{\overset{R'\,R''}{\underset{|}{C}}\!\!-\!Y} \longrightarrow \underset{R'''}{\overset{R'\,R''}{\underset{|}{C^+}}} + Y^- \quad \begin{array}{c} \xrightarrow{X^-} \underset{R'''}{\overset{R''}{\underset{|}{C}}\!\!-\!X} \quad \text{Retention} \\ \\ \xrightarrow{X^-} X\!-\!\underset{R'''}{\overset{R''\,R'}{\underset{|}{C}}} \quad \text{Inversion} \end{array}$$

The S_N1 and S_N2 mechanisms, in fact, represent the two extremes of a very involved process. Products from substitutions at optically active centres can vary greatly depending upon reagents, solvents and physical conditions.

Examples of inversions

Configurational inversions occur regularly in sugar chemistry, and a number of well-established instances have been described in detail by Baker[9]. The conversion of methyl-3-amino-3-deoxy-β-L-xylopyranoside (A) to methyl-3-acetamido-3-deoxy-α-D-ribopyranoside (B) with inversion of the hydroxyl groups at the 2 and 4 positions is an example

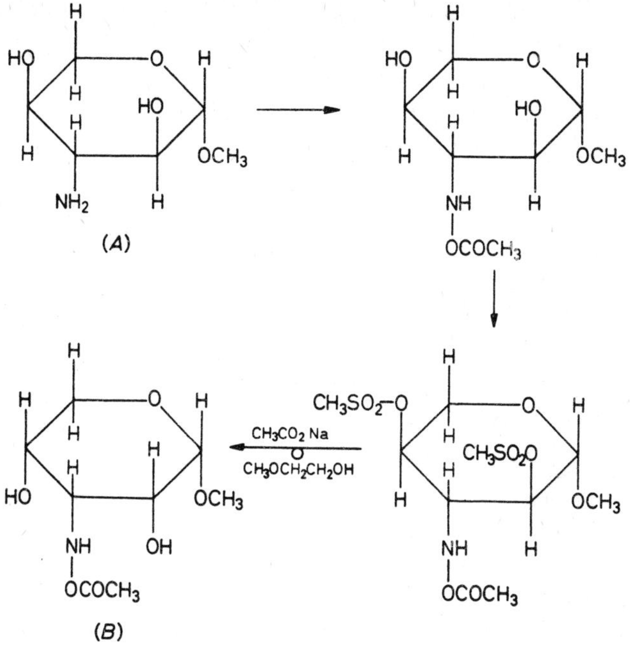

WALDEN INVERSION

Inversions are also frequently encountered in steroid chemistry. Thus, the action of phosphorus pentachloride on 5α-cholestane-3β,6α-diol leads to inversion at two points to give 3α,6β-dichloro-5α-cholestane[10]

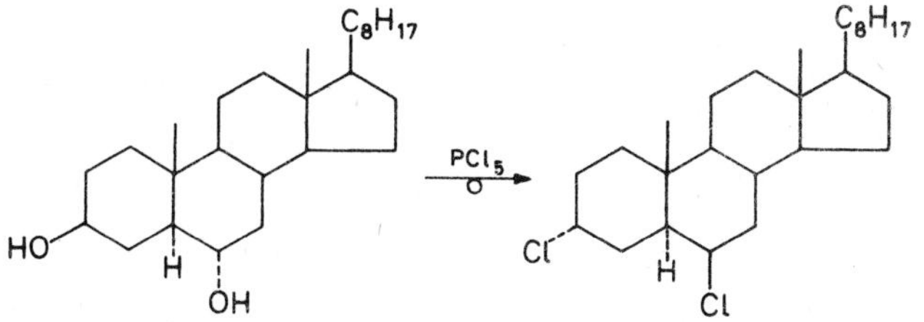

The studies on the Walden inversion have in recent years been extended to organosilicon compounds possessing an asymmetric silicon atom[11]. Brook and Warner[12] have described a complete cycle, starting with methylphenyl-α-naphthylchlorosilane, in which several inversion steps are involved.

NAMED ORGANIC REACTIONS

REFERENCES

[1] Walden, P. *Ber. dt. chem. Ges.* 26 (1893) 210; 28 (1895) 1287, 2766 [*J. chem. Soc.* (*A*) 64 (1893) [1] 250; 68 (1895) [1] 450; 70 (1896) [1] 139]
[2] Walden, P. *Ber. dt. chem. Ges.* 30 (1897) 3146; 32 (1899) 1855 [*J. chem. Soc.* (*A*) 74 (1898) [1] 178; 76 (1899) [2] 538]
[3] Fischer, E. *Ber. dt. chem. Ges.* 39 (1906) 2894; 40 (1907) 1051 [*J. chem. Soc.* (*A*) 90 (1906) [1] 808; *Chem. Abstr.* 1 (1907) 1546]
[4] Walden, P. *Salts, Acids and Bases, Electrolytes and Stereochemistry*, p. 353, New York (McGraw-Hill) 1929
[5] Lowry, T. M. *Inst. int. Chim. Solvay, Cons. Chim.* 2 (1925) 135 [*Chem. Abstr.* 20 (1926) 3620]
[6] Kenyon, J., Phillips, H., with Turley, H. G. *J. chem. Soc.* 127 (1925) 399; *Trans. Faraday Soc.* 26 (1930) 451
[7] Hughes, E. D. and Ingold, C. K. *J. chem. Soc.* (1935) 254; *et al.* (1937) 1252
[8] Ingold, C. K. *Structure and Mechanism in Organic Chemistry*, p. 372, Ithaca, N.Y. (Cornell Univ. Press) 1953
[9] Baker, B. R. *Meth. Carbohyd. Chem.* 2 (1963) 444
[10] Shoppee, C. W. and Lack, R., with McLean, B. *J. chem. Soc.* (1964) 4996; with Sharma, S. C. *C* (1968) 2083
[11] Sommer, L. H. and Frye, C. L. *J. Am. chem. Soc.* 82 (1960) 3796
[12] Brook, A. G. and Warner, C. M. *Tetrahedron Lett.* 8 (1962) 815

70

WILLGERODT–KINDLER REACTION

Nature of the reaction

The action of aqueous ammonium polysulphide, or sulphur, in the presence of an amine, leads to migration of ketonic carbonyl groups and formation of acid amides possessing the original number of carbon atoms

$$ArCO(CH_2)_n CH_3 \xrightarrow[210°]{(NH_4)_2 S_x / H_2O} Ar(CH_2)_{n+1} CONH_2$$

Historical development

Willgerodt found[1] that when alkyl aryl ketones were heated in sealed tubes with aqueous ammonium polysulphide, acid amides were formed. An unexpected feature was that, irrespective of the length of the alkyl chain, the amide was always formed without loss of any of the chain carbon atoms

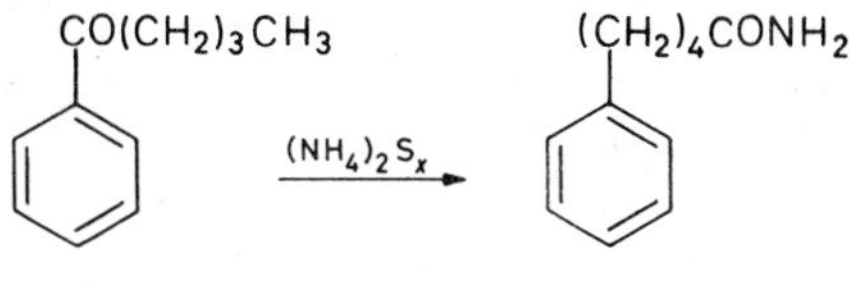

WILLGERODT–KINDLER REACTION

$$\text{PhCO(CH}_2)_3\text{CH}_3 \xrightarrow{(NH_4)_2 S_x} \text{Ph(CH}_2)_4\text{CONH}_2$$

Investigation into this reaction was carried out for many years by Willgerodt and his collaborators[2], but it was little used until later modified by Kindler[3]. The reaction conditions were changed by substituting equimolecular amounts of sulphur and an anhydrous amine for the aqueous ammonium polysulphide. The products from this latter method are thioamides which may be hydrolysed to carboxylic acids or reduced to amines

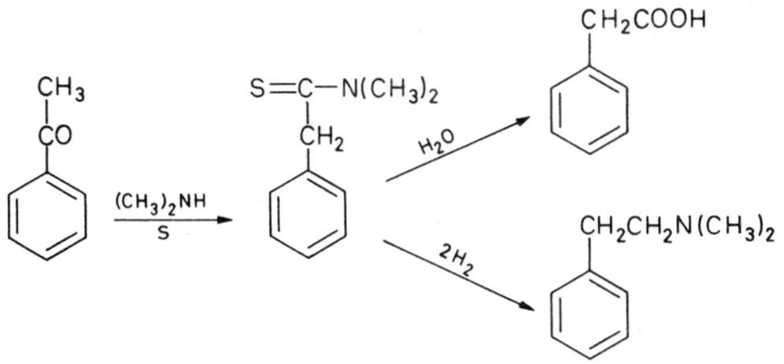

$$\text{PhCOCH}_3 \xrightarrow[S]{(CH_3)_2NH} \text{PhCH}_2\text{C}(=S)\text{N}(CH_3)_2 \begin{array}{c} \xrightarrow{H_2O} \text{PhCH}_2\text{COOH} \\ \xrightarrow{2H_2} \text{PhCH}_2\text{CH}_2\text{N}(CH_3)_2 \end{array}$$

Kindler's modification of the reaction has since been extended to aldehydes, alcohols and alkenes as starting materials[4].

Yields are often high, 50–80 per cent, but decrease with increased length of the alkyl chain. King and McMillan[5] showed that the reaction does not occur if there is a quaternary carbon atom in the chain.

The reaction has been reviewed by Carmack[6], and the preparation of thioamides by Kindler's modification has been covered by Hurd and Delamater[7].

Mechanism

A full explanation of the mechanism that can satisfy all experimental observations has yet to be given. It is generally believed that alkenes and/or alkynes are intermediates, as these too undergo the reaction. The work of De Tar and Carmack[8] indicated that this occurs initially at the carbonyl group, followed by migration along the alkyl chain

$$Ar-COCH_2CH_3 + HNR_2 \rightleftharpoons Ar\underset{NR_2}{\underset{|}{\overset{OH}{\overset{|}{C}}}}-CH_2CH_3 \overset{-H_2O}{\rightleftharpoons} Ar\underset{NR_2}{\underset{|}{C}}=CHCH_3$$

$$\rightleftharpoons ArCH=\underset{NR_2}{\underset{|}{C}}CH_3 \rightleftharpoons ArCH_2-\underset{NR_2}{\underset{|}{C}}=CH_2 \rightleftharpoons$$

$$ArCH_2-CH=\underset{NR_2}{\underset{|}{C}}H \overset{S}{\longrightarrow} ArCH_2CH_2-\underset{NR_2}{\underset{|}{C}}=S$$

Dauben and Rogan[9], however, gave evidence that the attack occurs initially on the methylene carbon atom α to the carbonyl group. A speculative route involving a series of oxidation–reduction cycles along the alkyl chain has been suggested by Wegler and co-workers[10].

General reaction conditions

The conditions used by Willgerodt necessitated preparing ammonium polysulphide by passing hydrogen sulphide into concentrated aqueous ammonia. The solution was then heated above 200° in a sealed glass tube with the alkyl aryl ketone for several hours. The cooled reaction mixture was treated with hydrochloric acid to decompose excess ammonium polysulphide before the amide was isolated.

Reactions under these conditions have been improved by the addition of 40 per cent dioxan to the mixture to increase homogeneity.

Kindler's modification is carried out by carefully refluxing a mixture of the ketone with sulphur and the amine (usually morpholine) for 6-20 h until no more hydrogen sulphide is evolved. After cooling, the mixture is hydrolysed with sodium hydroxide and the free acid obtained by acidifying with hydrochloric acid.

Several procedures of the type shown have been described in detail by Vogel[11]

p - Methoxyacetophenone → *p* - Methoxyacetic acid

WILLGERODT–KINDLER REACTION

Applications

The reaction occurs with a wide variety of compounds, most of the common alkyl aryl ketones giving satisfactory results.

Barrett[12] found that the Willgerodt–Kindler reaction with 3-acetyl-1-methylpyrrocoline gave two main products

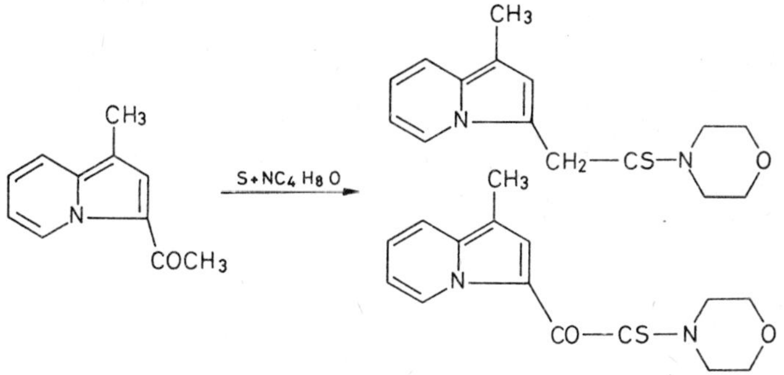

Similar results were obtained with *p*-dimethylamino-acetophenone.

REFERENCES

[1] Willgerodt, C. *Ber. dt. chem. Ges.* 20 (1887) 2467; 21 (1888) 534 [*J. chem. Soc.* (A) 52 (1887) 1045; 54 (1888) 476]
[2] Willgerodt, C., with Scholtz, T. *J. prakt. Chem.* 81 (1910) [2] 382; with Albert, B. 84 (1911) [2] 383 [*Chem. Abstr.* 4 (1910) 2283; 6 (1912) 80]
[3] Kindler, K. *Justus Liebigs Annln Chem.* 431 (1923) 187; with Li, T. *Ber. dt. chem. Ges.* 74 (1941) 321 [*Chem. Abstr.* 17 (1923) 2278; 35 (1941) 5102]
[4] McOmie, J. F. W. *Rep. Prog. Chem.* 45 (1948) 210
[5] King, J. A. and McMillan, F. H. *J. Am. chem. Soc.* 68 (1946) 525
[6] Carmack, M. and Spielman, M. A. *Org. React.* 3 (1946) 83
[7] Hurd, R. N. and Delamater, G. *Chem. Rev.* 61 (1961) 52
[8] De Tar, D. F. and Carmack, M. *J. Am. chem. Soc.* 68 (1946) 2029
[9] Dauben, W. G. and Rogan, J. B. *J. Am. chem. Soc.* 78 (1956) 4135
[10] Wegler, R., Kühle, E. and Schäfer, W. *Newer Methods of Preparative Organic Chemistry* (ed. Foerst, W.), 3, 1, New York (Academic Press) 1964
[11] Vogel, A. I. *A Textbook of Practical Organic Chemistry*, 3rd edn., p. 923, London (Longmans) 1956
[12] Barrett, P. A. *J. chem. Soc.* (1957) 2056

71

WILLIAMSON SYNTHESIS

Nature of the reaction
Ethers are formed by reacting alkyl halides with sodium alkoxides

$$RONa + XR' \longrightarrow ROR' + NaX$$

Historical development
Williamson's method for preparing ethers[1] was introduced at a time when the structures of acids, alcohols and ethers were uncertain. On the basis of his results, he proposed what are now known to be the correct structures for many of these compounds. In one particular case he established that an unknown compound with an empirical formula C_3H_8O was the unsymmetrical ethyl methyl ether, which could be prepared by two different approaches by his method

(1) $CH_3ONa + IC_2H_5 \longrightarrow CH_3OC_2H_5 + NaI$

(2) $CH_3I + NaOC_2H_5 \longrightarrow CH_3OC_2H_5 + NaI$

Williamson's approach incurred the displeasure of Kolbe[2] and led to a rather acrimonious exchange between them[3] about chemical structures. The method was, however, quickly accepted and found to be of general application. It is of particular value for forming unsymmetrical ethers incorporating secondary and tertiary alkyl groups. In these reactions the alkyl group with the greatest degree of branching is used as the alcoholate, and that with least branching as the alkyl halide. This is necessary because alkenes are formed as by-products from secondary and tertiary halides.

Yields from the reaction are high, generally above 60 and frequently >80 per cent. For the formation of methyl and ethyl ethers, alkyl sulphates are often used in place of the corresponding alkyl halides.

WILLIAMSON SYNTHESIS

Mechanism

An S_N2 mechanism is believed to operate in the reaction[4]

$$RO^- + \overset{\downarrow}{\underset{|}{C}}-X \longrightarrow RO---\overset{\downarrow}{\underset{|}{C}}---X \longrightarrow RO-\overset{\downarrow}{\underset{|}{C}} + X^-$$

Working with sodium eugenoxide and several different alkyl iodides, Woolf[5] showed that ether formation took place by second-order kinetics. Hughes and Ingold[6] at the same time reported that alcoholysis of 2-halogeno-octanes occurred with 95–100 per cent inversion, confirming that the mechanism is predominantly S_N2

$$C_2H_5O^- + \underset{(CH_2)_5CH_3}{\underset{|}{\overset{H \ CH_3}{\overset{\downarrow}{C}}-Br}} \longrightarrow \underset{(CH_2)_5CH_3}{\underset{|}{\overset{CH_3 \ H}{C_2H_5O-\overset{\downarrow}{C}}}} + Br^-$$

General reaction conditions

The simplest procedure is to convert the alcohol into the sodium alkoxide by careful addition of clean sodium metal until the required amount of alkoxide has been formed. The alkyl halide is added to the solution and the resulting ether may be distilled off.

Alternatively, the alkoxide is formed by adding the theoretical quantity of alcohol to powdered sodium suspended in xylene or toluene. The alkyl halide is then added gradually to the refluxed suspension. This method has been used to prepare benzyl phenethyl ether[7]

$$\text{C}_6\text{H}_5\text{-CH}_2\text{CH}_2\text{OH} \xrightarrow[\text{Toluene}]{\text{Na}} \text{C}_6\text{H}_5\text{-CH}_2\text{CH}_2\text{ONa}$$
$$\downarrow C_6H_5CH_2Cl$$
$$\text{C}_6\text{H}_5\text{-CH}_2\text{CH}_2\text{OCH}_2\text{-C}_6\text{H}_5$$

In the case of methylations with dimethylsulphate as applied to phenols, reactions are carried out in aqueous solution. The phenol is dissolved in aqueous sodium hydroxide solution, and the methylating agent is gradually added to the cooled mixture. The product often precipitates from solution. Several procedures of this form have been given by Vogel[8]

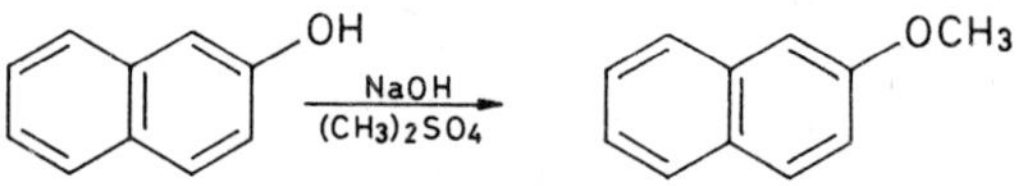

$$\text{naphthol-OH} \xrightarrow[(CH_3)_2SO_4]{NaOH} \text{naphthol-OCH}_3$$

Applications

The reaction is not limited to simple alkyl halides. Leffler and Calkins[9] reacted monochloroacetic acid with menthol, by the toluene–sodium powder procedure, to give 80 per cent yields of menthoxyacetic acid

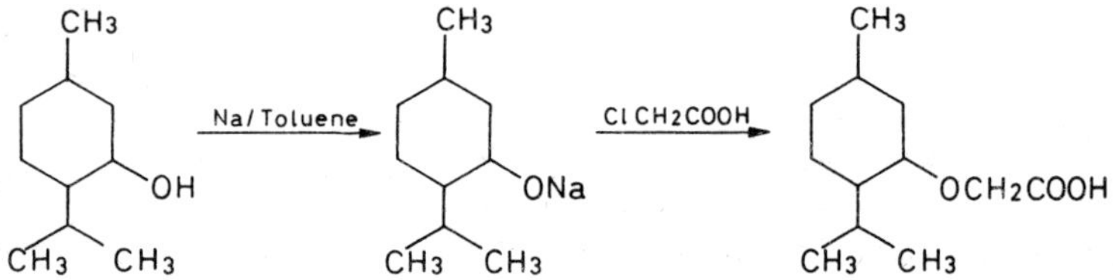

$$\text{menthol-OH} \xrightarrow{Na/Toluene} \text{menthol-ONa} \xrightarrow{ClCH_2COOH} \text{menthol-OCH_2COOH}$$

Intramolecular reactions have been used to give cyclic ethers. In many cases compounds possessing suitable halogen atoms and hydroxy groups cyclize by heating with solid sodium or potassium hydroxide. Gaylord et al[10] prepared 2-chloro-4-hexanol by this method, although a number of by-products were also formed

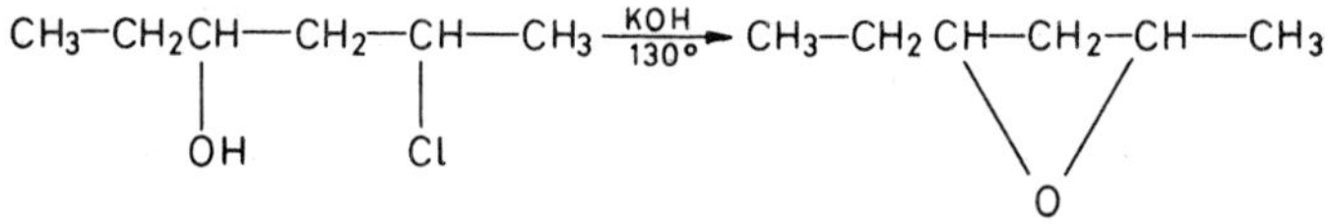

$$CH_3-CH_2CH(OH)-CH_2-CH(Cl)-CH_3 \xrightarrow[130°]{KOH} CH_3-CH_2CH-CH_2-CH-CH_3 \text{ (epoxide)}$$

REFERENCES

[1] Williamson, A. W. *J. chem. Soc.* 4 (1851) 106, 229
[2] Kolbe, H. *J. chem. Soc.* 7 (1854) 111
[3] Williamson, A. W. *J. chem. Soc.* 7 (1854) 122
[4] Alexander, E. R. *Principles of Ionic Organic Reactions*, p. 213, New York (Wiley) 1950
[5] Woolf, S. S. *J. chem. Soc.* (1937) 1173
[6] Hughes, E. D., Ingold, C. K. and Masterman, S. *J. chem. Soc.* (1937) 1196
[7] Baker, R. H. *J. Am. chem. Soc.* 70 (1948) 3857

[8] Vogel, A. I. *A Textbook of Practical Organic Chemistry*, 3rd edn., p. 670, London (Longmans) 1956
[9] Leffler, M. T. and Calkins, A. E. *Org. Synth.*, Coll. Vol. 3 (1955) 544
[10] Gaylord, N. G. et al., *J. Am. chem. Soc.* 76 (1954) 59

72
WURTZ–FITTIG REACTION (WURTZ REACTION)

Nature of the reaction

Hydrocarbons are formed by reacting aryl and/or alkyl halides together in the presence of metallic sodium

$$2\ RX + 2\ Na \longrightarrow R-R + 2\ NaX$$

$$ArX + RX + 2\ Na \longrightarrow ArR + 2\ NaX$$

Historical development

Wurtz[1] introduced his method for preparing alkanes from alkyl halides in 1855 by reporting the reactions occurring with the short-chain halides

$$2\ C_2H_5I + 2\ Na \longrightarrow C_4H_{10} + 2\ NaI$$

It was found that attempts to prepare chains with odd numbers of carbon atoms from mixtures of two alkyl halides led to a mixture of products

$$RI + R'I \xrightarrow{Na} R-R + R'-R' + R-R'$$

Fittig and his collaborators[2], however, applied the reaction to mixtures of aryl and alkyl halides and found that the unsymmetrical reaction predominates over the symmetrical

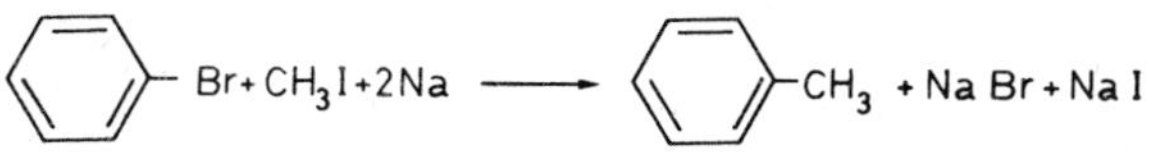

$$C_6H_5-Br + CH_3I + 2\ Na \longrightarrow C_6H_5-CH_3 + NaBr + NaI$$

The name 'Wurtz reaction' is applied to reactions in which alkanes are formed, and 'Wurtz–Fittig reaction' in cases where aryl halides are involved.

In the case of dihalogeno-benzenes, two alkyl groups may be introduced into the aromatic system. The disadvantage of the method lies in the formation of alkenes as by-products due to dehydrohalogenation.

Wagner and Zook[3] have discussed some of the literature on the reaction. Yields are moderate, between 30 and 60 per cent. The method is generally unsuitable for the preparation of diaryls, as yields are very low.

Mechanism

Evidence suggests that both ionic and free-radical mechanisms are involved[4]. Predominance of one over the other depends upon the nature of the alkyl halide and the preparative conditions. A simple free-radical mechanism for the reaction is

$$CH_3Br + Na^\bullet \longrightarrow NaBr + CH_3^\bullet$$
$$C_6H_5Br + Na^\bullet \longrightarrow NaBr + C_6H_5^\bullet$$
$$C_6H_5^\bullet + CH_3^\bullet \longrightarrow C_6H_5CH_3$$
$$2\ CH_3^\bullet \longrightarrow C_2H_6$$
$$2\ C_6H_5^\bullet \longrightarrow C_6H_5-C_6H_5$$

Morton and co-workers[5] considered the normal Wurtz–Fittig reaction to be ionic, involving the intermediate formation of organosodium compounds. Hine[6] has summarized the work on this and considers that, under ordinary reaction conditions, an ionic mechanism is most likely. This may be represented as

$$CH_3Br + 2Na \longrightarrow CH_3^-Na^+ + NaBr$$
$$C_6H_5Br + 2Na \longrightarrow C_6H_5^-Na^+ + NaBr$$
$$C_6H_5Br + CH_3^-Na^+ \longrightarrow C_6H_5CH_3 + NaBr$$
$$CH_3Br + C_6H_5^-Na^+ \longrightarrow C_6H_5CH_3 + NaBr$$
$$CH_3Br + CH_3^-Na^+ \longrightarrow C_2H_6 + NaBr$$
$$C_6H_5Br + C_6H_5^-Na^+ \longrightarrow C_6H_5-C_6H_5 + NaBr$$

Work on optically active alkyl halides has shown[7] that inversion of configuration occurs at the optically active centre due to an $S_N 2$ reaction. This lends support to the proposed ionic mechanism.

WURTZ–FITTIG REACTION (WURTZ REACTION)

General reaction conditions

An excess of clean sodium is placed in dry ether and cooled to 0°. The alkyl (or mixture of alkyl and aryl) halide is added gradually to the suspended sodium and the exothermic reaction commences almost immediately. Addition of the halides is controlled to maintain steady refluxing which is continued after addition is complete. On cooling, the solution is decanted and fractionally distilled.

This type of procedure has been used[8] to obtain n-butylbenzene with yields of 65–70 per cent

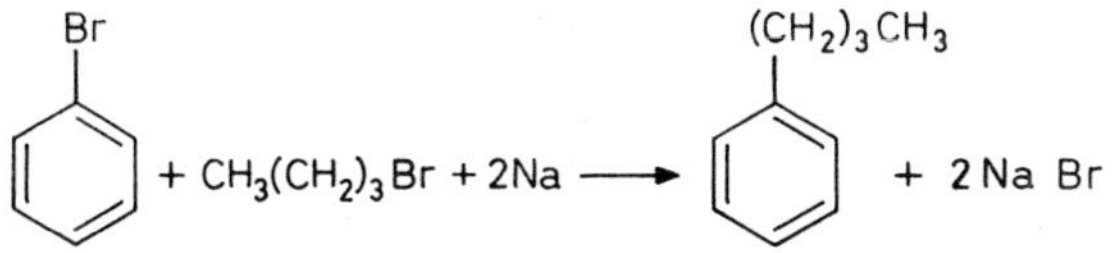

$$C_6H_5Br + CH_3(CH_2)_3Br + 2Na \longrightarrow C_6H_5(CH_2)_3CH_3 + 2\,Na\,Br$$

Reactions may be carried out using other solvents (benzene, toluene) and other metals (zinc).

Applications

There has been a recent revival of interest in the Wurtz–Fittig reaction as a method for preparing organosilicon compounds[9]. Work of this nature has been carried out by Eaborn and Walton[10] who reacted a number of substituted chlorobenzenes with chlorotrimethylsilane

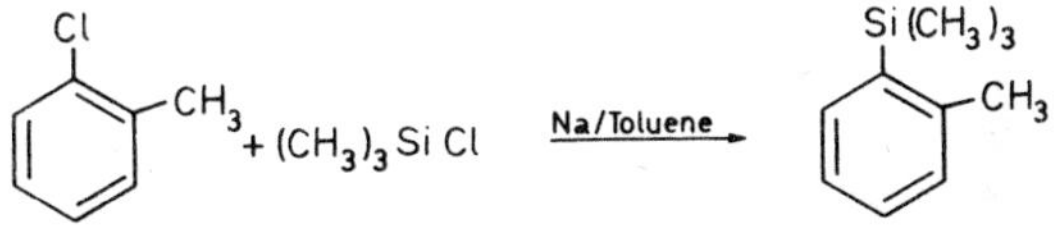

$$o\text{-}ClC_6H_4CH_3 + (CH_3)_3SiCl \xrightarrow{Na/Toluene} o\text{-}(CH_3)_3Si\,C_6H_4CH_3$$

The method has also been of value in the preparation of cyclic arsines[11]. Mann and his collaborators[12] have produced a number of interesting compounds of this type from dichloroarsines

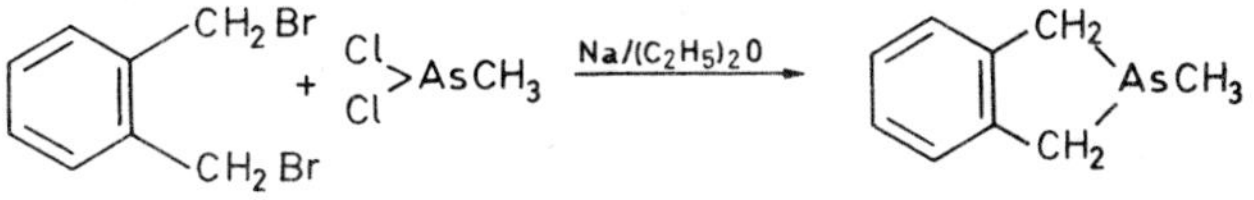

$$o\text{-}C_6H_4(CH_2Br)_2 + Cl_2AsCH_3 \xrightarrow{Na/(C_2H_5)_2O} \text{cyclic } o\text{-}C_6H_4(CH_2)_2AsCH_3$$

REFERENCES

[1] Wurtz, A. *Annls Chim. Phys.* **44** (1855) [3] 275; *Justus Liebigs Annln Chem.* **96** (1855) 364

[2] Fittig, R., with Tollens, B. *Justus Liebigs Annln Chem.* 131 (1864) 303; with König, J. 144 (1867) 277
[3] Wagner, R. B. and Zook, H. D. *Synthetic Organic Chemistry*, p. 10, New York (Wiley) 1953
[4] Waters, W. A. *The Chemistry of Free Radicals*, 2nd edn., p. 207, Oxford (Univ. Press) 1948
[5] Morton, A. A., Davidson, J. B. and Hakan, B. L. *J. Am. chem. Soc.* 64 (1942) 2243
[6] Hine, J. *Physical Organic Chemistry*, 2nd edn., p. 245, New York (McGraw-Hill) 1962
[7] Le Goff, E., Ulrich, S. E. and Denney, D. B. *J. Am. chem. Soc.* 80 (1958) 622
[8] Read, R. R. *et al.*, *Org. Synth., Coll. Vol.* 3 (1955) 157
[9] Emblem, H. G., Ridge, D. and Todd, M. *Chemy Ind.* (1955) 905
[10] Eaborn, C. and Walton, D. R. M. *J. organomet. Chem.* 3 (1965) 168
[11] Cullen, W. R. *Adv. organomet. Chem.* 4 (1966) 149
[12] Lyon, D. R. and Mann, F. G. *J. chem. Soc.* (1945) 30; with Cookson, G. H. (1947) 662

INDEX

Abietic acid, 24
Abietinol, 24
Acetaldehyde, 38, 217
Acetanilide, 21
Acetone, 29, 48, 57, 112
 acetyl, 42
Acetophenones, 25, 42, 48, 69, 122, 177
5-Acetyl-indan-4-ol, 179
3-Acetyl-1-methylpyrrocoline, 237
Acid
 anhydrides, 54
 chlorides, 2, 15, 28, 83, 144
Acids,
 dicarboxylic, 10, 72, 125, 127, 130, 134, 200
 monocarboxylic, 2, 8, 19, 34, 74, 77, 114, 157, 187, 200, 204, 235
 α,β-unsaturated, 50
Acrolein, 141
Acrylic acids, 22, 230
Adipic acid, 125, 127, 132
Adiponitrile, 135
Alanine, 217
 β-, 14
Alcohols,
 primary, 17, 23, 34, 86, 107, 111, 157, 180, 187, 226, 235, 238
 secondary, 86, 107, 111, 157, 180
 tertiary, 86, 114, 157
Aldehydes, 10, 19, 24, 28, 31, 37, 47, 50, 54, 57, 100, 107, 111, 115, 122, 137, 148, 151, 155, 160, 177, 207, 210, 214, 217, 235
Aldimines, 214
Aldohexoses, 90, 99, 106
Aldopentoses, 90, 99, 106
Aldol condensation, 47
Aldoses, 90, 93, 101, 104
Aldoximes, 4, 105
Alkaloids, 3, 45, 136, 194
Alkanes, 25, 72, 241
Alkenes, 61, 75, 81, 114, 144, 193, 235
Alkyl
 chlorides, 180, 200
 halides, 238, 241
Alkylidene succinic esters, 57

Alkynes, 186, 235
Alkynyl bromides, 212
Allyl ethers, 65
1-Allyl-2-naphthol, 65
Aluminium
 alkoxides, 37, 107, 111
 benzyloxide, 107
 t-butoxide, 112
 chloride, 68, 144, 151
 isopropoxide, 108, 113
Amides, 1, 4, 12, 19, 100, 234
Amines,
 primary, 8, 11, 19, 83, 86, 121, 140, 174, 190, 198, 207
 secondary, 83, 86
 tertiary, 83, 180, 193
Amino acids, 13, 22, 79, 190, 217
γ-Aminobutyric acid, 191
p-Aminoethyl benzene, 199
2-Amino-ethylphosphonic acid, 192
3-Amino-2-naphthoic acid, 175
4-Amino-1-phenylpyrazole, 142
4-Amino veratrole, 13
Aniline, 9, 197
Anthracene, 145
Anthraquinone dioxime, 7
γ-Apopicropodophyllin, 186
Arabinose, 53, 97, 101, 104
Arndt–Eistert synthesis, 1
Arsines, 243
Aryl halides, 220, 241
Ascorbic acid, 91
Aspartic acid, 22, 230
Azides, 8, 20

Baekeland polymerization process, 170
Baeyer reaction, 171
Bakelite, 171
Barbier–Wieland degradation, 114
Beckmann rearrangement, 4
Benzaldehyde, 28, 34, 38, 48, 50, 89, 102, 122, 153, 177, 210
Benzanilide, 19
Benzene
 diazonium chloride, 166

Benzene—*cont.*
  sulphonyl chloride, 83
Benzil, 112
3,4-Benzocarbazole, 139
Benzoic acid, 15, 34, 166, 223
Benzophenone, 19, 58, 61, 112, 144
  oxime, 4
p-Benzoquinone, 112
Benzothiadiole-1,1-dioxide, 183
Benzoyl
  azide, 9
  chloride, 15, 28, 86
  pyridine, 3-, 146
Benzyl
  amines, 192
  benzoate, 38
  chloride, 210
  phenethyl ether, 239
Betaine, 62
Biaryls, 220
2,2′-Bipyridine, 222
Blanc reaction (Blanc's rule), 125
Borneol, 227
Bornyl chloride, 226
  iso-, 228
Borodine–Hunsdiecker reaction, 200
Boron trifluoride, 153
Bouveault–Blanc reduction, 23
N-Bromoacetamide, 11, 80
Bromal, 39
Bromine, 11, 78, 201
Bromo-
  acids, 79
  benzene, 201
  cyclopropane, 202
  3-methyl benzoic acid, 2-, 225
  nitrobenzene, *p*-, 165, 223
  4-nitrotoluene, 2-, 225
  phthalimide, *N*-, 80
  pyridine, 2-, 222
  succinimide, *N*-, 80, 116
  toluene, *o*-, 164
Bucherer reaction, 174
Butadiene, 185
1-n-Butylpyrrolidine, 85
n-Butyraldehyde, 38
iso-Butyric acid, 78

Calcium
  gluconate, 97
  suberate, 130
  amphene, 227

Camphor, 18, 31, 126, 183, 226
Camphoric acid, 126
Cannizzaro reaction, 34
ε-Caprolactam, 7, 21
δ-Caprolactone, 10
Carbamates, 14
Carbanions, 41, 48
Carbohydrates, 53, 89, 93, 97, 100, 104
Carbon tetrachloride, 207
Carbonium ions, 38, 75, 115, 145, 211, 226, 231
Carotenoids, 63
Catechol, 119, 178
Chalcone, 48
Chloral, 39
Chlorine, 11
Chloro-
  aniline, *o*-, 168
  benzene, 166
  benzoic acid, *m*-, 225
  bromobenzene, *o*-, 168
  2,6-dinitrobenzene, 1-, 168
  4-hexanol, 2-, 240
  nitrobenzene, *p*-, 165, 225
  α-phenyl acetophenone, α-, 181
  sulphonic acid, 84, 153
  trimethylsilane, 243
Chloroform, 207
Cholanic acid, 115
Cholestanone, 27, 82
Cholestenone, 64, 111
Cholesterol, 27, 111, 117
Chromium trioxide, 115
Chromyl chloride, 122
Chrysene, 227
Cincholoiponic acid, 46
Cinnamaldehyde, 107, 113
Cinnamic acid, 54, 188
Cinnamide, 100
Cinnamoyl chloride, 88
Cinnamyl alcohol, 107
Citral, 162
Civetone, 27, 131
Claisen
  condensation, 40
  reaction, 47
  rearrangement, 65
  –Schmidt condensation, 47
Clemmensen reduction, 25
Coumarin, 54, 70
Creosol, 27
Crotonaldehyde, 51, 109
Crum–Brown Walker reaction, 72

INDEX

Cumaranone, 68
Curtius reaction (rearrangement), 8
Cumene, 146
Cyano-
　hydrins, 89, 218
　pyridine, β-, 216
Cyclization, 42, 46, 127, 130, 133, 137, 140, 146, 224, 240
Cyclo-
　heptanone, 130
　hexane, 31
　hexanol, 108
　hexanone, 21, 31, 62, 82, 108, 112, 131, 138
　hexene, 80
　hexyl carbinol, 157
　octadiene, 195
　octane, 133
　pentadiene, 183
　pentanone, 132
　pentenone, 184
　phanes, p-, 129
　propyl carbinol, 113

Dakin reaction, 177
Darzens' procedure, 180
Dean and Stark trap, 52, 189, 195
Decarboxylation, 51, 126
Dehydration, 17
2-Deoxy-D-ribose, 92, 99
Deuterium, 94, 185, 194
2,6-Dialkylhydroquinones, 178
Diarylidenes, 59
Diastereoisomers, 90
Diazo-
　ethane, 2
　methane, 1
Diazonium salts, 163, 166
Dibenzoyl
　methane, 40, 42
　peroxide, 81
ω,ω'-Dibromohydrocarbons, 76
Di-n-butylamine, 85
Dichalcone, 49
Dichloro-
　acetone, 1,3-, 38
　arsines, 243
　5α-cholestane, 3α, 6β-, 233
　methylene, 208
1,4-Dicyanobutane, 133
Dieckmann reaction, 127
Diels–Alder reaction, 183

Dienes, 183
Dienophiles, 183
Diethyl
　malonate, 43, 50
　octadecanedioate, 73
　succinate, 58
3,4-Dihydroxy-5-methoxyphenylacetic acid, 3
1,3-Diketones, 40
Dimethyl
　allyl ethers, α,γ-, 67
　amino-acetophenone, p-, 237
　bromosuccinate, 230
　butadiene, 2,3-, 185
　catechol, 3,4-, 179
　-8-nitroquinoline, 2,4-, 141
　octadecanedioate, 76
　sulphate, 195, 239
　terephthalaldehyde, 2,5-, 210
　tetrahydroanthraquinone, 185
3,5-Dinitroaniline, 21
2,2'-Dinitrodiphenyl, 221
Diphenyls, 220
Disaccharides, 106
Docosanedioic acid, 33

E 1, E 2 processes, 194, 227
Elbs persulphate oxidation, 118
Electrolysis, 72, 74
Electron spin resonance, 123, 145
Electrophilic reactions, 69, 115, 119, 172, 205
Elimination processes, 194, 227
Epicamphor, 18
Equilenin, 59
Erythrose, 101, 104
Esters, 23, 37, 40, 43, 50, 68, 72, 116, 127, 157, 187
Étard reaction, 122
Ethers, 65, 238
Ethyl
　acetoacetate, 41, 50, 128
　acetone dicarboxylate, 189
　benzene, 25, 122
　bromoacetate, 189
　α-bromopropionate, 78
　cinnamate, 43
　laurate, 24
　-o-methoxybenzoate, 42
　methyl ether, 238
　monochloroacetic acid, 159
　phenyl acetate, 23

NAMED ORGANIC REACTIONS

Ethyl—*cont.*
  phenyl cyanoacetate, 45
  sodiomalonate, 43
Ethylene
  chlorohydrin, 180
  diamine, 190
  dibromide, 190
o-Eugenol, 67

Fenton's reagent, 97
Ferric chloride, 144, 228
Fischer
  indole synthesis, 137
  –Speier esterification, 187
Flavones, 42, 88, 121
Fluorenone, 113
9-Fluorenylmethylacetates, 228
Fluoroalkenoic acids, 117
Fluorocarbons, 202
Formaldehyde, 36, 50, 157, 170
Formic acid, 153
Formimino chloride, 148
Formyl chloride, 152
Free radicals, 19, 67, 73, 75, 81, 98, 119, 145, 167, 201, 242
Friedel–Crafts reaction, 143
Fries rearrangement, 68
Fructose, 94
Furaldehydes, 30
Furans, 149
Furfural, 31, 36, 52
2-Furfuryl alcohol, 36
2-Furoic acid, 36
Furoyl chlorides, 30
Furylacrylic acid, 55

Gabriel Synthesis, 190
Galactose, 90
Gas chromatography, 32, 181
Gattermann
  aldehyde synthesis, 148
  reaction, 163
  –Koch reaction, 151
Geranic acid, 162
Glucose, 94, 101, 104
Glutaric acid, 50
Glycals, 99
Glycerol, 86, 140
Glycine, 86, 190
Grignard reagents, 154
Guiacol allyl ether, 67
Guinic acid, 202

α-Halogeno esters, 159
Hell–Volhard–Zelinsky reaction, 77
Hemiacetals, 94
Hendecanedioic acid, 33
Heptaldehyde, 25
n-Heptylamine, 20
2-Heptynal, 212
Hexahydrobenzaldehyde, 124
Hexamethylenetetramine, 210
Hinsberg's separation of amines, 83
Hippuric acid, 86
Hofmann
  exhaustive methylation, 193
  –Martius rearrangement, 197
  reaction, (rearrangement), 11
Homocamphoric acid, 126
Hunsdiecker reaction, 200
Hydrazine, 32, 191
Hydrazoic acid, 19
Hydrazones, 31, 137
Hydride ions, 35, 211
Hydrocarbons, 26, 31, 74, 109, 241
Hydrogen peroxide, 73, 97, 177
Hydrogenation, 29
Hydroxamic acids, 15
Hydroxy-
  acetophenone, o-, 88
  α-amino acids, β-, 219
  arylmethanes, 171
  3,4-dimethyl acetophenone, 2-, 179
  esters, β-, 159
  ketones, p-, 68
  3-methoxybenzaldehyde, 2-, 178
  5-methoxy-2-naphthoic acid, 1-, 205
  1-naphthaldehyde, 2-, 208
  phenyl benzyl ketones, 177
  propiophenone, 70
Hydroxylamine, 15

Imides, 13
Imines, 51
Imminium salts, 51
Indane-4,5-diol, 179
3-Indazolones, 224
Indoles, 82, 137, 149
Infra-red spectroscopy, 205
Inversion, 230, 242
Ion-exchange resins, 71, 95, 98
Ionanes, 33
Ionols, 113
Ionones, 33, 113
Irone, 113

INDEX

Isocyanates, 9, 11, 15, 101
Isomerization, 95, 184, 231
Isomers,
  cis, trans, 55, 62, 67
  syn, anti, 5, 21
Isoprene, 186
Isotope labelling, 94, 138, 156, 185, 188, 192, 194, 196

Ketals, 23
β-Keto-esters, 40, 127
Ketones, 7, 19, 24, 31, 47, 57, 68, 108, 111, 115, 122, 125, 130, 137, 144, 155, 160, 177, 218, 234
Ketoses, 93
Ketoximes, 4
Kiliani–Fischer synthesis, 89
Knoevenagel condensation, 50
Kolbe
  –Schmitt reaction, 204
  reaction, 74

Lactones, 7, 10, 89, 101
Lauryl alcohol, 24
Law of mass action, 189
Lederer–Manasse reaction, 171
Leucine, 219
Lewis acids, 5, 137, 143, 148, 151, 186, 228
Lignans, 146
Lithium aluminium hydride, 23
Lobry De Bruyn–Alberda Van Ekenstein transformation, 93
Lossen rearrangement, 15
Lycopene, 63

Magnesium, 154
Maleic anhydride, 183
Malonic acid, 51
Mandelamide, 102
Mandelic acid, 89
Mannose, 90, 94, 101, 106
Meerwein–Ponndorf–Verley reduction, 107
Menthol, 240
Menthoxyacetic acid, 240
Mescaline, 3
Mesitaldehyde, 30, 150
Mesitylene, 150
Methane, 158

Methionine, 219
Methoxy-
  acetic acid, *p*-, 236
  acetophenone, *p*-, 236
  amides, α-, 101
  1-naphthol, 5-, 205
  8-nitroquinoline, 6-, 142
Methyl
  abietate, 24
  amine, 11
  3-amino-3-deoxy-β-L-xylopyranoside, 232
  aniline, N-, 135, 197
  benzaldehyde, 4-, 36
  benzylalcohol, 4-, 36
  bromide, 62, 200
  5-bromovalerate, 202
  cyanoacetate, 52
  cyclohexane, 124
  ethyl ketone, 88, 112
  formate, 37
  heptanone, 162
  indole, 2-, 149
  iodide, 194
  magnesium bromide, 114
  oxindole, 3-, 137
  3-phenylhexanoic acid, 3-, 116
  phenyl-α-naphthylchlorosilane, 233
  quinoline, 6-, 140
  sebacate, 76
  silver adipate, 202
  thiophene, 3-, 81
  vinyl ketone, 44, 186
Methylene-
  benzoate, *p*-, 39
  cyclohexane, 62
  triphenylphosphorane, 61
Michael condensation (addition), **43**
Monochloroacetic acid, 240
Morphine, 46

Naphthaldehyde, 30, 212, 215
Naphthalene, 111
1,4-Naphthaquinone, 185
Naphthionic acid, 174
1-Naphthol-4-sulphonic acid, 174
Naphthols, 65, 139, 174
Naphthylamine, 174
β-Naphthyl benzoate, 86
Nicotinaldehyde, 216
Nicotinamide, 13
Nitriles, 19, 104, 133, 157, 214

NAMED ORGANIC REACTIONS

Nitrilium salt, 214
Nitro-
  aniline, p-, 164
  benzaldehyde, o-, 56
  benzene, 142, 222
    sulphonic acids, 225
  benzidine, 2-, 168
  diphenyls, 168, 220
  9-fluorenylmethyl acetate, 2-, 228
  methane, 44
  6-methoxybenzaldehyde, 2-, 124
  phenanthrene, 2-, 228
  phenol, o-, 118
  quinol, 118
Nucleophilic substitution, 180, 227, 231, 239, 242

Octadecanedioic acid, 72
Oestradiols, 110
Oestrone, 33, 110
Olefins, 61, 75, 114, 144, 193
Oleyl alcohol, 24
p-Oligophenylenes, 222
Oppenauer oxidation, 111
Optical activity, 6, 12, 94, 180, 230, 242
Organo-
  magnesium halides, 154
  silicon compounds, 233, 243
  zinc compounds, 159
Oxidation, 97, 111, 118, 122, 140
Oximes, 4, 6, 105
Oxindoles, 82

π complex, 70, 205
Palladium, 28
Paraconic ester, 58
Paraffins, 25, 72, 241
Perkin reaction, 54
Perylene, 222
Phenanthrenes, 56, 111, 228
Phenanthroline, 140
Phenols, 66, 118, 139, 148, 170, 174, 204, 207, 239
Phenyl-
  acetaldehyde, 100, 122
  acetate, 69
  acetic acid, 56
  alanine, 22, 192, 219
  benzene sulphonates, 71
  butyryl chloride, 1-, 146
  esters, 68

Phenyl- —cont.
  ethanol, β-, 180
  glycine, 219
  hydrazine, 138
  β-hydroxypropionate, β-, 161
  itaconic acids, 57
  3-methyl butanol, 1-, 155
  pentane, 1-, 144
  sulphamic ester, 10
Phosphonates, 60
Phosphorane, 62
Phosphorus
  pentachloride, 233
  pentoxide, 5, 228
iso-Phthalaldehyde, 213
Phthalic acid, 190
Pimelic acid, 125, 127
Pinacolones, 26, 227
Pinacols, 26, 227
Pinane, 227
Piperidine, 194
Plasmoquin, 142
Platinum, 31, 73, 75
Polyimides, 103
Polymerization, 133, 170
Polymers, 88, 171
Polyphosphoric acid, 6, 18, 21, 142, 146, 228
Polysaccharides, 103
Potassio-pyrrole, 206
Potassium
  acetate, 74
  t-butoxide, 57, 63, 112, 129
  ethyl suberate, 72
  phthalimide, 190
β-Propionylphenyl hydrazone, 137
Pschorr Synthesis, 56
Pseudopelletierine, 195
Pyrimidine-5-carboxaldehydes, 208
Pyrogallol 1-methyl ether, 178
Pyrrole-3-carboxylic acid, 206
Pyrroles, 149
Pyruvic acid, 137

Quaternary ammonium compounds, 193
Quinolines, 30, 140

Racemization, 231
Reduction,
  acid chlorides, 28
  carbonyl compounds, 25, 31, 107

INDEX

Reduction—*cont.*
  esters, 23
  lactones, 90
Reformatsky reaction, 159
Reimer–Tiemann synthesis, 207
Resins, 170
Resorcinol, 206
β-Resorcylic acid, 206
Retention of configuration, 231
Rosenmund–Saytzeff reduction, 28
Ruff degradation, 97
Ruthenium tetroxide, 117
Ruzicka ring synthesis, 130

S_N reactions, 180, 227, 231, 239, 242
Salicylaldehyde, 54, 178
Salicylic acid, 120, 204
Sandmeyer reaction, 166
Schmidt reaction, 19
Schotten–Baumann reaction, 86
Semicarbazones, 31, 102
Silver oxide, 2, 104, 195, 202
Skatole, 82
Skraup synthesis, 140
Sodium
  amalgam, 90
  borohydride, 91
  t-butoxide, 135
  ethoxide, 17, 40, 57, 127, 133, 238
  eugenoxide, 239
  hydride, 41, 57, 128
  N-methylanilide, 134
  phenoxide, 204
Sommelet reaction, 210
Sorbic acid, 52
Stannic chloride, 144, 186
Stannous chloride, 216
Stephen reaction, 214
Steric hindrance, 188, 218
Steroids, 3, 27, 33, 45, 59, 64, 82, 95, 110, 113, 115, 229, 233
Sterols, 110–11
Stilbene, 62
Stobbe condensation, 57
Strecker synthesis, 217
Suberic acid, 130
Suberone, 130
Substance S, 96
Succinic
  acid, 231
  ester, 57
Sulphobenzoic acids, 225

Sulphonamides, 83
Sulphones, 44, 71
Sulphonyl halides, 83

Tartaric acid, 97
Tautomerism, 16, 84, 174
Tazettine, 136
Teraconic acid, 57
Terephthalaldehyde, 39, 49
Terpenes, 27, 45, 126, 226
Tetrahydro-
  carbazole, 1,2,3,4-, 138
  furfuryl chloride, 181
  quinoline, 1,2,3,4-, 199
  sulphonic acids, 175
α-Tetralone, 146
Tetramethyl-*p*-xylylene, 195
2,3,4,5-Tetraphenylpentadienone, 186
1-Thiacyclo-octane-5-one, 129
Thioamides, 235
Thionyl chloride, 15, 17, 181
2-Thiophene-
  acetaldehyde, 103
  aldehyde, 212
Thiosemicarbazone, 216
Tishchenko reaction, 37
Titanium tetrachloride, 144
Tolualdehydes, 122, 153
Toluene, 122
  sulphonyl chloride, *p*-, 84
p-Toluidine, 16, 20, 140, 197
o-Toluonitrile, 214
3,4,5-Triacetoxycyclohexanone, 203
6-Trichloromethylpurine, 209
Trifluoroacetic acid, 42
2,3,5-Trihydroxybenzoic acid, 120
Trimethylamine, 195
Triphenyl-
  methyl sodium, 41
  phosphonium salts, 196
Tritium, 95, 196
Triton B, 45

Ullmann reaction, 220
Ultra-violet irradiation, 10, 81
Undecyl isocyanate, 10
Ureas, 8
Urethans, 8, 12, 102

Valine, 79
Vanillin, 27, 31

NAMED ORGANIC REACTIONS

Veratric amide, 13
Vitamin C, 91
von Richter reaction, 223

Wagner–Meerwein rearrangement, 226
Walden inversion, 230
Weerman degradation, 104
Willgerodt–Kindler reaction, 234
Williamson synthesis, 238
Wittig reaction, 61
Wohl
  degradation, 104
  –Ziegler reaction, 80

Wolff
  –Kishner reduction, 31
  rearrangement, 1
Wurtz (–Fittig) reaction, 241

Xylenes, 122

Ylides, 62

Zerewitinoff determination, 158
Ziegler–Thorpe ring synthesis, 133
Zinc, 25, 159, 164, 243
  chloride, 137, 144, 148, 175, 228
Zwitterion, 194